Science Communication

This book is accompanied by a companion website at:

www.wiley.com/go/bowater/sciencecommunication

Visit the website for:

- Figures and tables from the book
- Useful forms for downloading
- Examples of marketing material
- Ideas for activities

Science Communication

A Practical Guide for Scientists

Laura Bowater and Kay Yeoman
University of East Anglia, UK

WILEY-BLACKWELL

A John Wiley & Sons, Ltd., Publication

Library of Congress Cataloging-in-Publication Data

Bowater, Laura.
  Science communication : a practical guide for scientists / Laura Bowater and Kay Yeoman.
    p. cm.
  Includes bibliographical references and index.
  ISBN 978-1-119-99313-1 (cloth) – ISBN 978-1-119-99312-4 (pbk.)   1. Communication in science.
I. Yeoman, Kay.   II. Title.
  Q223.B69 2012
  501'.4–dc23
                                                                                    2012016771

A catalogue record for this book is available from the British Library.

Wiley also publishes its books in a variety of electronic formats. Some content that appears in print may not be available in electronic books.

Set in 9.5/12pt Optima by Aptara Inc., New Delhi, India

First Impression 2013

Contents

This book is accompanied by a companion website at:

www.wiley.com/go/bowater/sciencecommunication

Visit the website for:

- Figures and tables from the book
- Useful forms for downloading
- Examples of marketing material
- Ideas for activities

About the Authors

Laura Bowater

Laura is a Senior Lecturer at the Norwich Medical School, at the University of East Anglia (UEA), where she currently teaches microbiology, biochemistry and genetics to medical students. Laura previously worked as a microbiologist at the John Innes Centre – a BBSRC-funded research institute based on the Norwich Research Park – participating in research into the different roles of cupin molecules in bacteria and fungi. Laura has always had an interest in science communication and, while at the John Innes Centre, she took part in the Community X-change project, where members of the public were able to discuss issues relating to climate change during the 2006 British Science Festival that took place in Norwich in 2006. After Laura moved to the Norwich Medical School, she continued her work with the public by organising Café sessions, tackling such subjects as personalised medicine and microbial evolution. Laura has also been involved with the 'Workshop on User Involvement in Research', which has helped set up a new network of researchers and Service User groups, and she has designed a workshop to encourage and facilitate the development of a range of skills and expertise within research. During 2009, Laura also spearheaded a collaboration with Future Radio in Norwich and the UEA to produce a series of radio programmes about the life and work of Charles Darwin. Laura has also written articles about her engagement experiences in documents produced by Sciencewise for the Department for Innovations, Universities and Skills and the Research Councils UK. She is a member of the Society for General Microbiology (SGM), and is on the SGM committee for education. In recognition of her science communication work, she was awarded the Cue East Public Engagement Award in 2010.

Kay Yeoman

Kay is a Senior Lecturer in the School of Biological Sciences at UEA. She is a microbiologist with a particular interest in soil bacteria and fungi. She lectures in microbiology and molecular biology to both undergraduate and postgraduate students. In addition, she runs an undergraduate module in science communication and is the director of the Biology with Science Communication degree programme. Kay is involved with research looking at the public understanding of science and how students can gain crucial employability skills by taking part in community engagement. She is a member of the Society for General Microbiology, on the committee for education for the British Mycological Society and is a science advisor for the Norwich Castle Museum. Kay has run school and public science communication events for many years, and has been awarded funding from a variety of sources including The Royal Society and the Wellcome Trust. In 2007 she was awarded a Wellcome Trust Peoples Award for the 'Mobile Family Science Laboratory' (http://biobis.bio.uea.ac.uk/family/index.html), which travels throughout Norfolk and also goes to rural hard-to-reach communities. The events provide hands-on biomedical-related science activities for children and adults of all ages and abilities using family learning as the platform. During this project she developed a series of primary school clubs and activity days where children have been able to acquire a more in-depth experience of science, and now several hundred pupils have benefitted from these clubs. In 2007, she was awarded the UEA teaching excellence prize and was also awarded the Cue East Public Engagement Award in 2009.

About the Contributors

Dr Martyn Amos is a Reader in Novel Computation at Manchester Metropolitan University. His PhD, from the University of Warwick (1997), was one of the first in the field of molecular computing, and he has since worked to engage the public with his ongoing research into synthetic biology and complexity science. His 2006 book *Genesis Machines* was a well-received popular science 'biography' of this new area of science. He was the Principal Investigator of the EPSRC-funded 'Bridging the Gaps: NanoInfoBio' project, as well as running the Wellcome Trust-supported DIYbio Manchester initiative. He also heads several European Commission-funded projects, and is a contributor to the Speakers for Schools initiative. (Case study 8.3)

Janice Ansine is the Project Manager for the Biodiversity Observatory (iSpot). She works in the Department of Environment, Earth and Ecosystems, at The Open University. iSpot and the Biodiversity Observatory are supported by a grant from the Big Lottery Fund for England as part of the OPAL project. The iSpot team and the iSpot user community are also acknowledged for contributing to the success of the project to date. (Case study 8.5)

Dr Alison Ashby is a member of the Department of Plant Sciences at the University of Cambridge. She promotes 'Fungal Science' through public engagement. On behalf of the British Mycological Society, she has produced a set of primary school resources for 'engaging primary school children with the Kingdom Fungi', including the development of the BMS's 'Mycokids' pages. She held a Royal Society Research Fellowship and is affiliated through fellowships to Jesus College and Fitzwilliam College, Cambridge. Fun with Fungi was supported by the British Mycological Society, the Royal Society and the Department of Plant Sciences, Cambridge. (Case study 10.2)

Dr Stephen Ashworth is a Senior Lecturer in the School of Chemistry, where his science interests lie in high-resolution spectroscopy, and other spectroscopy applied to atmospheric chemistry, using a number of laser techniques. He is very active in the area of science communication and delivers interactive demonstration lectures to a wide range of audiences. Stephen also won a Public and Community Engagement Award from CUE East the Beacon of Excellence in the Eastern Region in 2009. (Case study 7.2)

Dr Richard Bowater is a Senior Lecturer in the School of Biological Sciences at the University of East Anglia (UEA). His research uses a range of contemporary techniques to study the relationship between the structure of DNA and cellular processes. He teaches across a broad range of biological science topics and has developed an interest in science communication activities with the general public. Richard has received funding from the Society of General Microbiology and the UEA's Annual Fund to support these activities. (Section 7.3.2)

Dr Tristan Bunn is the Inspiring Young Scientists Coordinator for the BBSRC and manager of the BBSRC School Regional Champions. He is a former biochemist, lecturer and science teacher who received his PhD from Edinburgh University for his work at the National CJD Surveillance Unit on molecular neuropathology of prions before obtaining a PGCE at Durham University. The 'Taste and Flavour' activities were delivered as part of the national Social and Emotional Aspects of Learning (SEAL) initiative. The programme for Pupil Referral Units (PRUs) and pupils with moderate learning difficulties in Norwich was organised by Norfolk Children's Services with funding from SEAL and CUE East. (Case study 9.2)

Annabel Cook works in science communications for the Mathematical, Physical and Life Sciences Division of the University of Oxford. She has an MSc in Microbial Marine Ecology from the University of Warwick and gained most of her initial science communication experience through volunteering while working full-time in an unrelated field. For five years she ran all online and offline, internal and external communication and public engagement for the Wellcome Trust Centre for Human Genetics (WTCHG) at Oxford. Volunteers from the WTCHG were involved in putting together and running the 'Five-a-day DNA' activity. (Case study 6.4)

Dr Sheila Dargan is a Lecturer in the School of Biosciences at Cardiff University. She is an active member of the Neuroscience and Mental Health Research Institute's public engagement committee, and a longstanding member of the Biochemical Society's Education Committee. Sheila supervises final year undergraduate students conducting bioscience engagement projects focused on curriculum enhancement in schools and universities. As a STEM ambassador, she has contributed to public engagement activities across the UK and internationally. In 2008 Sheila was selected to participate in a national 'SEARCH' project, using action learning sets to effect culture change with respect to public engagement in higher education and to maximise impact. (Case study 7.4)

Dr Darren Evans is a Lecturer in Conservation Biology at the University of Hull. His research focuses on the impacts of human-driven environmental change on the relationships between plants and animals. He is a co-founder of 'Conker Tree Science' and a newly established citizen science project with *The Observer* newspaper to map the spread of 10 highly invasive species in the UK. A regular speaker at conferences, workshops and Science Cafe events and contributor on radio and television, he is passionate about the need to communicate excellent science to the public, media and policymakers. (Case study 8.4)

Dr Ken Farquhar is a world juggling champion and soap bubble chemist. He founded the Inspirational Science Theatre company with a grant from the National Lottery in 1995.His science communication style has developed from on-the-job experience in many complimentary fields including working as a street performer, mime artist, actor, school teacher, television researcher and presenter. His science and maths shows and workshops have toured schools, museums, science and arts festivals across the UK and Europe. He provides bespoke programmes for schools at all Key Stages and development programmes for educational professionals, academics and businesses. (Case study 7.5)

Dr Sarah Field is a Senior Research Associate in the School of Biological Sciences at the University of East Anglia where her research focuses on the fermentation of straw waste by yeast to produce bioethanol. She has been involved in outreach activities for over 10 years. She is an active member of the Norfolk Teacher Scientist Network. Talking Science was supported by the Faculty of Science at UEA. (Case study 7.1)

Sara Fletcher A physicist by training, Sara joined Diamond Light Source in 2005 as Scientific Information Coordinator. The following year she completed an MSc in Science Communication and now writes about Diamond's science as well as looking after all the websites and social media. Growing up, she wanted to be a writer before turning to science, so as a science writer enjoys the best of both worlds! She is also addicted to social media and is interested in how the web is changing the practice and communication of science. (Case study 8.2)

Jaeger Hamilton is studying on the Wellcome Trust 4-year PhD programme at the University of Dundee where he investigates bacterial membrane biology as part of the Frank Sargent group in the division of Molecular Microbiology. He did his undergraduate studies in Microbiology (2009) at the University of East Anglia, Norwich. (Case studies 4.1 and 5.2; section 4.6 and Figure 4.5)

Tim Harrison is the Bristol ChemLabS School Teacher Fellow, Director of Outreach and the Science Communicator in Residence at the School of Chemistry, University of Bristol. He is the recipient of the Royal Society of Chemistry Secondary Education Award, the Royal Society's Hauksbee Award and the Engagement Award from the University of Bristol. The Bristol ChemLabS outreach work that he now directs has also won the Bank of America Merrill Lynch Education Award and the Big Tick Award for Business in the Community (2009), which was reaccredited in 2010 and 2011. Tim Harrison was on the RSC Committee for Schools and Colleges. He is currently an Associate Editor for *Chemistry Review* magazine, a member of the RSC's Western

Region of the Analytical Division and is the Education Officer for the Bristol and District Region of the RSC. (Box 3.3 and Case study 9.1)

Dr Adam Hart is a biologist, entomologist and Reader in Science Communication based at the University of Gloucestershire. He has a special interest in ants and bees, studying how they work together, although his varied research also includes birds, mammals, evolution and ecology. He is a Fellow of the Royal Entomological Society and Outreach Editor for their magazine *Antenna*. He is widely involved with science outreach, public engagement and broadcasting, including a newspaper column on scientific research, a column in *Cotswold Life* magazine and a regular science slot on BBC Radio Gloucestershire. He also writes and presents programmes for Radio 4, the World Service and BBC4. In 2010, he was named the Society of Biology's Science Communicator of the Year and was also awarded an HEA National Teaching Fellowship in recognition of his scientific and public engagement work. (Case study 10.5)

Sarah Holmes is currently teaching science, specialising in physics, in Norfolk, where she is also heavily involved in STEM activities. She completed her undergraduate studies in Ecology at the University of East Anglia (2010) where she took part in science communication activities focused on schools. Sarah has also completed a Postgraduate Certificate of Education at the UEA (2011). (Case study 10.7)

Josh Howgego is a Chemistry PhD student at the University of Bristol. During his three years in the lab he has designed and built artificial molecules which mimic natural carbohydrate-binding proteins. Away from the bench, Josh is a keen blogger and aspiring science writer. His blog was included in *Open Laboratory 2009* – an annual compilation of the best writing on science blogs. In 2011, he was awarded the Royal Society of Chemistry's Marriott Science Writing Internship and has since written for the *Times Higher Education* and *Chemistry World*. (Case study 8.6)

Dr Naomi Jacobs obtained her bachelor's degree in Biological Sciences from Nottingham Unviersity before moving to the University of Sussex for postgraduate work in Experimental Psychology. She was Project Co-ordinator on the EPSRC funded 'Bridging the Gaps: Nanoinfobio' project at Manchester Metropolitan Univeristy, designed to foster interdisciplinary research on the boundaries of biology, computer science and nanoscience. She has presented 'NanoInfoBio' at public community events such as 'Girl Geek Dinners' in Manchester, and has been involved in several high-profile events for the Manchester Science Festival, including, in 2011, delivering a series of events with the 'Monsters, Microbiology and Maths' team using zombies as a tool for engaging the public with disease, epidemiology and mathematical modelling. She is currently working at the University of Brighton supporting the management of European Commission Framework 7 Programme grants. (Case study 8.3)

Dr David Lewis is a Senior Lecturer in Neuroscience & Scientific Ethics within the Faculty of Biological Sciences at the University of Leeds where, in addition to running an active research laboratory interested in the brain's control of the gastrointestinal system, one of his principal responsibilities is the teaching of ethics to both undergraduate and postgraduate students across the Faculty. He also has significant involvement in outreach activities, regularly going into schools to discuss ethical issues in science with young people. (Case study 3.4)

Dr Niamh Ní Bhriain swapped postdoctoral for postnatal responsibilities when her first child was born and is now trying to enthuse and recruit the next generation of scientists! Her 'Aspects of Biology' course grew from her absolute conviction that schoolchildren are perfectly capable of grasping challenging scientific concepts if they are presented in an engaging and age-appropriate way. The course was developed with the help of the pupils and staff of St Attracta's Senior National School, Meadowbrook. The microbiology and microscopy components of the course are dependent on the goodwill and cooperation of Mr Gerard Dowd and his technical staff in the Moyne Institute, Trinity College Dublin and the financial support provided from the 'outreach' component of grant 07/IN.1/B918 from Science Foundation Ireland to Professor Charles J. Dorman. (Case study 10.1)

Professor Anne Osbourn is Associate Research Director of the John Innes Centre, Norwich. Her research focuses on plant natural products – function, synthesis and metabolic diversification. She is an author of over 100 peer-reviewed scientific publications and recently co-edited a comprehensive textbook on plant-derived natural products (Lanzotti, V. & Osbourn, A. (2009) *Plant-Derived Natural Products – Synthesis, Function And Application*. Springer, New York, USA). She has also developed and co-ordinates the Science, Art and Writing (SAW) initiative, a cross-curricular science education programme for schools (www.sawtrust.org). (Case study 10.6)

James Piercy has a degree in Chemistry and a MSc in Science Communication. He has been involved in writing, producing and delivering science shows, workshops and dialogue events for wide ranging audiences since 1995. He has been awarded the LAMDA Gold Medal in Public Speaking with Distinction, and was Chair of the British Interactive Group (BIG). Before joining *science made simple* James was Director of Inspire Discovery Centre, a small hands-on science centre in Norwich, where he developed the outreach and educational programmes. James has appeared on television numerous times to present scientific ideas and demonstrations, notably on *The Investigators* for Channel 4, which was nominated for a Children's BAFTA. Recent projects have involved the development of shows for National Museums Scotland, Butlins, and The Herschel Space Observatory. (Case study 5.1)

Dr Michael Pocock is an ecologist at the NERC Centre for Ecology and Hydrology, having previously held a NERC Fellowship at the University of

Bristol. He research is broadly to understand how 'we rely on nature and nature relies on us'. He is a co-founder of 'Conker Tree Science' and is developing more citizen science projects with *The Observer* newspaper, Open Farm Sunday and the Environment Agency. He is a member of the RCUK Public Engagement with Research Advisory Panel. Enthusiastically promoting public engagement with science through the media, speaking at conferences and training students, he still also participates in hands-on engagement in schools and at public events. (Case study 8.4)

Dr Jenni Rant is a scientist from the John Innes Centre, Norwich, interested in plant pathology and metabolic biology. She is a keen science communicator, working with the SAW Trust, a science education charity, bringing together science, art and writing to explore scientific themes in schools using a cross-disciplinary approach (www.sawtrust.org). (Case study 10.6)

Dr Dee Rawsthorne is Outreach Coordinator for the Norwich BioScience Institutes responsible for the schools and public programmes for the John Innes Centre and the Institute of Food Research. Dee is also founder and coordinator of Science Outreach in Norfolk and organises the annual Science in Norwich Day. 'Blooming Snapdragons' was an event held during the John Innes Centenary Celebrations funded by the John Innes Centre from materials uncovered in the John Innes Foundation archives by Dr Sarah Wilmot. Play production was by Sue Mayo, performed by Liz Rothschild and Syreeta Kumar, stage management by Tristan Bunn with additional support provided by the John Innes Centre Communications Team. (Case study 7.6)

Professor Dudley Shallcross is the Professor of Atmospheric Chemistry at the School of Chemistry, University of Bristol, and is a course coordinator for two modules within the MSc in Science & Education programme. He was appointed Schools Liaison Officer for Chemistry in 2000, later to become Bristol ChemLabS Outreach Director (until 2010).

In 2004, Dudley was awarded a National Teaching Fellowship by the Higher Education Academy (HEA) under the Rising Star category, for his excellence and innovation in teaching. In 2005, Dudley won the Faculty of Science Teaching Prize and the Higher Education Teaching Award from the RSC. In 2006, he was the first ever recipient of the SCI (Society of Chemical Industries) International Chemical Education Award. In 2007, he received the University of Bristol's Engagement Award for the Faculty of Science. In 2008, he was awarded the RSC Tertiary Education Award in recognition for his contributions to the promotion of the chemical sciences through outreach which have impacted on all ages both in the UK and overseas and, in particular, for work which addresses the school–university transition to include those with physical disability. In 2009, Dudley Shallcross received the Royal Meteorological Society's Michael Hunt Award which is given biennially for excellence in increasing the understanding of meteorology or its applied disciplines among members of the general public.

Most recently, in 2010, Dudley was appointed as the new Director of the AstraZeneca Science Teaching Trust (AZSTT) which promotes Science in Key

Stages 1–3. Dudley is also a member of the steering committee for the Science Learning Centre South West (SLCSW) University of Bristol subgroup. (Box 3.3 and Case study 9.1)

Dr Kenneth Skeldon is Head of Public Engagement with Research at the University of Aberdeen. He has held personal Research Fellowships from the Royal Society of Edinburgh and STFC and more recently, was the recipient of a NESTA Fellowship for Public Engagement. He has created and produced award-winning bespoke exhibits, outreach and education resources for science centres and museums around the world. In Aberdeen, he founded and currently coordinates one of the UK's biggest Cafe science programmes. He also played a leading role in securing the British Science Festival back to the city in 2012, the first time to return to Scotland for over a decade, and to Aberdeen for some 50 years. (Case study 7.7)

Dr Phil Smith MBE trained as a plant pathologist at the John Innes Centre where he worked in research for over 10 years. During this time, he began a long-term partnership with a primary school teacher, Mrs. Maxine Woods (through the Teacher Scientist Network). Their partnership, which continues to thrive, has worked at several schools across Norfolk and Essex. They have been successful recipients of both National Science Week awards and the Royal Society partnership grants for a range of primary school projects: 'Science in the Home' (2001); 'Seed to Sandwich – where does your food come from?' (2004); 'The Mobile Microbe Roadshow (MMR!)' (2005). Phil now runs the Teacher Scientist Network, based at the John Innes Centre, on the Norwich Research Park. He has been a reviewer of BBSRC Public Engagement awards and RCUK's NSEW awards and was fortunate enough to be awarded an MBE in the birthday honours list of 2008, for 'services to science education'. (Case studies 3.3 and 10.8)

Dr Nicola Stanley-Wall is a Lecturer in the College of Life Sciences at the University of Dundee. She is the recipient of the Brian Cox Senior Investigator Prize for Public Engagement awarded by the College of Life Sciences (2010), the Society for General Microbiology Outreach Prize (2011) and the Royal Society for Edinburgh Beltane Prize for Public Engagement (2012). 'Magnificent Microbes' was supported by staff and students from the Division of Molecular Microbiology at the University of Dundee and enjoyed financial support from many sources including the Society for General Microbiology and the Society for Plant Pathology. (Case study 6.2, Box 6.2)

Dr Elizabeth Stevenson is the Public Engagement Manager and a Teaching Fellow in the College of Science and Engineering at the University of Edinburgh. She is the recipient of the Royal Society of Chemistry Award for the Promotion of Chemistry. The event 'In Your Element' was supported by the National Museum of Scotland, the University of Edinburgh and Craft Reactor and was one event in a full programme of activities to celebrate International Year of Chemistry, 2011. (Case study 7.3)

Professor Joanna Verran is Professor of Microbiology in the School of Healthcare Science at Manchester Metropolitan University. She is a National Teaching Fellow, and recipient of the Mike Pitillo Award for Biomedical Science education, the Society for Applied Microbiology Communication Award and the Society for General Microbiology Peter Wildy Award for Microbiology Education. She leads public engagement activities in the Faculty of Science and Engineering at MMU, and encourages the participation of her postgraduate and undergraduate students. Jo has developed a variety of events for a range of audiences, focusing on microbiology, and has enjoyed support from the Society for General Microbiology, the Society for Applied Microbiology, the Manchester Beacon for Public Engagement and other sources. (Case studies 6.3, 7.8 and 8.3)

Dr Robert D. Wells is the Welch and Regents Professor Emeritus at the Institute of Biosciences and Technology in the Texas Medical Center, Houston, USA. His research career has focused on biochemical studies of DNA structure and DNA metabolism. Dr Wells' postdoctoral studies began in the laboratory of Dr H. Gobind Khorana, as part of the team that solved the genetic code, for which Khorana shared the Nobel Prize in Physiology or Medicine in 1968. Dr Wells has held academic and research posts at the University of Wisconsin-Madison, the University of Alabama at Birmingham, and Texas A&M University. He was Chairman of the Department of Biochemistry in the Schools of Medicine and Dentistry at the University of Alabama at Birmingham and Founding Director of the Albert B. Alkek Institute of Biosciences and Technology in Houston whilst also serving as the Head of the Department of Biochemistry and Biophysics, Texas A&M University in College Station, Texas. Dr Wells has acted as an advisor to the White House on healthcare reform and he served on the Scientific Advisory Council of the National Institutes of Health, National Institute of Environmental Health Sciences. Dr Wells has also served as the President of the American Society for Biochemistry and Molecular Biology and the Federation of American Societies for Experimental Biology. (Case study 7.9)

Dr Michael Wormstone is a Senior Lecturer in the School of Biological Sciences at the University of East Anglia. His research interests include the use of human tissue to study human eye disease and the wound-healing events that follow cataract surgery, which lead to a secondary visual loss. The Fight for Sight charity has supported research into blindness and eye disease for more than 40 years. Michael has worked with this charity to successfully fundraise for their research programmes. (Case studies 3.2 and 6.1)

Julie Worrall FRSA joined UEA in 2005. She instigated and now coordinates UEA's Annual Community Engagement Survey and in 2007 she co-authored UEA's successful bid to become a Beacon for Public Engagement. Julie is a member of UEA's Engagement Executive and is currently researching a doctorate on the culture of higher education in relation to public and community engagement, 'The academy and community: seeking authentic voices inside higher education'. (Case study 3.1)

Foreword

Science – and how it is done – has never been more important. Public interest and appetite remain high and science underpins almost every aspect of our daily life. Yet suspicion, prejudice, misunderstanding and ignorance about science still remain, even in some cases with scientists. Too few of our politicians have a science background and the same is true for many other professions, while something of the old C.P. Snow dichotomy still festers. A more scientific approach to countless policy matters is surely desirable and evidence-based decision-making certainly seems preferable to ignorance.

So perhaps it is not surprising that the top-down promotion of science-related public engagement activity is on the rise. Universities in particular are increasingly being asked to get their academics and other staff out of their ivory silos to talk to and listen to the various publics, individuals and organisations in the communities that they are embedded in. Talk of the *engaged university,* the *civic university* and *paths to impact* resonate around the corridors of academe, and the pressure is increasing on grant holders to engage more effectively with the public.

What does this mean in practice? The landscape is complex and effective science communication is not a straightforward matter. There is both scientific and sociological jargon to get to grips with. TLAs (Three Letter Acronyms!) abound and mathematicians find it odd that, for a biologist, multiplication means the same thing as division. And yet there is now enough case history and best practice on which to draw to know that some things work while others don't. This book is the repository for what has been learned – an all-in-one-volume synthesis of how to get involved. As the multiple ways through which science engagement can be mediated grow – cafés, exhibitions, books, articles, TV, radio, social media – this book enables you to get engaged effectively through their best-practice and evidence-based success stories. Such practical information, in a relatively new arena, has inevitably been scattered and is hard to access easily. So the success of our two authors in bringing together all you need to know into one handy volume – and in particular the many wide-ranging case studies – is a remarkably useful and pragmatic achievement.

We need more young people – students, technicians and young researchers – to be involved in science engagement activities in addition to the more secure academics who currently dominate. And we need more scientists from business, industry and management to get involved in addition to those from academia. There are many, well-recognised barriers to engagement activity, but no longer can it be said that being unsure of *how* to get engaged is a valid reason for inactivity. This book has removed that barrier at a stroke and I hope it becomes the turn-to manual of choice that it deserves to be.

Professor Keith Roberts OBE

Prologue

Practical science communication is an aspect of our job as scientists that we have both valued and enjoyed. However, it wasn't until we began to attend science communication conferences that we discovered the breadth and depth of research that surrounds and underpins the science communication field. We became much more aware of the impact that research in this area has had on the development of science communication in the UK and beyond. This realisation allowed us to develop our own personal understanding of both the context and the importance of scientists beginning to engage with the public. We realised that from a scientist's perspective, there are limited interactions between scientists and social scientists. We wanted this book to begin to bridge this gap by describing how science communication theory and activity has developed and changed. In addition we chose to describe some of the different issues, drivers and debates, familiar to many scientists, that have shaped the modern frameworks for communicating science to the public. Over recent years there have been many initiatives that have encouraged scientists to step out of the safety of the research environment and to face outwards towards the wider community. We have highlighted and described some of these initiatives, the impact they have on the day-to-day life of a working scientist and how they influence the way that scientists communicate with the public. We also provide insight and detail about how, why and where scientists fall into the wide spectrum of different communication activities.

This science communication book has been written from our perspective: two scientists who really enjoy and feel personally enriched by communicating science to the public. At the same time that we were developing our own communication skills, we recognised that there was a need for a modern practical guide that could offer sensible and straightforward advice to other scientists starting out on their own science communication career or seeking to diversify and widen their portfolio of communication experiences. We wanted to provide a book for scientists that had a familiar format. Initially we had the idea of developing a laboratory type manual for science communication, but we soon realised that we wanted to produce something that was more comprehensive. The result is a book that is intended to provide useful hints and tips to help you start to undertake science communication activities and events. It also gives examples of pitfalls that should be avoided and points you towards additional reading material to allow you to develop your own background knowledge and understanding of science communication. This book also provides scientists with the essential theories and models that underpin the role of a science communication practitioner. However, it is not intended to provide a detailed discourse of science communication as a discipline. There are already books written by experts that provide this type of information and knowledge and we have signposted them within our text.

We have used a series of case studies provided by colleagues who are scientists, students training to be scientists, scientists who have trained as teachers and also scientists who have established themselves as professional communicators in the science communication arena. The case studies highlight the depth and breadth of activities being undertaken by scientists who are passionate about communicating science to a wider audience. An unexpected outcome of the book, on a personal and professional level, was that it confirmed our belief that there is a whole army of scientists producing innovative and exciting science communication projects. We recognise that we have had an opportunity to highlight just a few of these initiatives but they clearly demonstrate the wealth of good practice and expertise that we wanted to use as an inspirational resource for peers and colleagues.

Acknowledgements

This book is built around the case studies that have been supplied by our colleagues. We are grateful for their generosity and support and allowing us to share their inspirational work with the wider world. As well as the authors of the case studies, we thank Dr Richard Bowater, Dr Stephen Ashworth, Dr Ian Gibson, Professor Robert Watson and Professor Joanna Verran for their generous help and support in discussing chapter material, commenting on draft chapters and providing useful suggestions and resources. A special thank you has to go to Alistair McWalter for being incredibly generous with his time, his graphic artist talents and his ability to listen to our abstract concepts and ideas and turn them into a visual reality. We love the images and the illustrations that he has produced for the book including the front cover – thank you so much (Figures 2.1, 6.1, 6.2, 6.3, 9.1 and 10.1). We must also thank Dr David Waterhouse who created the majority of the artwork that beautifully illustrates the top of each case study. We also thank our commissioning editors Liz Renwick and Lucy Sayer for all their help, encouragement and support with getting this project off the ground and keeping it on track, Fiona Seymour our project editor who has handled our uncompleted and completed manuscript with calmness and care, Jasmine Chang and Baljinder Kaur for helping us tidy up the final bits and pieces. We appreciated it hugely.

Finally we would like to thank our families, and in particular Richard, Jon, Charlotte, Ellie, Alex and George, for putting up with our distraction over the past two years – we couldn't have done it without your support.

CHAPTER ONE

A Guide to Science Communication

One can hardly believe that modern science is almost included within the present century. All before then, except astronomy, was more or less speculation. Scientists had only been playing, like children, in the vestibule of the great Temple. It may be that we ourselves have not advanced far within the precincts at least, those who study these subjects 100 years hence may think so.

—Dr J.E. Taylor (*The Playtime Naturalist*, 1889)

1.1 Introduction

The issue of science communication has risen globally in its importance in recent years, not least due to a belief that science and technology are the basis of a knowledge economy. Science and technology are an integral part of our culture and heavily influence our everyday lives. The knowledge and applications produced from science are powerful and exciting and it's reasonable to suggest that the public should know about these new advances because of the questions they raise for our society. Public money also pays for a substantial amount of research undertaken in many universities and government institutes, although we must also acknowledge that the ratio of private to public funding for scientific research and development has dramatically increased over the past 50 years (OECD, 2004). However, regardless of how research is funded, its impacts must be communicated to citizens, even if the strategies used and the motivations are different for research and development funded by private as opposed to public money (Bauer, 2010).

Communication by scientists to the public is not a new phenomenon. Even before the term scientist was first used (not coined until 1834; Hannam, 2011), Humphrey Davy and Michael Faraday were engaged in the popularisation of science and Joseph Priestly was even encouraging active science experimentation by the public (Broks, 2006). Twenty-first century examples of talented communicators include among others, the physicist Brian Cox and anatomist Alice Roberts, whose enthusiasm for and knowledge about their own subject and science in general has underpinned their willingness to communicate with the public.

Science Communication: A Practical Guide for Scientists, First Edition. Laura Bowater and Kay Yeoman.
© 2013 John Wiley & Sons, Ltd. Published 2013 by John Wiley & Sons, Ltd.

1.2 The influence of science societies, charities and organisations

1.2.1 Science societies

Science communication in the UK has been shaped by historical institutions such as the Royal Society, as they have commissioned influential reports that have described the relationship between science and society. The committees producing these reports have often been chaired by eminent and respected scientists and the reports have affected the way that science has been communicated to the public within the UK and across the world. The Royal Society was one of the first science societies to be established and has been in continuous existence for the longest. It was founded in 1660 by a group of well known individuals that included Robert Boyle, Robert Hooke and Christopher Wren. The Royal Society was granted a royal charter by Charles II in 1662 and the society maintained itself with dues from its members (McClellan and Dorn, 2006). The French established the Academie des Sciences in 1663, but it differed from the Royal Society in one key aspect, it was a government institution, with patronage from Louis XIV (Gribbin, 2002). Other countries also saw the value of a science society and by the end of the eighteenth century there were approximately 200 societies across Europe and North America (Fara, 2009). The Royal Society was not established to facilitate communication to a public audience, but it did begin the concept of the 'scientific paper' with the publication of the *Philosophical Transactions of the Royal Society* from 1666, enabling communication between individuals interested in science. This was published by Henry Oldenburg, first secretary to the Royal Society from his own private funds (Gribbin, 2002). Since then the phenomenon of the scientific paper has grown in importance. It can be equated to the 'unit of productivity' of science (McClellan and Dorn, 2006) and it forms a substantial part of the criteria used to judge scientists in the twenty-first century. This is epitomised by the 'scientific paper' being used as a major criteria within the UK's Research Excellence Framework (previously Research Assessment Exercise); a process used to judge research output from universities in order to determine the level of block governmental research funding (HEFCE, 2011).

Over a hundred years after the establishment of the Royal Society, the Royal Institution (RI) was founded in 1779 as a research laboratory. It also had a role in public education, specifically to educate young workmen (RIGB, 2011). The RI was intended to be different from the Royal Society; the science was meant to be sustainable, although in reality its activities were maintained by annual subscriptions. One of the original goals of the RI was to try to apply the latest scientific techniques to improve agricultural practices and reduce the level of poverty (Berman, 1978). This philanthropic goal was soon superseded by the use of science for entrepreneurial and professional purposes to improve and advance society (Berman, 1978; Broks, 2006). Notable scientific advances by the RI include the discovery of new elements calcium, magnesium, boron and barium by Humphrey Davy, confirmation of the structure of benzene in 1925 by Kathleen Lonsdale and the structure of the enzyme lysozyme in

1965 by David Phillips. The RI also popularised science and developed the public demonstration lectures first started by Humphrey Davy in 1802. Skilled workers would attend these lectures to gain knowledge they could use to advance their careers. The format of these demonstration lectures still exists today; the RI Christmas Lectures, first started by Michael Faraday in 1825, polled 0.86 million viewers when aired on BBC 4 in 2011 (Barb, 2011). These modern lectures have covered a wide range of scientific disciplines, and have been delivered by experts in their field.

Throughout the eighteenth and nineteenth centuries, subject-specific societies began to emerge in England, notably including:

- The Linnaean Society (1788);
- The Geological Society of London (1807);
- The Zoological Society of London (1826);
- The Royal Astronomical Society (1831);
- The Chemical Society of London (1841).

These societies began to publish their own subject-specific journals. The process of peer review materialised as one of the secretaries of the Geological Society, a certain Charles Darwin, developed a system of sending papers out for scrutiny prior to publication. This process is now standard practice among academic journals across all disciplines. Science became a common amateur pursuit in the nineteenth century and in America it became common for even small towns to have a 'science society'. Similarly in the UK, towns and cities were also hubs of amateur scientific activity. Case study 1.1 'The Playtime Naturalist', highlights such a society and pays tribute to its founder Dr John Ellor Taylor.

Case Study 1.1

The Playtime Naturalist
Kay Yeoman

By knowledge, by humour, by rare and excellent gifts of speech, he opened the eyes of many to the order, variety and beauty of nature.

—Memorial to Dr J.E. Taylor

While doing some reading on the history of science, I came across a reference to a British Science Association meeting held in Norwich in 1868. At this time, the president of the British Science Association was Joseph Hooker, the first Darwin supporter to hold this post.

The Darwinians minus Darwin assembled at Norwich for the Association jamboree. From far and wide they came, a rallying call of evolutionary pilgrims of every persuasion.

—Desmond and Moore (1991)

At that meeting, Thomas Huxley, staunch Darwin supporter, gave an address at the Norwich Drill Hall to a group of working men. The lecture was entitled 'On a Piece of Chalk' and described what could be learnt about the geological history of the Earth and the passage of time by examining not only the structure of the chalk, but also the fossil remains of plants and animals that lay within it. The lecture provided a vivid description of animal and plant life at Cromer on the Norfolk coast.

Thus there is a writing upon the wall of cliffs at Cromer, and whoso runs may read it. It tells us, with an authority which cannot be impeached, that the ancient sea bed of the chalk sea was raised up, and remained dry land, until it was covered with forest, stocked with the great game the spoils of which have rejoiced your geologists. How long it remained in that condition cannot be said; but, "the whirligig of time brought its revenges" in those days as in these. That dry land, with the bones and teeth of generations of long-lived elephants, hidden away among the gnarled roots and dry leaves of its ancient trees, sank gradually to the bottom of the icy sea, which covered it with huge masses of drift and boulder clay. Sea-beasts, such as the walrus now restricted to the extreme north, paddled about where birds had twittered among the topmost twigs of the fir-trees. How long this state of things endured we know not, but at length it came to an end. The upheaved glacial mud hardened into the soil of modern Norfolk. Forests grew once more, the wolf and the beaver replaced the reindeer and the elephant; and at length what we call the history of England dawned.

Joseph Hooker the president of the British Science Association had strong links to Norwich; his grandfather was a Norwich merchant and his father, Sir William Jackson Hooker, was born in Norwich in 1785. His father was a keen botanist who began the herbarium which eventually became the herbarium at Kew Gardens. This local link to these eminent past scientists caught my imagination and being interested in people, history and science, I began to delve into the science of Victorian Norwich. I was delighted to find several references to the 'Norwich Science Gossip Club', the records for which still exist today. With a mounting level of excitement (equal to unveiling a perfect Southern blot) I set off for Norfolk County Hall and asked to view the records of the Science Gossip Club. I was astonished to find a beautifully kept set of records detailing the activity of the club that started in 1870 and ended just after the Second World War. While reading and making notes on these records, I found a connection to a man, for whom I developed an enormous admiration, Dr John Ellor Taylor, naturalist, founder of both the Norwich and Ipswich Science Gossip Clubs, editor of the *Science Gossip Magazine*, prolific author, curator of Ipswich Museum and consummate science communicator.

J.E. Taylor was the son of a Lancashire cotton-factory foreman; he had a rudimentary education, but he was motivated and he learnt through private study. He was employed in the railway works at Crewe, but he attended evening classes at the Manchester Mechanics Institute. He became fascinated with geology and published his first work *Geological Essays* in which he described the geology of Manchester. He secured a position as a subeditor at the *Norwich Mercury* in 1863 and he devoted his leisure time to science and in 1864 he co-founded the Norwich Geological Society with John Gunn. I know he attended the 1868 British Science Association meeting in Norwich, as his name appears on a list of contributors (records kept at the Dana Centre in London). As well as being a talented scientist, J.E. Taylor was a natural communicator and gave many popular lectures. He published many books and one of them, *The Playtime Naturalist*, describes a fictional natural history club for boys at Mugby School. It's a beautifully illustrated book, full of hints and tips about collecting and classifying plant and animal species, all intended for a lay audience.

Taylor founded the Norwich Science Gossip Club in 1870 with the following objective:

The object of this society is the promotion among its members of a spirit of enquiry and investigation of scientific and literary knowledge by means of fortnightly papers on such subjects, and occasional excursions for open air study.

In my mind this objective is full of ideas surrounding self-improvement, at which Taylor himself was a master. Many members of the Gossip Club were listed in the records. Using the 1871 census I established that they came from a variety of different professions. The members presented papers, discussed new ideas and displayed specimens. Mr Manning P. Squirrell, a corn merchant gave a talk entitled 'Gleanings about Ostriches and Elephants' and Mr Thomas Bayfield, an ironmonger addressed the club on the subject of the *Lamellibranchiates*. In one particular meeting, Mr C.W Ewing displayed the fossilised remains of a tortoise (*Emyslutaria*) he had found at Mundesley, on the Norfolk coast. This specimen was later described by Dr E.T. Newton in the *Geological Magazine* of 1897. You can still see the specimen on display at Cromer Museum.

Taylor became curator of the Ipswich Museum in 1872, only nine years after his arrival in Norfolk, and he also took over the editorship of Hardwicke's *Science Gossip Magazine*. The previous editor and founder had been Mordecai Cubitt Cooke, one of Britain's first mycologists, also from Norfolk. J.E. Taylor was an immensely curious man, but at times this led him into trouble. He contracted smallpox whilst investigating a severe outbreak of the disease in Norwich and he was scarred for life. A contemporary of Charles Darwin, he greatly admired Darwin's work and on the 25th June 1878 wrote a letter to him presenting him with a copy of one of his books, *Flowers, their Origins, Shapes, Perfumes and Colours*.

> *Dear Sir,*
> *I have taken the liberty of forwarding to you for your acceptance a copy of my new book on "Flowers, their Origins, Shapes, Perfumes and Colours" in which I have freely referred to your various invaluable books. Please accept the volume as a sincere and humble tribute of respects from one of your most ardent students I have the honour to be, dear Sir,*
> *Yours sincerely*
> *J.E. Taylor*
>
> —Letter from John Ellor Taylor to Charles Darwin 1878. By permission of the Syndics of Cambridge University Library

Like Darwin, Taylor was a brilliant observer and a meticulous keeper of notes and records, but unlike Darwin, he did not come from a privileged background. Nevertheless, he managed to gain a doctorate and a career in science, which was an incredible achievement considering his lack of formal education.

Ill health forced him to leave Ipswich Museum in 1893 and sadly he died bankrupt in 1895. He was survived by his wife and four daughters.

I think Dr J.E. Taylor would have revelled in today's science and embraced the means for its promotion through the internet. I am convinced that were he alive today he would have been involved in citizen science projects such as iSpot and he would have produced the most amazing blog full of his ideas, observations and tips for the budding amateur naturalist!

It is fair to say that Norwich is not unique in having its own science society and with a little bit of digging there is a good chance that you could unearth similar science clubs and societies in your local area.

In 1830 the Cambridge mathematics professor, Charles Babbage, published his work on *Reflections on the Decline of Science in England and Some of its Causes*. This is still an interesting publication and many of his observations and reflections still apply today. Babbage was concerned that British science was lagging behind the rest of the world because of a lack of public interest. He wanted to see the establishment of a modern profession composed of paid and properly funded researchers. In response to this publication, the British Association for the Advancement of Science (formerly the BAAS, then the BA, now the British Science Association) was founded in 1831. It had a specific remit: to facilitate communication not only to the public but also to government. A similar organisation, the Association of German Researchers had already been in existence for nine years. The first meeting of the British Science

Association was held in York in 1831 and since then it has met annually in different provincial cities, but always avoiding London. Several years later in 1848, the American Association for the Advancement of Science (AAAS) was established with a mission 'To advance science, engineering, and innovation throughout the world for the benefit of all people'. The AAAS still has a strong commitment to public communication (Daley, 2000). It was originally modelled on the British Science Association, but has developed into a well-funded, highly influential society that also publishes the eminent weekly journal *Science*.

More recently, the British Science Association established the British Association of Young Scientists (BAYS), which at its height had 8000 individual members (Briggs, 2003). BAYS days became an established feature of the BA calendar, but it was replaced by the National Science and Engineering Week in 1994 which is still held during March every year. The concept of a science day or week is also seen in other countries, for example Australia, Denmark and Norway have a National Science Week (Riise, 2010), Sweden and Poland run science festivals, and other science communication events occur in Asia and Africa. These events are funded through different organisations and can be on a local, regional or national scale. The US has Public Science Day, founded by the AAAS, which also coordinates Project 2061 started in 1985, after the publication of the 'Science for all Americans' report. Project 2061 is a long-term, ambitious programme aimed at helping all Americans to become literate in science, mathematics and technology. Initiatives aimed at improving science education have included benchmarking for scientific literacy, which provides specific learning goals used to inform curriculum design in schools (Project 2061, 2011).

1.2.2 Charitable trusts

Charitable trusts and Institutes have also been founded by companies. Henry Wellcome with his partner Silas Burroughs established the pharmaceutical company Burroughs Wellcome and Company in 1880. This company introduced the concept of selling medicine in tablet form in England, and it also established several research laboratories. The Wellcome Trust was set up at the behest of Henry Wellcome in 1936 and it has become the UK's largest charity focused on improving human and animal health. It is also the largest non-profit funder of research in Europe and in 2007–8 gave away £620 million to fund research in and outside the UK (Stephan, 2010). The Wellcome Trust commissions reports and funds a substantial amount of work in the area of public engagement, aimed at raising the awareness of the medical, ethical and social implications of biomedical science. It has several funding streams for engagement between scientists and the public, including Peoples awards, Broadcast awards and larger Science and Society awards.

The Salters' Institute was founded in 1918 by the UK-based Salters' Company, with the initial aim of getting young people back into their chemistry studies after the Second World War. It now has a major role in supporting chemistry education in schools.

In the US, the Rockefeller Foundation established in 1913 promotes the well-being of humanity around the world (Bauer, 2010). More recently, the Bill and Melinda Gates Foundation, started in 1994, provides funding and resources to support people to lead healthy and productive lives. In the developing world this has focused on issues surrounding health and the foundation has supported work into fighting and preventing diseases such as malaria, HIV/AIDS and tuberculosis (Gates Foundation, 2011).

1.2.3 Organisations

There were also less formal organisations that influenced change in scientific culture. In 1864, eight eminent men from the world of science, including Thomas Huxley, John Tyndall and Joseph Hooker, met for dinner at St George's Hotel in Albemarle Street, in central London. Over dinner they founded the X-Club, a club that despite having no specific aim or rules, managed to have a significant influence over the professionalisation of Victorian science (Barton, 1998). Between them, at some point, members of this club held the presidency of the Royal Society, The Royal Institution and the British Science Association. The X-club seemed to function as a mentoring group for its members. It was relatively short lived and dissolved after their deaths. One of the lasting impacts of the X-Club was the support given to Tyndall and Huxley to establish the journal *Nature*, recognised today as a premier place for scientific publication.

1.3 Modern societies and organisations

Since the turn of the twenty-first century, several organisations and societies have emerged with science communication at their core. The European Science Events Association (EUSCEA) was founded in 2001 as a non-profit scientific society with a membership drawn from across Europe. The aims of EUSCEA are:

- to share good communication practice;
- to provide a forum for marketing communication events;
- to enable people to collaborate;
- to enable participation in EU funded projects.

Another network was established in the US in 2006, the Coalition on the Public Understanding of Science (COPUS). This organisation grew out of a concern about the state of science in the US, and unites universities, science societies, media, science educators, science advocacy groups, business and industry to work towards a better public understanding of science (COPUS, 2011). Throughout the world there are many other science communication societies and some of these are detailed in Table 1.1.

1.4 Science communication as a discipline

As a discipline, science communication faces several challenges and one of the biggest is its multidisciplinary nature; it can encompass communication

Table 1.1 Science communication societies.

Society	Description	Website
Indian Science Communication Society	Non-governmental organisation committed to bringing science to the public	http://www.iscos.org/
Australian Science Communicators	Supporting making science accessible	http://www.asc.asn.au/
Science Communicators association of New Zealand	Dedicated to improving science communication	http://www.scanz.co.nz/
South African Agency for Science and Technology Advancement	Aims to advance public awareness and appreciation of science	http://www.saasta.ac.za/
Danish Science Communicators	Non-profit organisation devoted to increasing public awareness and understanding of science and technology	http://www.formidling.dk/sw15156.asp
Coalition on the Public Understanding of Science	Network of organisations dedicated to improving public understanding of science	http://www.copusproject.org/

studies, sociology, education, philosophy, history, political science, ethics and, of course, science itself. Science communication is continuing to develop and it is important that scientists appreciate that it is emerging as an academic field of study in its own right with:

- theories and models;
- peer-reviewed journals that publish research and also practical case studies which attempt to bridge the gap between theory and practice (see Table 1.2);
- international conferences, e.g. Public Communication of Science and Technology (PCST) held biannually;
- university courses at both undergraduate and postgraduate levels (Yeoman *et al.*, 2011; Mulder, 2008).
- science communication societies, a few of which are detailed in Table 1.1.

The case studies presented in this book cover the practical side of science communication, the majority being designed and delivered by scientists. As Gregory and Miller (1998) point out, practical science communication is often done by scientists, but the reflection on its worth and effectiveness is most often undertaken by social scientists. The result can be a tension and a lack of common language between these two fields. There is an argument that *practical* science communication is separate from research on the *process* of science communication and more would be gained by practitioners learning about good narrative, communication and design (Davis, 2010). As practitioners ourselves, we have some sympathy with this view, but we feel that the evidence-based practice from investigating the process cannot be entirely ignored. As an introduction to this evidence base, in the rest of this chapter

Table 1.2 Science communication journals.

Journal	Access	Website
Public Understanding of Science	Subscription	http://pus.sagepub.com/
Journal of Science Communication	Subscription	http://scx.sagepub.com/
Journal of Science Education, Part B: Communication and Public Engagement (IJSE (B))	Subscription	http://www.tandf.co.uk/journals /RSED
Journal of Science Communication (JCOM)	Open	http://jcom.sissa.it/archive/06/02/
Indian Journal of Science Communication	Open	http://www.iscos.org/ijsc.htm
Journal of Higher Education Outreach and Engagement	Open	http://openjournals.libs.uga.edu /index.php/jheoe/index

we cover the phases of science communication which have been marked by key reports, underpinned by research and have shaped the approach to science communication in the UK and other countries. A first step is to begin to understand the language used by social scientists researching science communication. To help with this, Table 1.3 provides definitions of terms that are often found in the science communication literature.

These definitions are surprisingly hard to pin down, and this Table includes our own more simplified suggestions. You will find that there are alternative, more complex definitions described in the social science literature (Burns *et al.*, 2003, NCCPE 2011). Terminology differs within and also between countries for example 'outreach' is often used interchangeably with 'public engagement' in the UK. Outreach is often a term used in UK universities to describe their engagement with primary and secondary schools. Universities often have outreach offices, which employ people to specifically engage with schools. These offices tend to have a widening participation and a more general university admissions agenda. Many European nations use 'scientific culture' to mean Public Understanding of Science (PUS)(Burns *et al.*, 2003) but in the US they use scientific literacy to describe this. The US also uses the term the Public Understanding of Science and Technology (PUST) and the Public Appreciation of Science (PAS) (Daley, 2000), but PAS is also used to mean Public Awareness of Science (Stocklmayer, 2002) and can be abbreviated to PAWS. Public Engagement with Science and Technology (PEST) is also used and Holliman and Jensen (2009) also suggest the term SCOPE for Science Outreach and Public Engagement.

1.5   Phases of science communication

Science communication has gone through three phases: scientific literacy, public understanding of science (PUS) and public engagement with science and technology (PEST). There was considerable overlap between these phases and many of the terms are still used interchangeably (Section 1.4). Each

Table 1.3 Definition of terms often used in the science communication literature.

Term	Definition	Reference
Science communication	The popularisation of science	Davis, 2010
Public	Every person in society	Burns *et al.*, 2003
Lay public	People, including other scientists who are non-expert in a particular field	Burns *et al.*, 2003
Scientific literacy	Knowledge and understanding of science facts and processes	This book
Public engagement	Communication and discussion with a public audience	This book
Outreach	A meaningful and mutually beneficial collaboration with partners in education, business, public and social service	Abridged from Ray 1999
Public understanding of science	A knowledge of science and how it applies to everyday life	This book
Communication	Social interaction through symbols and message systems	Gerbner, 1966
Deficit model	Where the public is seen as lacking knowledge and understanding, which can only be remedied by imparting facts	This book
Dialogue model	Scientists and public in conversation	This book
Upstream engagement	Discussion takes place with the public before any new scientific developments and technology become reality	This book
Citizen science	Lay public participation in research	This book

of these phases had important reports and surveys associated with them, which often spurred a change of strategy for public science communication. Figure 1.1 gives a flow diagram of the models, movements and reports which have influenced science communication phases in the UK.

Science communication has also developed differently in different countries. For example, the US still maintains a strong scientific literacy and educational approach (Gregory and Miller, 1998; Miller, 2011).

1.5.1 Scientific literacy

The first phase of science communication was tied to ideas surrounding scientific literacy. Jon D. Miller (1983) identified the four components of scientific literacy as:
- a knowledge of basic text book facts of science;
- an understanding of scientific methods, e.g. experimental design;
- an appreciation of the positive outcomes of science and technology;
- a rejection of superstitious beliefs (Gregory and Miller, 1998).

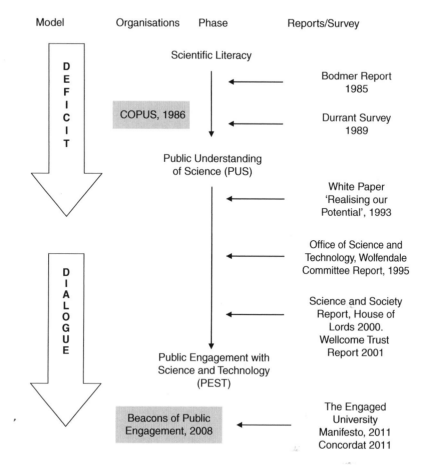

Figure 1.1 Models, organisations, phases and reports associated with the development of science communication in the UK.

Rightly or wrongly, scientific literacy suggests that the public should have a stock of scientific knowledge akin to literacy and numeracy (Bauer *et al.*, 2007). The implication is that this stock of knowledge can be tested, in the same way as you test for literacy and numeracy. The testing of public 'science knowledge' has led to many reports of an 'ignorant' public and highlights a knowledge 'deficit' which scientists need to fill with facts. The work commonly cited in this area was undertaken by Durrant *et al.* (1989). Their paper published in *Nature* in 1989 showed that citizens from Britain and the US were lacking knowledge and understanding of science, e.g. only 34% of the British public knew that the Earth went round the Sun once a year and only 17% spontaneously referred to experimentation and/or theory testing when asked what it means to study something scientifically. Similar studies have also been done more recently in the US by the Science Board and in Europe by the special Eurobarometer science and technology survey in 2005. The 13 questions asked in these surveys are given in Box 1.1 and are similar to the ones asked by Durrant *et al.* in 1989. In Europe, the level of scientific literacy has increased since 1992 and recent results indicate that Europeans actually have a fairly good, although not outstanding, knowledge of science. The average percentage of correct answers was 66%, although it must be noted that there was some degree of variation. Sweden had the highest rate of correct answers at 79% and Turkey, a non-member EU state, had the lowest at 44%. Despite this more positive data, there are still concerns about the level of scientific literacy, even though Russell (2010) has pointed out that there is a problem in defining how much factual scientific knowledge is needed to be deemed literate. For example, while I have a detailed knowledge of certain aspects of molecular biology, I am sadly lacking knowledge in physics, with a superficial understanding at best. On this theme, the Australian National Centre for Public Awareness of Science ran workshops for scientists on public communication, where the idea of scientific literacy was explored. The scientists were asked to complete a section of the Durrant survey; 193 scientists have taken part and it has highlighted that many scientists were unsure of answers to questions not directly related to their discipline. In addition, there were no questions that all scientists answered correctly and they were critical of the questions (Rennie and Stocklmayer, 2003). Perhaps an important point is not what is known at any one time, because we can't know everything, but instead it is the motivation to look for and the skills required in accessing and analysing information when it's needed.

Box 1.1   Eurobarometer Quiz

Which statements are true and which are false?

The Sun goes round the Earth

The centre of the Earth is very hot

The oxygen we breathe comes from plants

Radioactive milk can be made safe by boiling it

Electrons are smaller than atoms

The continents on which we live have been moving for millions of years and will continue to move in the future

It is the mother's genes that decide whether the baby is a boy or girl

The earliest humans lived at the same time as dinosaurs

Antibiotics kill viruses as well as bacteria

Lasers work by focussing sound waves

All radioactivity is man-made

Human beings, as we know them today, developed from earlier species of animals

It takes one month for the Earth to go round the Sun

The most important aspect of scientific literacy was the educational agenda and this phase increased the efforts in science education that are still seen today. Presently in the UK, all children up to the age of 16 have compulsory science education. On the negative side, it suggested that an ignorant public is disqualified from participating in science policy decisions. The approach of trying to fill the 'knowledge gap' between scientists and the public by imparting factual information has become known as the 'deficit model' of communication. Science and learning in schools is explored in more detail in Chapters 9 and 10.

1.5.2 Public understanding of science

The second phase of science communication was PUS. In the mid 1980s concerns were raised over the public attitudes towards science similar to those raised by Charles Babbage in 1829. These concerns were marked by an influential report by the Royal Society produced by a committee chaired by Sir Walter Bodmer (currently principal at Hertford College Oxford and former director general of the Imperial Cancer research fund), which has become known as the Bodmer Report (Bodmer, 1985). The ramifications of this report in establishing the new paradigm of PUS across the world cannot be underestimated. This report has been highly cited within the science communication literature and is now regarded as a key publication when describing the 'deficit' model of science communication.

The Bodmer report led directly to the foundation of the Committee on the Public Understanding of Science (COPUS – not to be confused with the current COPUS organisation in the US), where the three major UK historic institutions came together, The Royal Society, The Royal Institution and the British Science Association. The UK COPUS aimed to interpret scientific advances and make them more accessible to non-scientists. Several schemes for science promotion were initiated; including a fund for speakers to talk to organisations, e.g. Women's Institute. They also funded an annual book prize. In addition, they had a direct role in bringing about the highly successful National Science and Engineering Week (NSEW), which still operates today (Bodmer, 2010). Indeed many of the case studies in this book came from a desire on the part of scientists to take part in this UK event.

The Bodmer Report has been much criticised in the literature for what it seemed to represent, i.e. a deficit of knowledge and understanding of science within the public. When the original report is examined, one of the main themes is on improved education within the formal school system.

> A proper science education at school must provide the ultimate basis for an adequate understanding of science.
>
> —*Bodmer (1985, p. 6)*

The National Curriculum introduced in 1989 ensured that science was a core subject from the ages of 5 to 16. The Bodmer Report also suggests that *quality* of choice is better when an understanding of the issues is improved.

> Better overall understanding of science would, in our view, significantly improve the quality of public decision making, not because the right decisions would then be made, but because decisions made in the light of an adequate understanding of the issues are likely to be better than decisions made in the absence of such understanding.
>
> —*Bodmer (1985, p. 9)*

We interpret this as 'it is fine to disagree with the science, but by being better informed, your choice is built on more secure foundations'. We consider the Bodmer Report to have been somewhat misrepresented in the science communication literature. There were many incredibly positive outcomes of the report. It encouraged scientists to get involved with the education process of science at all stages. COPUS enabled scientists to take science communication with the public seriously. It removed the stigma associated with the popularisation of science and it became a more mainstream activity (Bodmer, 2010). This change was partly because COPUS provided a funding stream for engagement projects. In 2002, COPUS disbanded in the UK as more organisations became involved with PUS, but the individual founding organisations have remained committed to providing funds for engagement. For example, The Royal Society funds Partnership grants with schools and scientists and Case study 10.5 by Adam Hart on the Bee Guardian Foundation (BGF), is an example of engagement first established through such a Partnership grant.

The public understanding of science was a key issue in the 1993 science and technology White Paper 'Realising our Potential' which clearly stated the importance of the understanding and application of science to wealth creation and quality of life (British Council, 2001). In 1995, the Wolfendale Committe in the UK (chaired by former Astronomer Royal, Sir Arnold Wolfendale), also concluded that scientists receiving public funding had a duty to engage citizens with their research (Pearson, 2001; Poliakoff and Webb, 2007). The recommendation of this committee was the inclusion of a statement in research grants on how the public should be informed about the findings from the funded scientific research. The restructuring of the research councils, as a result of this White Paper, made it explicit that PUS was part of their responsibility. At the present time, all research councils in the UK require scientists to write impact statements as part of their research proposals. These impact statements are examined in more detail in Chapter 2.

1.5.3 Problems with public understanding of science

The PUS phase was not without its problems. The Economic and Social Research Council (ESRC) established a programme of research to investigate the relationship between science and society (Lock, 2011). As research projects progressed and papers were published it became clear that social scientists were critical of the PUS movement as:

- all the knowledge and expertise lay with the scientists;
- it implied that more knowledge of science on the part of the public would bring about a greater appreciation of science (Gregory and Miller, 1998).

While scientific literacy was seen as a deficit of knowledge, PUS was a deficit of attitude. The crisis surrounding both bovine spongiform encephalopathy (BSE) and genetically modified (GM) food in the UK are often cited in the science communication literature as perfect examples of the failure of the deficit model. The BSE crisis identified a need to try and communicate the ideas of risk and also highlighted the presence of different publics, e.g. consumers, activists, government and farming communities, all of whom had their own knowledge and stance on the issues (Irwin, 2009). There was also a crisis of trust, as the link between BSE and variant Creutzfeldt–Jacob disease (vCJD) became apparent, despite earlier assurances from the government that there was no link. In the case of GM, the campaign to raise awareness and a positive public attitude towards the technology had a negative effect instead, as the public became more sceptical (Irwin, 2006). In the wake of other crises, such as BSE, the public simply did not trust the government to make the right decisions for them and in the UK there is still a moratorium on the commercial growing of GM crops. In a recent meta-analysis, Allum *et al.* (2008) showed that there is only a weak correlation between science knowledge and attitude and sometimes a negative correlation when associated with specific issues, such as GM food. What surprised many supporters of the PUS movement was that their success at increasing the level of scientific literacy ultimately lead to a more sceptical public. Although this was an unexpected outcome, Bauer (2010) suggests that this should not be viewed as a negative result but rather regarded as an asset as it represents a public that is more critically aware of issues.

Nisbet and Scheufele (2009) argue that ignorance of the facts is not the reason why there are conflict issues between science and society. This is an interesting point; citizens are influenced by their own experiences as well as a variety of cultural and religious views (Davies, 2009). This is addressed in the contextual model of science communication, put forward by Falk and Dierking in 2000. This model takes into account the knowledge and experiences that the lay public have built up over time within different contexts. Scientists shouldn't ignore lay knowledge. They should consider that these experiences could be pertinent to science and scientists can learn from them (Irwin, 2009). A classic example from the literature is the work of Brian Wynne (1992), who looked at the knowledge built up by Cumbrian hill-sheep farmers. This group of individuals had considerable experience and knowledge about hill-farming management, sheep behaviour and also fell ecology. Being close to Windscale/Sellafield nuclear power station, they also had experience of grazing sheep on contaminated grassland after the disaster at Windscale in 1957. Thus after the Chernobyl nuclear accident in 1986, and the fallout of radioactive caesium which occurred over Cumbria, these farmers had specialist knowledge which could and should have been immensely useful in determining a response to the crisis. However, scientists chose to ignore the experience of the hill farmers, which left the farmers feeling belittled and threatened.

1.5.4 Public engagement with science and technology

The third and current phase of science communication is PEST, also referred to as Science and Society. The House of Lord's *Science and Society* report which came out of a committee chaired by Lord Jenkins in 2000 stated that the PUS movement was arrogant and outdated and there was only a 'top-down' one-way communication from the science community to the public. PEST has less emphasis on the one-way dissemination of facts, and focuses instead on dialogue, or two-way engagement between the scientists and the public. Simply talking to the public about science is not sufficient. Instead scientists should listen to the public, enter into a conversation with them and record their views. This is essential in terms of public involvement in policy, as it allows democracy and increased trust and confidence in the regulation of science and the decisions that are subsequently taken by the government (Haste *et al.*, 2005).

The idea of dialogue isn't new. There are two good examples of dialogue occurring prior to the PEST movement, firstly in the 1970s and then again in the 1990s. The first example is the Genetic Manipulation Advisory Group (GMAG) established in 1976. This was a highly unusual government advisory committee as it included representatives of the 'public interest' (Bauer *et al.*, 1998). The second example was in 1994 when the Biotechnology and Biological Sciences Research Council (BBSRC) sponsored a UK National Consensus Conference on Plant Biotechnology (Trench, 2010). This was an example of a citizen jury, where a panel of 16 lay public volunteers set the agenda for the conference, chose the expert witnesses, conducted the questions and then delivered the verdict (NCBE, 2011).

1.5.5   Problems with the dialogue model

Examples of twenty-first century dialogue events include café scientifique, scenario workshops, deliberative opinion polls, citizen juries, people's panels and in the US, consensus conferences (Russell, 2010). On the surface these seem to be good examples of dialogue events, but closer scrutiny has revealed some problems with a dialogue-focused approach. In 2009, Sarah Davies examined informal public dialogue events at the Dana Centre in London (a purpose built centre, part of the Science Museum). These were panel events, where expert panel members spoke and then the public audience were able to comment and ask questions. What she discovered was that this format of comments, questions and responses, was not a simple dialogue event, instead these panel events had elements of both deficit and dialogue. This research indicates that a pure dialogue event is often difficult to achieve. In addition, it isn't clear how these examples of informal dialogue actually feed into government policy.

GM nation was an example of a formal dialogue event with a larger audience and it took place in the UK between 2002 and 2003. This was an ambitious public consultation project costing £1million, where the government promised to take into account both public and expert opinion prior to making any policy decisions about the commercialisation of GM technology. After examining the findings of this event, it became clear that there was a need for upstream engagement, i.e. a discussion that takes place with the public before any new scientific developments and technology become a reality. This enables reflective practice, to discuss ethical issues and risks before the public become polarised in their views (Haste *et al.*, 2005). The emerging area of nanotechnology was seen as an excellent opportunity to practice and experiment with upstream engagement. One example was undertaken in the UK by DEMOS (an independent political think-tank) and researchers at Lancaster University in collaboration with the BBSRC and the Engineering and Physical Sciences Research Council (EPSRC). This experiment was a dialogue event, called Nanodialogues, run over three sessions covering public values, concerns, aspirations and also the role of public engagement in influencing scientific research. There were two groups of citizens involved, the first group consisted of full-time mothers and the second, professional men and women. The evaluation, conducted by Chilvers (2006), showed that the events were successful because:

- access to specialists was provided;
- multiway dialogue was observed, with scientists talking to each other as well as to the public.

However, this dialogue process did have a problem – public retention. Only four people attended the last session (out of a total of 14) and all participants claimed that the money offered for taking part was their strongest motivation for attending. It was suggested that while the citizens involved had learnt about nanotechnology and something about the operation of the research councils, the real value in the event was the influence upon the research councils, as the BBSRC/EPSRC learnt and reflected upon the role that citizens could play in shaping the research agenda (Chilvers, 2006).

Another recent example was a synthetic biology dialogue event organised by the BBSRC and EPSRC with support from the Department for Business, Innovation and Skills Sciencewise-ERC programme. The event took place in 2009 with 12 deliberative workshops, 160 members of the public, and it was held three times in four different locations across the UK. The evaluation findings from the event showed that the public were appreciative of the process and felt that their views were valued and listen to. However, they were less clear about how this would feed directly into policy decisions, a point Davies (2009) also mentions in the Dana Centre activities. The participants also indicated that they wanted a continuation of dialogue and the term, 'long stream engagement' was introduced.

The evaluation reports described above highlight that two-way engagement events have their own shortcomings:

- they can only involve a limited number of people;
- participants don't usually have a role in shaping the agenda;
- there is no direct responsibility of the organisers to feed the findings into policy;
- participant expectations need to be managed in terms of continued dialogue;
- the citizens taking part are unrepresentative of the public as a whole – those who take part are likely to be well informed and have strong views on the issues being discussed (Nisbet and Scheufele, 2009);
- there are difficulties in translating a dialogue model into real practical science communication events for large audiences. Most events are likely to be a mixed approach of deficit and dialogue, suggesting that despite the rhetoric of dialogue, a deficit approach is still common.

We agree with the suggestion by Brake and Weitkamp (2010) that it is not necessary for all science communication events to be dialogue oriented, as long as there is the opportunity for citizens to take part in discussion or in policy decisions. Science events which inform and excite the public about science are still very important. The recent Ipsos MORI poll on public attitudes to science for the Department for Business Innovations and Skills (PAS, 2011) suggested that the public were quite cynical about public consultation events, with 50% of respondents agreeing with the statement 'consultation events are just public relations activities and don't make any difference to policy'. People feel that consultation is important, but don't necessarily want to get involved in it themselves. The Danes have recognised the importance of public consultation for many years. In 1995, the Danish Parliament established the Danish Board of Technology (DBT), an independent body committed to the dissemination of knowledge about technology. Its central mission is 'to promote the technology debate and public enlightenment concerning the potential, and consequences of technology'. The DBT advise the Danish Parliament and Government and report to the Parliamentary Committee on Research.

In an article looking at the democratisation of science, Turney (2011) points out that an area missing from public involvement is the setting of the actual research agenda. Whilst there are isolated examples of this happening, it's not universal. One example mentioned in the article is the Medical Research Council (MRC) that had a panel involving the lay public who were specifically involved in assessing grants for the third phase of the Lifelong Health and Wellbeing initiative. Another good example of public involvement in agenda setting is the UK Alzheimer's Society. In 2000, they established a network called Quality Research in Dementia (QRD), patients and carers have involvement in research priorities, they review research proposals and also have a role in assessment and monitoring of research grants (Stilgoe and Wilsdon, 2009). A good example of where consultation events can work to ultimately influence the research agenda is the EPSRC-funded SuScit project which is Citizen Science for Sustainability. This project is coordinated by Brunel University, the Centre for Sustainable Development at the University of Westminster and Capacity Global. The aim of the project was to provide local communities with a voice in environmental and sustainability research. They particularly worked with hard-to-reach groups, including older citizens, people with disabilities and those from ethnic minority backgrounds. SuScit used a mix of panels, focus groups, community videos and deliberative workshops to develop a research agenda and recommendations for the EPSRC. As a result of the project, researchers, practitioners and residents are now working together on local initiatives and future research projects.

1.6   Recent initiatives

In 2008, the Beacons for Public Engagement were established in the UK. This project was the biggest investment of money into public engagement to date and was funded by the Research Councils UK (RCUK), Higher Education Funding Councils and The Wellcome Trust. The investment is to help universities engage better with the public, not just in science, but across all disciplines. Six university partnerships were awarded Beacon status, and they are located in Edinburgh, Cardiff, Newcastle, Manchester, London and Norwich (UEA), with a National Co-ordinating Centre for Public Engagement (NCCPE) at Bristol University.

In the UK there has been a manifesto for public engagement, *The Engaged University*, drawn up by the NCCPE where universities and research institutes have been asked to sign up to 'celebrate and share their public engagement activity, and to express their strategic commitment to engaging with the public'.

The funders of research in the UK have also recently drawn up a set of principles for engaging the public with research: the Concordat. 'The signatories of the Concordat recognise the importance of public engagement to help maximise the social and economic impact of UK research' (Concordat, 2011). More details about these recent initiatives can be found in Chapter 2.

1.7 A way forward

The Ipsos MORI poll on public attitudes to science (PAS, 2011) suggests that the public attitude towards science in the UK is really positive, 86% are 'amazed by the achievements of science' and 82% agree that 'science is such a big part of our lives we should all take an interest'. This is mirrored by other studies in Europe, the US and Australia (Wilkinson, 2010). As scientists wishing to communicate our science to the public, we should be encouraged by these findings. There are many exciting and entertaining ways to communicate science through a variety of different media: face-to-face (e.g. science cafés), exhibitions, popular books, magazines, television programmes, web sites and social media. We also have to acknowledge that we are individuals with our own strengths, experiences and different personalities, and might prefer using some approaches more than others. The case studies contained within this book give marvellous examples of the many different forms of engagement with a variety of audiences. Although we have used this chapter to highlight different models of communication in terms of deficit and dialogue, we mustn't get too hung up on a 'one approach' fits all. We want to use this book and the case studies it contains to demonstrate that it is perfectly acceptable to use different approaches at different times, in different situations and with different audiences. This will lead to a dynamic and vibrant community of scientists communicating effectively with the public.

References

Allum, N., Stugis, P., Tabourazi, D. and Brunton-Smith, I. (2008) Science knowledge and attitudes across cultures: a meta-analysis. *Public Understanding of Science* 17 (1), 35–54

Barb (2011) http://www. Barb.co.uk (accessed January 2011)

Bauer, M.W., Durant, J., Gaskell, G., Liakopoulos, M. and Bridgman, E. (1998) United Kingdom, in *Biotechnology in the Public Sphere. A European Handbook* (eds Durant, J., Bauer, M.W and Gaskell, G.). The Science Museum

Bauer, M.W, Allum, N. and Miller, S. (2007) What can we learn from 25 years of PUS survey research? Liberating and expanding the agenda, *Public Understanding of Science*, 16, 79–95

Bauer, M. (2010) Paradigm change for science communication: commercial science needs a critical public, in: *Communicating Science in Social Contexts* (eds Cheng, D., Claessens, M., Gascoigne, T., Metcalfe, J., Schiele, B and Shi, S.). Springer

Barton. R. (1998) "Huxley, Lubbock and half a dozen others": Professionals and Gentlemen in the formation of the X-Club. *History of Science Society* 89 (3), 410–444

Berman, M. (1978) *Social Change ad Scientific Organisation. The Royal Institution 1799–1844.* Heinemann Educational Books

Biosphere (2011) http://www.biosphere-expeditions.org/english.html (accessed December 2011)

Bodmer, Sir Walter (1985) I. Royal Society, London

Bodmer, W. (2010) Public Understanding of Science: The BA, the Royal Society and COPUS. *Notes and Records of the Royal Society* S151–S161

Brake, M.L. and Weitkamp, E. (2010) *Introducing Science Communication.* (eds Brake, M.L. and Weitkamp, E.). Palgrave Macmillan

Briggs, P. (2003) The BA at the end of the 20th Century: A personal account of the 22 years from 1980 to 2002. http://www.britishscienceassociation.org/web/AboutUs/OurHistory/BA20thCentury.htm (accessed 3 August 2011)

British Council (2001) Briefing Sheet 6, Public Understanding of Science, UK Partnerships. http://www.britishcouncil.org

Broks, P. (2006) *Understanding Popular Science. Issues in Cultural and Media Studies* (ed. Allan, S.) Open University Press

Burns, T.W., O'Connor, D.J. and Stocklmayer, S.M. (2003) Science communication: a contemporary definition. *Public Understanding of Science* 12, 183–202

Chilvers, J. (2006) Engaging Research Councils? An evaluation of a Nanodialogues experiment in upstream public engagement. http://www.bbsrc.ac.uk/web/FILES/Workshops/nanodialogues_evaluation.pdf (accessed 2 August 2011)

Concordat for Engaging the Public with Research (2011). A set of principles drawn up by the Funders of Research in the UK. http://www.researchconcordat.ac.uk/ (accessed 11 May 2011)

COPUS (2011) www.copusproject.org/ (accessed 18 November 2011)

Davies, S (2009) Learning to engage; engaging to learn; the purposes of informal science-public dialogue, in *Investigating Science Communication in the Information Age* (eds Holliman, R., Whiteleg, E., Scanlon, E., Smidt, S. and Thomas, J.). Oxford University Press, The Open University

Davis, L. (2010) Science communication: a "down under" perspective. *Japanese Journal of Science Communication* 7, 66–71

Darwin Correspondence project, Charles Darwin papers, Cambridge University Library, MS. DAR.202:124. http://www.darwinproject.ac.uk/

Daley, S.M. (2000) Public Science Day and the public understanding of science in America. *Public Understanding of Science* 9, 175–181

DEMOS http://www.demos.co.uk/ (accessed 28 December 2011)

Desmond, A. and Moore, J. (1991) *Darwin.* Michael Joseph, the Penguin Group, London

Durrant, J.R., Evans, G.A. and Thomas, G.P (1989) The public understanding of science. *Nature* 340, 11–14

Eurobarometer Survey (2005) Europeans, Science and Technology. Special Eurobarometer 224, http://ec.europa.eu/public_opinion/archives/ebs/ebs_224_report_en.pdf (accessed 3rd May 2012)

Eurobarometer Survey (2010) Employers' perception of graduate employability N° 304. http://ec.europa.eu/public_opinion/index_en.htm (accessed 11 May 2011)

European Science Events Association (EUSCEA) http://www.euscea.org/ (accessed December 2011)

Evolution MegaLab (2011) http://www.evolutionmegalab.org (accessed December 2011)

Fara, P. (2009) *Science: A Four Thousand Year History.* Oxford University Press

Falk, J.H. and Dierking, L.D. (2000) *Learning from Museums. Visitor Experiences and the Making of Meaning.* Altmira Press

Gates Foundation http://www.gatesfoundation.org/Pages/home.aspx (accessed, 11 December 2011)

Gregory, J. and Miller, S. (1998) *Science in Public Communication Culture and Credibility*. Basic Books

Gerbner, G. (1966) An institutional approach to mass communications research, in *Communication: Theory and Research*, Thayer, L, Springfield, IL: Charles C. Thomas p.429-445

Gribbin, J. (2002) *Science a History*. Penguin Books

Hannam, J. (2011) Explaining the world: communicating science through the ages, in *Successful Science Communication* (eds Bennett, D.J. and Jennings, R.C.). Cambridge University Press

Haste, H., Kean, S., Peacock, M., Russell, C. and Whitmarsh, L. (2005) *Connecting Science: What We Know and What We Don't Know About Science in Society*. The British Science Association

HEFCE (Higher Education Funding Council for England) (2011) http://www.hefce.ac.uk/research/ref/ (accessed 10 May 2011)

Holliman, R. and Jensen, E. (2009) (in)authentic sciences and (im)partial publics: (re)constructing the science outreach and public engagement agenda, in *Investigating Science Communication in the Information Age* (eds Holliman, R., Whiteleg, E., Scanlon, E., Smidt, S. and Thomas, J.). Oxford University Press, The Open University

Irwin, A. (2006) The politics of talk: coming to terms with the 'new' scientific governance. *Social Studies of Science* 36, 299–320

Irwin, A. (2009) Moving forwards or in circles? Science communication and scientific governance in an age of innovation, in *Investigating Science Communication in the Information Age* (eds Holliman, R., Whiteleg, E., Scanlon, E., Smidt, S. and Thomas, J.). Oxford University Press, The Open University

Lock, S.J. (2011) Deficits and dialogues: science communication and the public understanding of science in the UK, in *Successful Science Communication* (eds Bennett, D.J. and Jennings, R.C.). Cambridge University Press

McClellan, J.E. and Dorn, E. (2006) *Science and Technology in World History, an Introduction*, 2nd edn. The John Hopkins University Press

Miller, J.D. (1983) Scientific literacy: a conceptual and empirical review. *Daedalus*, Spring: 29–48

Miller J.D. (2011) To improve science literacy, researchers should run for school board. *Nature Medicine* 17 (1), 21

Mulder, H.A.J., Longnecker, N. and Davis, L.S. (2008) The State of Science Communication Programs at Universities Around the World. *Science Communication* 30 (2), 277–287

National Co-ordinating Centre for Public Engagement (2010) http://www.publicengagement.ac.uk/ (accessed 11 May 2011)

National Science Board http://www.nsf.gov/nsb/about/

NCBE (2011) http://www.ncbe.reading.ac.uk/ncbe/gmfood/conference.html (accessed 5 June 2011)

Nisbet, M.C. and Scheufele, D.A. (2009) What's next for science communication? Promising directions and lingering distractions. *American Journal of Botany* 96 (10), 1767–1778

OECD (Organisation for Economic Co-operation and Development) (2004) *Basic R&D Statistics*. OECD, Paris:. http://www.oecd.org/home/0,2987,en_2649_201185_1_1_1_1,00.html (accessed 18 November 2011)

Pearson, G. (2001) The participation of scientist in public understanding of science activities: The policy and practice of the UK research councils. *Public Understanding of Science* 10, 121–137

Poliakoff, E. and Webb, T.L. (2007) What factors predict scientists' intentions to participate in public engagement activities? *Science Communication* 29 (2), 242–263

Project 2061 (2011) http://www.project2061.org/ (accessed 15 May 2011)

Public Attitudes to Science (2011) Ipsos Mori, http://www.ipsos-mori.com/Assets/Docs/Polls/sri-pas-2011-summary-report.pdf

Ray, E. (1999) Outreach, engagement will keep academia relevant to twenty-first century societies. *Journal of Public Service & Outreach* 4, 21–27

Rennie, L.J. and Stocklmayer, S.M. (2003) The communication of science and technology: past, present and future agendas. *International Journal of Science Education* 25 (6), 759–773

Riise, J. (2010) Bringing Science to the Public. In: *Communication Science in Social Contexts* (eds Cheng, D., Claessens, M., Gascoigne, T., Metcalfe, J., Schiele, B. and Shi, S.). Springer

Royal Institution for Great Britain (RIGB) (2011) http://www.rigb.org/registrationControl?action=home (accessed 5 June 2011)

Russell, N. (2010) *Communicating Science*. Cambridge University Press.

Salters' Institute http://www.salters.co.uk/institute/ (accessed 2 August 2011)

Stilgoe, J. and Wilsdon, J. (2009) The new politics of public engagement with science, in *Investigating Science Communication in the Information Age* (eds Holliman, R., Whitelegg, E., Scanlon, E., Smidt, S. and Thomas, J.). Oxford University Press, The Open University

Stephan P.E. (2010) The Economics of Science Funding for Research. International Centre for Economic Research. Working Paper no. 12

Stocklmayer, S.M. (2001) The background to effective science communication by the public, in *Science Communication in Theory and Practice* (eds Stocklmayer, S.M., Gore, M.M. and Bryant, C.). Kluwer Academic Publishers

Stocklmayer, S.M. (2002) New experiences and old knowledge: towards a model for the personal awareness of science and technology. *International Journal of Science Education* 24 (8), 835–858

Taylor, J.E. (1889) *The Playtime Naturalist*. Chatto and Windus, Piccadilly, London

The Engaged University, A Manifesto for Public Engagement (2011) http://www.publicengagement.ac.uk/why-does-it-matter/manifesto (accessed 16 May 2011)

Trench, B. (2010) Towards and analytical framework of science communication models, in *Communicating Science in Social Contexts* (eds Cheng, D., Claessens, M., Gascoigne, T., Metcalfe, J., Schiele, B. and Shi, S.). Springer

Turney, J. (2011) Democratic Experiments, Times Higher Education, http://www.timeshighereducation.co.uk/story.asp?storyCode=417989&sectioncode=26 (accessed 12 December 2011)

Wilkinson, C. (2010) Science and the citizen, in *Introducing Science Communication* (eds Brake, M.L. and Weitkamp, E.). Palgrave Macmillan

Wynne, B. (1992) Misunderstood misunderstanding: social identities and public update of science. *Public Understanding of Science* 1, 281–304

Yeoman, K.H, James, H.A. and Bowater, L. (2011) Development and Evaluation of an Undergraduate Science Communication Module, beej, 17-7. http://www.bioscience.heacademy.ac.uk/journal/vol17/beej-17-7.pdf

CHAPTER TWO

Scientists Communicating with the Public

Nothing in science has any value to society if it is not communicated, and scientists are beginning to learn their social obligations.

—Anne Roe, *The Making of a Scientist* (1953, p.17)

2.1 Introduction

Public engagement can be seen as the philosophy which overarches all communication to the public in both the sciences and humanities. Science and humanities communication can occur in the school environment (Chapters 9 and 10), with the wider public (Chapters 7 and 8) and also with policymakers (Chapter 7). It can be applied to both the practice of public engagement and the research into the process (Figure 2.1).

As described in Chapter 1, we are experiencing the third and current phase of science communication: 'Science and Society'. The call for scientists to be more engaged with the public has never been clearer. As scientists, our first response should be to ask:

- Has 'Science and Society' changed the working environment of scientists?
- Who are the science communicators?
- What are the current levels of science communication among scientists?

At the present time, there are a limited but growing number of studies in the primary literature that address these questions, but research in this area exists. It is found in several seminal reports commissioned by The Wellcome Trust, The Science for All Expert Group, The Beacons for Public Engagement, The Royal Society and The Royal Academy of Engineering with the Department for Business, Innovation and Skills (BIS). A summary of these reports can be found in Table 2.1.

2.2 What does 'Science and Society' mean for scientists? The changing environment

To answer this question we need to examine the changes that have taken place over the past few decades that have made the working environment

---

Science Communication: A Practical Guide for Scientists, First Edition. Laura Bowater and Kay Yeoman.
© 2013 John Wiley & Sons, Ltd. Published 2013 by John Wiley & Sons, Ltd.

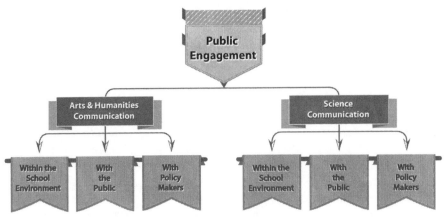

Figure 2.1 Hierarchy of public engagement. Public engagement can be seen as the philosophy which overarches all communication to the public in both the sciences and humanities. Science and humanities communication can occur in the school environment, with the wider public and also with policymakers. It can be applied to both the practice of public engagement and the research into the process of public engagement.

of scientists more conducive to science communication activities. What are these changes from the UK perspective?

2.2.1 The UK perspective

The publication of several seminal reports, including the Bodmer Report (1985) and the 1993 science and technology White Paper 'Realising our Potential' led to the removal of the stigma associated with the popularisation of science. Since then, engaging the public through science communication has become a more mainstream activity (Chapter 1). In addition, the restructuring of the research councils resulted in a new role for the Research Councils UK (RCUK): to encourage researchers to engage with the public and embed the philosophy of public engagement across *all* disciplines in the higher education (HE) and the research sectors (Figure 2.1). Evidence of the commitment of UK funding bodies to encourage public engagement was clearly seen when the Beacons for Public Engagement were established in the UK. This project was the biggest investment of money into public engagement to date. It was funded by RCUK, Higher Education Funding Council for England (HEFCE) and The Wellcome Trust. In February 2007, HEFCE invited applications from Higher Education Institutions (HEIs) to set up six collaborative centres and one coordinating centre under this 4-year pilot project. The sum of money was not trivial; funders provided just over £9 million to support and encourage universities to engage more effectively with the public, not just in science, but across all academic disciplines. The Beacons are university-based collaborative centres working to support, recognise, reward and build capacity for public engagement. Over 100 HEIs applied, and there were six successful applicants.

- **Beacon North East:** a partnership between Newcastle University, Durham University and the Centre for Life.

Table 2.1 A summary of the published reports that explore the science communication environment.

The Report	Funder	Aims and objectives	Website
The Role of Scientists in Public Debate	Report of the Research study conducted by Ipsos MORI for the Wellcome Trust December 1999 – March 2000	Research aims to investigate whether scientists consider themselves most responsible for and best equipped to communicate their scientific research and its implications to the public, what benefits and barriers they see to a greater public understanding of science, and what needs to change for scientists to take a greater role in science communication	http://www.wellcome.ac.uk/stellent/ groups/corporatesite/@msh_peda/ documents/web_document/ wtd003425.pdf
Science Communication	The Royal Society June 2006	Survey of factors affecting science communication by scientists and engineers	http://royalsociety.org/policy/ publications/2006/science-communication/
Reward and Recognition of Public Engagement	Report for the Science for All Expert Group November 2009	The two main objectives are to pull together findings and recommendations from previous studies to explore: • the barriers to recognition of effective public engagement within academia, industry and health and public services; and • the role of reward and recognition as a factor in influencing decisions to participate	http://www.britishscienceassociation .org/NR/rdonlyres/B5899730-F2D5-481B-9047-577C1A600A10/0/ RewardandrecognitionFINAL.pdf
A Qualitative Baseline Report on the Perceptions of Public Engagement in University of East Anglia Academic Staff	Report of the Research conducted by The Research Centre, City College Norwich, for CUE East December 2008	The overall goal of the research was to explore academic attitudes towards public engagement and the cultural and institutional factors that affected their involvement	http://www.uea.ac.uk/polopoly_fs/ 1.134441!Baseline%20research %20report.pdf

(Continued)

Table 2.1 (*Continued*)

The Report	Funder	Aims and objectives	Website
Public Culture as Professional Science	Scientists on public engagement (ScoPE) produced by BIOS Funded by the Wellcome Trust September 2009	Data-led, sociological analysis of the understandings, views, perspectives, judgements and experiences of scientists working in the life sciences with respect to public engagement and public dialogue	http://eprints.kingston.ac.uk/20016/1/ScoPE_report_-_09_10_09_FINAL.pdf
Towards a Professional Development Framework for Scientists Involved in Public Engagement Work	Prepared for the Wellcome Trust Sanger Institute March 2010	The document reports on the research design and initial testing of a framework for improving professional development in science communication	http://www.sanger.ac.uk/about/engagement/docs/professionaldevelopmentframework.pdf
Embedding Public Engagement in Higher Education	Final report of the national action research programme on behalf of the NCCPE September 2011	Its aim is to support the embedding of public engagement in the Higher Education sector of the UK by generating insight into key issues and building an evidence-based change programme from the issues that were raised	https://www.publicengagement.ac.uk/sites/default/files/Action%20research%20report_0.pdf
Business Motivations for Engaging the Public in Science and Engineering	The Royal Academy of Engineering and the Department for Business, Innovation and Skills September 2011	The aims are to understand the motivations and rewards for, and barriers to, public engagement within science, technology, engineering and mathematics based businesses. To provide the wider science and engineering communication and public engagement community with a resource that will enable them to develop more fruitful relationships and partnerships with the business community	http://raeng.org.uk/news/publications/list/reports/engaging_the_public_in_science_and_engineering.pdf
Science and the Public Interest: Communicating the Results of New Science to the Public	The Royal Society (May 2006)	This report is designed to help researchers whose imminent publication might merit broader communication (Appendix 1 is particularly useful)	http://royalsociety.org/policy/publications/2006/science-public-interest/

- **CUE East:** led by the University of East Anglia (UEA), has the largest number of community partners including the Norwich Research Park, Norwich University College of the Arts, the Sainsbury Centre, Norwich City Council, BBC East, the Teacher Scientist Network and the SAW trust (Science, Art and Writing).
- **Edinburgh Beltane:** is one of the larger Beacon partnerships, bringing together the expertise from five Higher Education Institutions in Edinburgh and the prospective University of Highlands and Islands (UHI).
- **Manchester Beacon:** partners are the University of Manchester, Manchester Metropolitan University, University of Salford, Manchester Museum of Science and Industry, and Manchester: Knowledge Capital.
- **University College London:** partners are the British Museum, the South Bank Centre, Birkbeck College, Cheltenham Science Festival, Arts Catalyst, and City and Islington College.
- **The Beacon for Wales:** is a partnership between Cardiff University, University of Glamorgan, Techniquest, AmgueddfaCymru – National Museum Wales and BBC Cymru Wales.

As well as the six Beacons, a National Co-ordinating Centre for Public Engagement (NCCPE) was established at Bristol University[1]. To summarise, the Beacons for Public Engagement state that its strategic aims are to:

1 inspire a shift in culture;
2 increase capacity for public engagement;
3 build effective partnerships to encourage partners to embed public engagement in their work.

—NCCPE (2010)

Recently, the NCCPE has drawn up a manifesto[2] for public engagement; 'The Engaged University'. This manifesto invites universities and research institutes to sign up to: 'celebrate and share their public engagement activity, and to express their strategic commitment to engaging with the public.'

Currently the manifesto is attracting a growing number of signatories from a variety of HEIs from across the UK. As of January 2012, 39 HEIs had signed this manifesto[3]. In 2010, the Concordat[4] for Engaging the Public with Research was launched in the UK. The Concordat outlines the expectations and responsibilities of research funders with respect to embedding public engagement across *all* disciplines in universities and research institutes. The Concordat's key principles are:

- UK research organisations have a strategic commitment to public engagement;
- researchers are recognised and valued for their involvement with public engagement activities;
- researchers are enabled to participate in public engagement activities through appropriate training, support and opportunities;

---

[1] www.publicengagement.ac.uk/

[2] http://www.publicengagement.ac.uk/why-does-it-matter/manifesto

[3] http://www.publicengagement.ac.uk/why-does-it-matter/manifesto/signatories

[4] http://www.rcuk.ac.uk/Publications/policy/Pages/perConcordat.aspx

- the signatories and supporters of this Concordat will undertake regular reviews of their and the wider research sector's progress in fostering public engagement across the UK.

—Reproduced by permission of the RCUK

The Concordat has been endorsed by a number of signatories that include the RCUK, HEFCE, the Royal Society and the Department for Environment, Food and Rural Affairs (DEFRA). In addition, it is supported by Government departments, scientific academies and societies and universities in the UK. A signatory to the Concordat recognises that: 'engaging with the public is an approach that can be integrated across HEIs (for example in teaching and learning)'.

The funding provided for the 4-year Beacon pilot project ends in 2012. However, the RCUK has recognised that there is still a need to:

- support scientists and researchers to embed public engagement within their career portfolio;
- integrate public engagement activities into the core strategies of HEIs.

Without this political motivation, its financial underpinning and the institutional support for public engagement, researchers will struggle to continue to undertake public engagement within their workplace. As a response to this concern, in September 2011 the RCUK announced a new funding stream, the Catalysts[5], designed to support a formal culture of public engagement within HEIs. Higher Educational Institutes that did not take part in the Beacons project were eligible to bid for funding. The successful applicants will build on the experiences of the Beacons project to embed public engagement within their institution. Researchers will also be incentivised to participate in public engagement activities through reward and recognition mechanisms, and institutions will be encouraged to celebrate and share good practice.

In the UK, examples of this change in culture towards an environment that encourages researchers to engage with the public and embed public engagement in the higher education and the research sectors are already visible. Statements encouraging public engagement can be found in HEIs, for example the University of East Anglia states that two of its key strategies outlined in its Corporate Plan (2008–2012)[6] are to:

- support active engagement with international agencies, governments, businesses and charities over global and national challenges, especially those in the cross-faculty research priority areas of climate change, global health and poverty reduction;
- make CUE East an effective Beacon for Public Engagement, contributing to the public understanding of major issues and building on our existing work with teachers, parents and pupils.

—Reproduced by permission of the University of East Anglia

Similarly, when the corporate plan of The John Innes Centre, a plant science research institute strategically funded by the Biotechnology and Biological

---

[5] For more information: http://www.rcuk.ac.uk/per/pages/catalysts.aspx
[6] Corporate plan; http://www.uea.ac.uk/polopoly_fs/1.74259!corporateplan.pdf

Sciences Research Council (BBSRC) and located on the Norwich Research Park, is examined, it states that

> Our mission is to generate knowledge of plants and microbes through innovative research, to train scientists for the future, to apply our knowledge to benefit agriculture, the environment, human health and well-being, and engage with policymakers and the public.[7]
>
> —*Reproduced by permission of The John Innes Centre*

Clearly different HEIs and research institutes display different levels of strategic and practical support towards embedding public engagement within their mission statements, corporate plans and culture. Within the UK, there are HEIs that have embraced this philosophy and have already firmly embedded public engagement within the institutional make-up. Progress is variable; although all HEIs within the UK are continuing to move towards this goal, some are moving more slowly than others. This aspiration to embed engagement in HEIs and research institutes is not unique to the UK. A recent paper has examined the public engagement activities among 40 science research institutes throughout Europe (Neresini and Bucchi, 2010). It found that very few of the 40 institutes surveyed have embedded engagement within their core activities. This lack of institutional support was conceptualised within this report as:
- a lack of institutional funding;
- a lack of support for the design, organisation, delivery and evaluation of activities;
- a lack of training to support engagement through science communication activities.

Since the early 1990s all RCUK funding bodies have promoted science communication to the public. The RCUK now have pathways to impact, a mandatory submission which is made alongside the research grant. It covers the economic and societal impact of scientific research and this can include public engagement through science communication (Figure 2.2). The primary assessment criteria used to judge whether funding for research should be awarded for grant applications is still 'science excellence' (RCUK, 2011);[8] however,

> Pathways to Impact are one of a number of other criteria taken into account during the peer review process and by panels in prioritising applications.
>
> —*Reproduced by permission of the RCUK*

These pathways to impact statement submissions are peer reviewed and form part of the secondary assessment criteria of grant applications. In addition, many research councils offer small-scale funds to allow development of science communication activities. A range of these can be found on the

---

[7] JIC, http://www.jic.ac.uk/corporate/index.htm

[8] This statement is found on the RCUK website, http://www.rcuk.ac.uk/documents/impacts/RCUKImpactFAQ.pdf

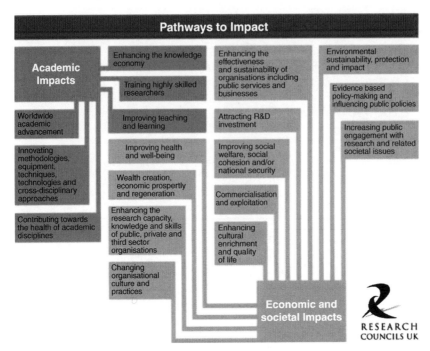

Figure 2.2 Research Councils UK Pathways to Impact. Reproduced by permission of the RCUK.

website which accompanies this book. The inclusion of the impact statement was in part, as a result of the Wolfendale Committee (Chapter 1).

> In principle, all who receive grants from public funds should accept a responsibility to explain to the general public what the grant is enabling, or has enabled them, to do and why it is important and how it fits into the broader area of knowledge.
> —*Wolfendale Committee (1995)*

Currently, the RCUK describes how grant applications are to be assessed for their Impact Summary and Pathways to Impact. The first thing you should notice is who will review these proposals:

> ...your Impact Summary and Pathways to Impact will be looked at by peer reviewers who may be non-academics (i.e. potential users and beneficiaries) as well as academic reviewers...
> —*Reproduced by permission of the RCUK*

This means that you need to consider whether the language you are using to convey the science and its potential impact is suitable for a lay audience. For some general advice on scientific writing for the public see Section 4.7.

The RCUK also provides guidance about what reviewers are looking for when they screen the Impact Summaries and the Pathways to Impact. They state that they should be:

...appropriate and justified in terms of the nature of the proposed research project and sufficient attention has been given to who the beneficiaries might be and appropriate ways to engage with them throughout the project.[9]

— Reproduced by permission of the RCUK

This statement means that you need to consider which part of the public you are trying to reach. The public is a wide and disparate group, populated with different personalities, attitudes, interests, ages, sexes and socioeconomic status (see Section 6.2.2). Once you have decided on your audience, you also need to consider the mechanism you are going to use for your engagement activity, for example will you engage directly or indirectly, will your activity be information driven or conversation oriented (see Table 6.1)?

In 2014, the Research Excellence Framework (REF) will review the quality and outputs of academic research at UK academic institutions. There is a huge amount at stake for all UK HEIs taking part in this activity. The results of this exercise will dictate the level of Quality-Related[10] (QR) funding provided by HEFCE to support the research across all disciplines (including science) undertaken in these institutions. It is absolutely clear that the quality of the research output from each institution will be the main driver for decisions concerning the level of funding provided, but other factors are to be taken into consideration. The website for REF 2014 also has a section entitled 'Decisions on assessing research impact.'[11]: It states that:

Case studies may include any social, economic or cultural impact or benefit beyond academia that has taken place during the assessment period, and was underpinned by excellent research produced by the submitting institution within a given timeframe.

— HEFCE (2011)

This research impact highlighted in the REF statement reflects the RCUK pathways to impact in Figure 2.2. This has the potential to provide researchers with an environment that supports and encourages public engagement activity if it can be shown to provide a 'social, economic or cultural impact or benefit'. Nevertheless, the inclusion of impact has been controversial. The concerns raised include the issues surrounding:

- the lack of equivalency between research disciplines when it comes to assessing 'impact';
- how 'impact' and the benefit of research can be successfully judged with transparency and equality across research disciplines;
- the relative merits of 'enterprise' impact vs. 'engagement' impact.

---

[9] Further information can be found on the RCUK website, including the following document. http://www.rcuk.ac.uk/documents/impacts/RCUKImpactFAQ.pdf

[10] The Research Excellence Framework (REF) is the new system for assessing the quality of research in UK higher education institutions (HEIs). It replaces the Research Assessment Exercise Research (RAE) and it will be completed in 2014. The REF will be undertaken by the four UK higher education funding bodies. http://www.hefce.ac.uk/research/ref/

[11] Information is available at, http://www.hefce.ac.uk/research/ref/pubs/2011/01_11/

From the perspective of a scientist who is also a science communicator, the inclusion of 'impact' within the REF assessment provides a policy that allows a culture of science communication and public engagement to be developed within HEIs. It will be interesting, post 2014, to examine the ratio of enterprise to engagement included in the impact case studies.

2.2.2   Policy changes beyond the UK

Clearly in the UK, science communication has moved up the political agenda in terms of its importance. In 2009, Lord Drayson, the UK's Science Minister made a statement: 'scientists have a duty – particularly when they are funded by tax payers to engage in the public arena.' This has been echoed globally, with other policy statements outlining the importance of a society engaged with science, facilitated by scientists who engage with the public.

It is not just scientists working within the UK who must consider how they can demonstrate the impact of their research to secure funding. This is shown in impact statements from the international grant awarding bodies. Pace *et al.* (2010) report that in the US 'the National Science foundation established a "broader impacts" criterion in 1997 as part of its merit review process.'[12]

Similar criteria were introduced in research granting agencies in the European Union such as the European Commission Funding Frameworks.[13] In addition, clear, politically driven policy decisions designed to encourage a society that engages with science can be found in other countries. For example:

- The Inspiring Australia strategy aims to build a strong, open relationship between science and society, underpinned by effective communication of science and its benefits. As a significant contribution to this national strategy, the Government has provided $21 million over three years in the 2011–12 budget towards an Inspiring Australia programme (Inspiring Australia Report, 2011).
- The China Association for Science and Technology (CAST) was founded in1958 to promote and popularise science. CAST coordinates interactions between the scientific community, the Communist Party and the Chinese Government. In addition it has a key role in encouraging scientists to communicate science to the wider public, including remote communities. In 2002, China issued a law for science popularisation and in 2006, CAST, the Ministry of Science and Technology and other government departments ran popularisation campaigns on topics such as SARS (severe acute respiratory syndrome) and H5N1 bird flu as part of the Scheme for the Advancement of Chinese Scientific Literacy.[14]
- The South African Government's Department of Science and Technology has recently launched a Public Understanding of Biotechnology (PUB)

---

[12] Further details on the National Science foundation's '*broader impacts*' criterion can be found on the following website: http://www.nsf.gov/pubs/gpg/broaderimpacts.pdf. http://www.csid.unt.edu/topics/NSF-broader-impacts-criterion.html

[13] As highlighted in the following website: http://ec.europa.eu/governance/impact/key_docs/key_docs_en.htm

[14] english.cast.org.cn/

programme. Its mandate is 'to ensure a clear, balanced understanding of the scientific principles, related issues and potential of biotechnology and to stimulate public debate around its applications in society'. The aims of PUB are to provide awareness of and provide opportunities for public dialogue and debate around biotechnology and its applications.[15]

As a result of such policies, scientists on a global scale are being asked to engage in conversations with the public.

2.2.3   Practical steps to assess your institution's commitment to public engagement

There are straightforward, practical steps that you as an individual working within the UK and beyond can undertake to better assess your institution's 'publicly voiced' commitment towards public engagement. These steps will allow you to understand how your institution compares to similar, alternative institutions within the sector. These steps are described in Box 2.1.

Box 2.1   Practical Steps to Assess your HEIs Commitment to Public Engagement

- Scrutinise the mission statements, corporate plans, vice chancellor statements or addresses as well as the public facing website(s) of your institution. Ask yourself whether they contain any statements detailing their commitment to public engagement? Do they provide examples of any public engagement activities?
- Explore the NCCPE website that provides details about the Manifesto for Public Engagement. Is your institution listed as a signatory and have other members of your University group signed up as signatories. In the UK your Higher Education Institution is a member of one of the following groups;
  a the Million+ group;
  b the UKADIA (UK Arts and Design institutions);
  c the 1994 Group;
  d the Russell Group;
  e the University Alliance that share ideas and resources towards higher education.
  The website for the manifesto can be found at the following web address. In addition, details about how to sign up to the manifesto are provided as well as the potential benefits of becoming a signatory.
    http://www.publicengagement.ac.uk/why-does-it-matter/manifesto/signatories.
- The supporters of the Concordat for Engaging the Public with Research can be found at the following website. http://www.rcuk.ac.uk/per/Pages/supporters.aspx
  It will give you an insight into the number of HEIs as well as the funding bodies and science societies that have signed up to this agreement. Taking time to read the Concordat will also reveal the depth and breadth of the commitment towards public engagement that your institution has publically agreed to support.
  http://www.rcuk.ac.uk/Publications/policy/Pages/perConcordat.aspx

---

[15] More information can be found at http://www.pub.ac.za/index.php?option=com_content&view=frontpage&Itemid=93

2.2.4 The EDGE tool

A simple, quick and effective way to assess how well your school, department, faculty, institution or place of work supports public engagement is to use the EDGE tool (Figure 2.3). This tool is one of the outcomes of the work undertaken by the Beacons for Public Engagement project. It is designed to allow you to:

- consider the level of strategic support at a departmental, faculty or institutional level;
- identify areas where you can support and encourage positive changes towards public engagement.

You can use the tool yourself or you might choose to use it with a group of colleagues as part of an informal workshop session. Alternatively, you could invite several colleagues to complete the tool individually and then compare the responses from different departments or across faculties. Comparing results in this way can potentially provide powerful insights into the different perceptions of the support offered by your institution towards public engagement. The tool provides descriptive summary statements of the different levels of support for public engagement in an institution's people, day-to-day processes, strategies, mission and vision. It can be found in Figure 2.3, but further information and resources regarding the EDGE tool can be found on the website of the NCCPE at www.publicengagement.ac.uk/support/planning-change.

2.3 Are academics involved in public engagement?

In 2008, a qualitative baseline report on the perception of public engagement among academic staff at the University of East Anglia was published (McDaid, 2008). A cross section of 55 academic staff were interviewed, 14 from the Faculty of Science, 14 from the Faculty of Social Science and 12 from the Faculty of Health with a high proportion of senior academics. This report showed that 84% of the academics who were interviewed had been involved in 'some form of self-defined public engagement'.

Another exploratory study was recently published by Glass et al. (2011). This study was undertaken at the research-intensive Michigan State University, which has a historical mandate to serve the public good. The researchers examined the promotion and tenure documents of 173 faculty members across all disciplines who underwent promotion or tenure review from 2002 to 2006 to identify the different types of public engagement activities listed by staff. The researchers determined the frequency levels and categorised the different types of activities:

- **Publicly Engaged Research and Creative Activities**, which includes community-driven and funded research opportunities and arts projects;
- **Publicly Engaged Instruction**, which includes teaching (may be credit bearing or non-credit bearing) for, or in association with community partners. It also includes scholarly resources that are available for the general public;
- **Publicly Engaged Service**, which includes expert testimony, technical assistance, clinical and diagnostic provisions;
- **Publicly Engaged Commercialised Activities**, which includes the 'for profit' translation of new knowledge generation into the public domain through the commercialisation of discoveries, such as technology transfer and licenses.

beacons
for public
engagement

National Co-ordinating Centre
for Public Engagement

EDGE self-assessment matrix

Support

National Co-ordinating Centre for Public Engagement   Self Assessment Matrix – SUPPORT   www.publicengagement.ac.uk

Figure 2.3 The EDGE tool. Reproduced by permission of the National Co-ordinating Centre for Public Engagement.

EDGE tool: Support

By reflecting on the work of the six beacon projects, and drawing on the experiences of other institutions, projects and research, we have identified nine dimensions which are critical to building a supportive culture for public engagement:

- mission
- leadership
- communication
- recognition
- support
- learning
- staff
- students
- public

This document provides a self-assessment matrix to help you review your support for public engagement. It identifies five key challenges, and maps each against a scale:

E *Embryonic: activity is currently patchy or non-existent for this challenge*

D *Developing: Some activity is underway, but in a relatively unsystematic and non-strategic fashion*

G *Gripping: The institution is taking steps to develop a more systematic and strategic approach*

E *Embedding: The institution has put in place strategic and operational support for this challenge, resulting in a supportive culture for engagement.*

Instructions for use

You can use the matrix in a variety of ways, for example:

- You could fill it in individually, relying on your own knowledge of your institution;
- You could use it as part of a workshop with colleagues and other stakeholders;
- Or you could invite a number of people to fill it in individually and then bring them together to compare their perspectives. Comparing different departments across an institution can be a powerful exercise.

While the levels presented here assume that embedding engagement brings benefits to an institution, some may choose not to seek to 'embed' support in all of the areas identified in the tool. In some instances informal and emergent approaches may be preferred to formalised and embedded ones.

Elsewhere on our website, in the Planning for Change section of our website, you can explore an in depth account of how the UCL Beacon, and others, have tackled the challenge of supporting public engagement. This includes short case studies and reflections from staff about the lessons learned: www.publicengagement.ac.uk/support/planning-change/support.

National Co-ordinating Centre for Public Engagement   Self Assessment Matrix – SUPPORT   www.publicengagement.ac.uk

Figure 2.3 (*Continued*)

Focus	Embryonic	Developing	Gripping	Embedding
Investment in expert support	There are no staff members with responsibility for supporting and embedding PE on the campus. There may be individuals in a few departments with PE roles.	There are some staff who are tasked with supporting and embedding PE; however their appointments are temporary / not core funded and PE is only one of their responsibilities.	Staff are employed in the institution with explicit responsibility for supporting and embedding PE. Some appointments are permanent but most are temporary / not core funded	The institution core funds staff members with expertise in public engagement, who take responsibility for supporting and embedding PE across the organisation.
Effective networks and co-ordination	There is no attempt to co-ordinate public engagement activity or to network learning and expertise across the institution.	There are some informal attempts being made to co-ordinate public engagement activities, but there is no strategic plan for this work. Some self-forming networks exist, not supported by the institution.	Oversight and co-ordination of PE has been formally allocated (e.g. to a working group or committee) but there is minimal support and resource to invest in activity. There are some subject or career-level specific networks of engaged staff.	The institution has a strategic plan to focus its co-ordination, a body (or bodies) with formal responsibility for oversight of this plan, and resources available to assist the implementation and embedding of PE. There are a number of recognised and supported networks.
Opportunities for staff and students	There are few if any opportunities for staff to get involved in public engagement. Staff find their own external opportunities.	Several departments provide some opportunities for staff and student involvement, but there is no systematic support. Central brokerage may provide some details of external opportunities.	The majority of departments have made some provision to facilitate opportunities for staff and students to get involved in public engagement activities	The institution actively facilitates and communicates opportunities to get involved, and provides practical support measures (e.g. brokerage; bursaries; fellowships; secondments). It also invests in institution-wide programmes that provide first steps.
Evaluation of activity	There is no organized, institution-wide effort underway to evaluate the quantity and quality of public engagement activities taking place, nor any recognition of the value of formative evaluation.	A few departments attempt to evaluate the number and quality of public engagement activities. There are no efforts across the institution. Evaluation is focussed on monitoring.	A systematic effort to evaluate the number and quality of public engagement activities has been initiated. Summative evaluation is common.	An ongoing, systematic effort is in place to evaluate the number and quality of public engagement activities that are taking place throughout the institution. Evaluation feedback is being used to inform future activity and strategy. Formative evaluation is an expected part of engagement activities.
Brokerage and partnership working	There is little or no attempt made to facilitate public access to information, advice or expertise within the institution. There is little or no support for staff outreach e.g. access to training, writing grant proposals for outreach projects.	Some basic 'signposting' is in place – e.g. web pages – to describe the institution's public engagement offer and facilitate contacts. There is some support for staff outreach.	Effective 'signposting' is in place, and there are some attempts being made to broker partnerships with external organisations. The organisation has active 'front doors' which will respond to new requests from outside. Staff are supported to initiate their own outreach projects.	The institution has invested in signposting to facilitate contact with the community, provides some dedicated brokerage and is taking a strategic approach to partnership development. It is involved in long-term partnerships with local community neighbours.

This study showed that overall, 94% of faculty members reported at least one type of publicly engaged scholarship activity. Closer examination focusing on the four different categories revealed that:

- 72% of faculty members reported at least one type of Publicly Engaged Research and Creative Activities;
- more than 88% reported at least one type of Publicly Engaged Instruction;
- 71% of faculty members undertook Publicly Engaged Service;
- only 15% of faculty members indicated that they had undertaken Publicly Engaged Commercialised Activities.

An interesting result of this study from the scientist's perspective, is the difference in public engagement among faculty members from different disciplines. The paper reports that faculty members in the science and health disciplines were more likely to report different types of public engagement activities compared with faculty members from the arts and humanities (although physical and biological sciences faculty didn't fare too well either!). Public engagement is clearly a term and a philosophy that applies across all disciplines. Within this book we define Science Communication as public engagement applied to science, with communication happening between different groups including policymakers, the wider public and the school environment as shown in Figure 2.1.

2.3.1   Who are the science communicators?

The Wellcome Trust Sanger Institute is one of the key institutes involved in the Human Genome Project. It is another example of a research institute that has supported science communication and it is currently working to encourage more scientists to become actively involved in communicating with the public. In 2010, a report was published by Alison MacLeod for the Sanger Institute, 'Towards a Professional Framework for Scientists Involved in Public Engagement Work'. Its aim was to 'look at science communication skills and how they develop.' It describes the research, design and initial testing of a framework for improving professional development in science communication. As a first step, the author recognised that science communication is undertaken for different reasons by different groups of communicators. Science communication is also cross-disciplinary and many people can claim to be involved with it. These different groups of communicators are defined in Box 2.2.

Box 2.2   Who are the Science Communicators?

- **Professional Science Communicators** – This group of science communicator may be employed by science centres or museums, research institutes or universities or they may be self employed. They have often trained as scientists but they no longer have 'scientist' as their primary occupation.
- **Academic Science Communication Experts** – (who may also be involved in science communication activities). This group is composed of academic specialists *who may have a background in science or social science [. . .] conduct their own research and run masters degree courses on science and society'*. (MacLeod 2008)

- **Science Popularisers** – who gain popularity with the public through writing, lecturing, journalism, or broadcasting: this group are scientists who are motivated to enthuse new audiences about science.
- **Science Defenders** – whose work is either important or controversial, and who seek to explain their work: this group of scientists are heavily involved in science communication but define themselves as scientists first and foremost. Their involvement in science communication has stemmed from their research areas which are either '*controversial or newsworthy*'.
- **Scientists** – who are involved with science communication in schools or science events, as a 'sideline to their main work': this final group are motivated to engage with the public because they want to 'give back to the community' by explaining their work or the work of other scientists to an audience that is outside the research or academic community.

Adapted from 'Towards a professional framework for scientists involved in public engagement work', a report prepared for the Wellcome Sanger Institute by A. Macleod (2010). Reproduced by permission of the Wellcome Trust Sanger Institute.

These definitions highlight the disparate nature of the groups currently placed under the science communication umbrella, but it is also clear that scientists can participate in each group (MacLeod, 2010). As a scientist who is reading this book, you are probably interested in engaging the public through science communication activities and you can probably recognise the group you belong to. It is worth bearing in mind that as your career progresses, you might start off in one group but move into another as your reasons to become involved in science communication activities change or develop.

2.4   What is the current level of science communication by scientists?

2.4.1   'The Role of Scientists in Public Debate' report

By the end of the twentieth century, the Wellcome Trust (Section 1.2.2), was keen to understand scientists' views towards public engagement. In particular it wanted to:

> ...investigate whether scientists consider themselves to be the people most responsible for and best equipped to communicate their scientific research to the public, what benefits and barriers they see to a greater public understanding of science, and what needs to change for scientists to take a greater role in science communication.

In 2000 they published a report, 'The Role of Scientists in Public Debate' based on a survey they had commissioned from Ipsos MORI, a leading market research company in the UK and Ireland. Ipsos MORI conducted interviews with research scientists from HEIs (1540) and various research institutes (112) across the UK. This was one of the first reports to reveal the views and attitudes of scientists towards science communication. The report identified that out of the pool of scientists who took part in the survey:

- 84% agreed they had a duty to communicate their research findings to the lay public;
- 91% believed that all scientists have a responsibility to communicate the social and ethical implications of their research to policymakers and the public;
- 97% recognised the benefits associated with the lay public having a greater understanding of science;
- 56% had undertaken a science communication activity in the preceding year;
- 69%, a surprisingly high percentage, felt that they should have the main responsibility for communicating their science to a wider audience.

This survey shows that 15 years on from the Bodmer Report (Section 1.5.2), which called scientists into action, the majority of scientists feel responsible for, and see the benefits associated with, communicating their research to a lay audience.

2.4.2 Are scientists communicating with the public?

By 2006, the funding organisations for scientific research in the UK, had embraced the concept of a more dialogic approach to public engagement and the current phase of science communication, Science and Society, was underway (Section 1.5.4). Funding organisations were keen to encourage this dialogic approach to science communication between scientists and society. To support this, The Royal Society established a study with the support of Research Councils UK and the Wellcome Trust.

The outcome of this study was 'The Science Communication Report' (The Royal Society, 2006). Its aim was to provide the funders of scientific research with the evidence they needed to encourage scientists to undertake science communication. The report's objectives included, 'exploring the factors that may facilitate or inhibit science communication' but it also sought to provide evidence about 'how universities and other research institutions and funders can promote effective science communication'. The report summarised the findings of an initial web-based survey undertaken on a representative group of scientists and researchers (1485) at various stages in their career, who were working in HEIs. A selection of scientists who responded to the initial survey, were then interviewed for their views on science communication. A key finding of this report was that 74% of scientists stated they had taken part in at least one science communication activity in the year prior to the survey (not including institutional open days). The implication was that in the UK, between 2000 and 2006, the percentage of scientists actively taking part in science communication over this time frame had increased.

2.4.3 The career divide

Bauer and Jensen (2011) decided to characterise the groups of scientists taking part in science communication activities. They reanalysed the data of The Royal Society's science communication report and concluded that senior researchers were three to four times more likely to take part in these activities than their junior colleagues. They also discovered that scientists with a mainly research role were 55% less likely to engage than the scientists who balance research with teaching. 'The Role of Scientists in Public Debate Report' commissioned by the Wellcome Trust also showed that 'senior scientists ... are also more likely to have participated' (Wellcome Trust, 2000). A similar finding was also reported in a study looking at French scientists working in the Centre National de la Recherche Scientifique (CNRS) where 'activity strongly increases with increasing seniority' (Bauer and Jensen, 2011).

If we turn our attention to the US, a study published by Andrews *et al.* in 2005 surveyed the outreach activities ('outreach' is often used interchangeably with 'public engagement' in the US) of 73 scientists from the University of Colorado at various career stages. This study also showed that the career stage of scientists had an impact on their involvement in outreach activities. However, unlike the British scientists surveyed in 2000 and 2006 and the French scientists tracked over the 6-year period from 2004 to 2009, scientists at the University of Colorado were much more likely to take part in outreach activities if they were at an early career stage; graduate students spent twice as much time on outreach activities compared to scientists with faculty positions (Andrews *et al.*, 2005).

2.4.4 The gender divide

If we compare the levels of public engagement activities across the gender divide, the evidence is conflicting.

Mostly male: In the UK, 'The Role of Scientists in Public Debate Report' (Wellcome Trust, 2000) showed that there was a difference in the number of science communication activities between male and female scientists: 59% of male scientists stated they had participated in communication activities compared to 48% of female scientists.

No difference: The more recent 'Science Communication Report' (The Royal Society, 2006) does not show any differences between male and female participation. A reanalysis of the data contained within this report was undertaken by Bauer and Jensen (2011) and they report that 'Considering the factors that make a difference for being intensely active in public engagement seniority stands out. Sex ... makes no difference.'

Mostly female: However, in the US the study undertaken by Andrews *et al.* (2005), showed that in contrast to the British Scientists surveyed in 2000, female scientists at the University of Colorado were more likely than men to take part in engagement activities. Data received from a statistical study of 7000 French scientists involved in science communication activities over a 6-year period also suggested the likelihood that female scientists were more inclined to take part in these activities compared to their male counterparts: '60.9% have been active compared to 57.4% of men (*p*-value<0.0001)' (Jensen 2011).

This data indicate that there are different patterns of science communication activity and the gender effect is probably influenced by multiple, different factors.

Table 2.2 Levels of engagement of scientists with different science communication activities. Adapted from the 'Science Communication' report (2006) established by the Royal Society with the support of the RCUK and the Wellcome Trust.

Public engagement activity	None	Once	More than once (up to five times)
Institutional Open Day	44%	36%	21%
Public Lecture (include being part of a panel)	60%	21%	20%
Engaged with policymakers	67%	16%	17%
Worked with teachers/schools	70%	15%	15%
Interviewed by newspaper journalist	77%	13%	11%
Engaged with non-governmental organisations (NGOs)	77%	9%	14%
Written for non-specialist public	75%	15%	10%
Public dialogue event	80%	13%	7%
Worked with science centres/museums	87%	6%	7%
Interviewed on radio	88%	7%	6%
Judged competitions	89%	8%	11%

2.4.5   Levels of science communication activities among scientists

The 2006 'Science Communication Report' asked scientists to report whether they had taken part in a science communication activity within the last 12 months. The report categorised science communication activities into the 11 separate types listed in Table 2.2. In the year prior to the survey, 74% of scientists stated they had taken part in at least one of ten science communication activities (only ten types were used because institutional open days were not included in this calculation). If we examine the report more closely, a clearer picture emerges of the level of communication activities of this representative group of UK scientists. Closer scrutiny shows that for each separate activity, more scientists had *not* taken part than had actually *taken* part. The most popular type of communication activity that scientists had taken part in were institutional open days which, at most Universities, tend to have an 'admissions and marketing' driven agenda.[16] (The open day figures were not included in the science communication activity statistics highlighted in the 'Science Communication Report' and described below.) The report also revealed that:

- although 74% of scientists had engaged with at least one of the ten activities, 63% had a low to medium level of activities (1–10 a year, median of two), with 10% of scientists reporting they had high levels of activities (>10 a year);

---

[16] Scientists supporting Open Days were provided as an option for a science communication activity. The majority of Open Days at most universities and research institutes have an Admission or Marketing Agenda. For this reason the figures for the open days are not included in the figures produced by the report in assessing the levels of science communication activities.

- there were 26% of scientists who had no communication activity but over half (53%) of these scientists stated that 'they would like to spend more time engaging with the non-specialist public, about science'. However this leaves a proportion of scientists who clearly don't want to spend time taking part in science communication activities.

This survey also asked scientists which groups they felt it 'important or very important' to engage with about their research. Their responses showed that policymakers (60%) were the front runners, closely followed by teachers (50%), industry/business communities (not directly concerned with funding the research)(47%), journalists (45%), the non-specialist public (39%), young people outside school (38%), non-governmental organisations (NGOs) (34%), others in the media such as writers, documentary and other programme makers (33%) and, finally, general journalists (31%). This survey reveals a disparity between the groups targeted by the two most frequent public engagement activities undertaken by scientists – open days and public lectures – (Table 2.2), and the groups that scientists feel it is important or very important to engage with – policymakers and teachers.

When Bauer and Jensen (2011) undertook a secondary analysis of the data contained in the 'Science Communication Report' they found the separate science communication activities listed in the report could be grouped. They discovered that scientists:

- who engage with primary and secondary teachers also tended to participate in public lectures and open days;
- who are actively involved with policymakers were also active with NGOs;
- who work with museums and science centres are also likely to engage in public debates and serve as jury members for competitions;
- who work with newspapers also work with radio and in popular science book publishing.

The suggestion is that as scientists we feel certain affinities towards different types of engagement activity, which suit our interests and personalities (Chapter 6). This suggests that one opportunity can lead to another, for example engaging with schools and getting involved with teachers will lead to more, similar opportunities.

2.5 Concluding remarks

Clearly, there is a political commitment to realise the 'Science and Society' vision. Financial incentives have been provided and mechanisms have been put in place to encourage HEIs, research institutes and individual scientists to demonstrate their willingness to become 'engaged' with the public. Although the direction of change is constant, there are different levels of activity towards and commitment to this vision throughout the science sector. The reasons for embracing the 'science and society' vision are also varied and disparate as are the perceptions of what 'science and society' means to individuals, institutes and institutions (Burchell *et al.*, 2009). This current phase of science communication is still taking shape and forming its own identity. Its achievements and its successes are visualised as:

- a more committed scientific workforce willing to embrace this philosophy and consider the societal impact of their work;
- a public that is more engaged with science and scientific issues. Only time will tell if this vision is successfully becoming a reality.

References

Andrews, E., Weaver, A., Hanley, D. *et al.* (2005) Scientists and public outreach: participation, motivations, and impediments. *Journal of Geoscience Education* **53**, 281–293

Bauer, M.D. and Jensen, P. (2011) The mobilization of scientists for public engagement. *Public Understanding of Science* **20** (1), 3–11

Bodmer, W. (1985) Government White Paper; Realising our Potential.

Burchell, K., Franklin, S. and Holden, K. (2009) Public culture as professional science: final report of the ScoPE project – Scientists on public engagement: from communication to deliberation? BIOS, London School of Economics and Political Science

Glass, C.R., Dobernec, D.M., Schweitzer, J.H. *et al.* (2011). Unpacking faculty engagement: the types of activities faculty members report as publicly engaged scholarship during promotion and tenure. *Journal of Education Outreach and Engagement* **15** (1), 7–30

Inspiring Australia (2011) A National Strategy for Engagement with the Sciences. A Report to the Minister for Innovation, Industry, Science and Research Science Communication and Strategic Partnerships Questacon – the National Science and Technology Centre Department of Innovation, Industry, Science and Research. http://www.innovation.gov.au/Science/InspiringAustralia/Documents/InspiringAustraliaReport.pdf (accessed 5 May 2012)

Jensen, P. (2011) A statistical picture of popularization activities and their evolutions in France. *Public Understanding of Science* **20** (1), 26–36

McDaid, L. (2008) A Qualitative Baseline Report on the Perceptions of Public Engagement in University of East Anglia Academic Staff. Report No. Rs7408, the Research Centre, Ccn, Norwich. http://www.theresearchcentre.co.uk/files/docs/publications/rs7408.pdf (accessed 7 May 2012)

MacLeod, A. (2010) Towards a Professional Framework for Scientists Involved in Public Engagement Work' Report. The Wellcome Trust Sanger Institute. http://www.sanger.ac.uk/about/engagement/docs/professionaldevelopmentframework.pdf (accessed 5 May 2012)

Neresini, F. and Bucchi, M. (2011). Which indicators for the new public engagement activities? An exploratory study of European research institutions. *Public Understanding of Science* **20** (1), 64–79

Pace, M.L., Hampton, S.E. *et al.* (2010). Communicating with the public: opportunities and rewards for individual ecologists. *Frontiers in Ecology and the Environment* **8** (6): 292–298

Steering Committee for a National Science Communications Strategy (2011) Inspiring Australia: A National Strategy for Engagement with Sciences. A report to the Minister for Innovation, Industry, Science and Research Science Communication and Strategic Partnerships. http://www.innovation.gov.au/

The Wellcome Trust. (2000) The Role of Scientists in Public Debate Report. http://www.wellcome.ac.uk/stellent/groups/corporatesite/@msh_peda/documents/web_document/wtd003425.pdf

The Royal Society (2006) The Science Communication Report. http://royalsociety.org/uploadedFiles/Royal_Society_Content/Influencing_Policy/Themes_and_Projects/Themes/Governance/Final_Report_-_on_website_-_and_amended_by_SK.pdf

Science for All Expert Group (2009) Reward and Recognition of Public Engagement; Report for the Science for all Expert Group- People Science & Policy Ltd. http:// interactive.bis.gov.uk/scienceandsociety/site/all/files/2010/02/Reward-and-recognition- FINAL1.pdf

Wolfendale Committteee (1995) Report from the Committee to Review the Con- tribution of Scientists and Engineers to the Public Understanding of Science, Engineering and Technology; section 3.2.2. https://legacy.ueaac.uk/owa/redir. aspx?C=7b64d9295bcd4031bced7b5bab13592c&URL=http%3a%2f%2fcollections. europarchive.org%2ftna%2f20060215164354%2fhttp%3a%2fwww.dti.gov.uk%2fost%2 fostbusiness%2fpuset%2freport.htm (accessed 4 May 2012)

Useful websites

Beacons Project. http://www.publicengagement.ac.uk/about/beacons

China Association for Science and Technology (CAST) (2006) http://english.cast.org.cn/ (accessed 7 May 2012)

European Commission Funding Frameworks. http://ec.europa.eu/governance/impact/key_ docs/key_docs_en.htm (accessed 7 May 2012)

National Science and Technology Centre Department of Innovation, Industry, Science and Research. http://www.innovation.gov.au/Science/InspiringAustralia/Documents/Inspiring AustraliaReport.pdf

Public Understanding of Biotechnology (PUB). http://www.pub.ac.za/index.php?option= com_content&view=frontpage&Itemid=93

Research Council UK. http://www.rcuk.ac.uk/

The Research Excellence Framework 2014 (REF) http://www.hefce.ac.uk/research/ref/

CHAPTER THREE

Encouraging Scientists to Communicate with the Public

Our most direct and urgent message must be to the scientists themselves: learn to communicate with the public, be willing to do so and consider it your duty to do so.

—Bodmer (1985, p.36)

3.1 Introduction

Clearly there is a move towards a more engaged scientific community and encouraging scientists to buy into this vision is vital. Scientists are beginning to respond to this call and are starting to take a more active role in science communication activities (Chapter 2) but barriers still remain. This chapter will address three questions.

1 Should all scientists 'learn to communicate with the public, be willing to do so and consider it their duty to do so'?
2 What are the barriers to science communication activities?
3 How can we remove these barriers and provide the incentives that will encourage scientists to become more involved in science communication activities?

3.1.1 Distributing scientists on the spectrum of science communication with the public

The result of recent surveys highlighted in the 'Reward and Recognition of Public Engagement' report released by the Science for All Expert Group states that:

> Researchers will fall on a scale of willingness to participate that extends from those who will simply never take part to those whose interest/desire is such that they will overcome any barriers to participate.
>
> —*(Science for All Expert Group, 2009)*

The report clearly states that this 'duty to' communicate with the public should not extend to all researchers. There are scientists who would be happy

---

Science Communication: A Practical Guide for Scientists, First Edition. Laura Bowater and Kay Yeoman.
© 2013 John Wiley & Sons, Ltd. Published 2013 by John Wiley & Sons, Ltd.

to take part in science communication activities but even with training, they are unable to develop the skills or competence needed to be effective communicators. The report suggests that these scientists should be discouraged from taking part in face-to-face activities with the public. Burns and Squires (2011) also identified this group as 'academics that shouldn't be allowed anywhere near the general public'. There are other members of the research population who may be willing to take part in science communication activities but lack confidence in their skills or abilities. An obvious solution is to train and support these scientists to develop both their skills and their confidence. In contrast, we have the scientists, who have the natural ability to be effective science communicators but have no desire to take part. These scientists need to be motivated and supported because they clearly have the potential to contribute. The report suggests this group could be encouraged by offering incentives and rewards. Finally we have the group of scientists who will never take part in science communication activities even if they are offered support, reward and recognition. Poliakoff and Webb (2007) also identified this group of scientists. They reported 'some scientists did not intend to participate despite recognising potential career benefits.'

We can distribute scientists along a communication spectrum as illustrated in Figure 3.1. The upper right-hand quadrant represents scientists who are willing and able to become effective communicators. These scientists should be rewarded and their achievements should be recognised, but it is just as important that these scientists are not discouraged from taking part. The bottom

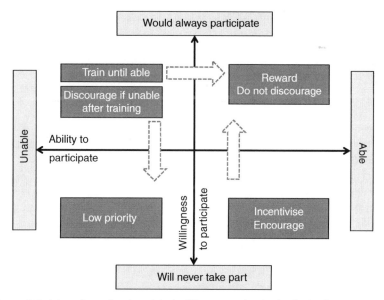

Figure 3.1 A two-dimensional model of willingness and aptitude of scientists to participate in Public Engagement Activities. Adapted from the 'Reward and Recognition of Public Engagement' report (Dyball and King (2009)). Reproduced with permission of the RCUK.

right-hand quadrant represents the scientists with natural talents who need to be incentivised and encouraged to take part. The top left quadrant represents the scientists who are not natural communicators. The scientists in this quadrant fall into two clear groups. The first group can become effective science communicators if they receive training and support. In contrast, the second group are the scientists who even after training should not take part in face-to-face activities with the public. Finally the last quadrant, the bottom left, represents the scientists who have no natural ability and are unwilling to take part in science communication activities. There is little impetus to offer this group of scientists encouragement, incentives and training. They are a low priority.

3.2   Science communication: the barriers

As the results from The Royal Society's 2006 'Science Communication Report' indicate Figure 3.1 highlights, there is a group of scientists who are unwilling to take part in public communication activities. Like any new activity, there are barriers that hinder scientists from taking part. These barriers can be divided into issues surrounding time, peer support, discomfort, exposure and vulnerability.

3.2.1 Time as an inhibiting factor

One of the questions scientists were asked in the 'Science Communication Report' (Chapter 2), was whether they 'would like to spend more time, less time or about the same amount engaging with the non-specialist public about science?' The responses were:

- 45% of scientists said that they would like to spend more time;
- 41% stated that they were content with the amount of time spent on this activity;
- 3% reported that they would like to spend less time;
- 11% said that they didn't know.

The implication is that science communication activities suffer because scientists lack the time needed to get involved. Globally, scientists from different research backgrounds such as Higher Education Institutes, Research Institutes or industry and at all stages of their career have many different and competing demands on their time. For example, if we look at research scientists currently employed in the HE sector in the UK, a typical job role may include:

- preparing, delivering and assessing teaching;
- enhancing and ensuring student satisfaction with the student experience;
- various administrative roles essential for the smooth running of a department, school or faculty within HE;
- initiating, negotiating, stimulating, promoting and delivering enterprise and money generating opportunities;
- ensuring research output levels that include supervising research students and post-doctoral workers, writing papers and securing grant income;

- sitting on public panels or research council panels, reviewing academic papers and/or editorial duties for scientific journals, organising conferences;
- undertaking clinical work.

Recently the demands and pressures on UK scientists working in HE have increased due to a growing, market-driven economy within this sector. There have been three key factors that are responsible for this change:

1 increased student fees;
2 pressure from the Research Excellence Framework (REF);
3 a requirement to generate third-stream income (enterprise).

The distribution and source of university funding is being radically over-hauled, with core funding from the government being replaced through a substantial hike in student fees (£3000 per year up to a maximum of £9000). Students are going to demand 'value for money' and an excellent student experience is going to be firmly placed at the top of every student's wish list. In order for universities to charge these increased student fees, they need to deliver a 'product' that students are prepared to pay for. Inevitably this will affect academic time as teaching loads and staff–student contact time increases to satisfy demands. Burns and Squires (2011) also point out that staff who spend time undertaking science communication activities can im-pact on their colleagues' time. This is because staff with a heavy teaching load can resent having to take on more teaching responsibility while their colleagues are spending time outside the university communicating with the public.

The second key factor is ensuring that research output levels are satisfactory (or above) as judged by the REF due to take place in 2014. This exercise will review the quality of outputs of academic research at UK academic institu-tions. The results of this exercise will dictate the level of Quality-Related[1] (QR) funding to these HE institutes. Clearly there is intense pressure on scientists to be publishing world-renowned research in high impact journals to secure QR funding for their institutions. But achieving this level of research success will impact negatively on the amount of time that can be spent on other activities, including science communication.

Finally, there is a growing tacit understanding that in addition to teaching and research income, researchers should be generating revenue from enter-prise activities. Although these changes are affecting established scientists, early career scientists (e.g. graduate students) are not immune to demands on their time. Andrews et al. (2005) discussed the time constraints experienced by graduate PhD students at the University of Colorado. Graduate students don't just have the pressure to successfully complete their research on time, but they often encounter financial pressure that they ameliorate by undertak-ing paid research or teaching activities. This means they have little time left

---

[1] The Research Excellence Framework (REF) is the new system for assessing the quality of research in UK higher education institutions (HEIs). It replaces the Research Assessment Exercise Research (RAE) and it will be completed in 2014. The REF will be undertaken by the four UK higher education funding bodies. http://www.hefce.ac.uk/research/ref/

to undertake science communication activities. Early career researchers who are further up the career ladder, also encounter competing time pressures. They are trying to secure a successful scientific career which includes gaining tenure or a permanent position. This is still judged by the time-intensive demands of demonstrating research potential which is evaluated using research outputs such as publishing papers and writing successful grants. As such we need to ask, is it ethical to expect early career scientists to undertake science communication activities when this could jeopardise the likelihood of establishing a successful career?

In 2007, Poliakoff and Webb also asked what predicts whether scientists intend to participate in science communication activities. They distributed 1000 questionnaires to scientists at different stages of their career at the University of Manchester and received 169 completed questionnaires in return.[2] Their research suggested that a lack of time had little influence on a scientist's intention to participate in science communication activities. Instead they found the perceived lack of time was 'associated with more negative perceptions of participating in public engagement' and that 'time constraints may be used as an excuse to mask other concerns about participating in public engagement activities.' These other concerns were not clearly defined by the authors, but for those of us currently pursuing a career in science it is quite likely that these concerns include the balancing of administrative duties with teaching, research and enterprise. Negotiating these different, competing demands required for a successful scientific career is challenging, without the additional requirement for science communication.

3.2.2   Peer support as an inhibiting factor

It is not just the time constraints that can prove a barrier to science communication; a lack of support from peers and colleagues also inhibits these activities. Early career scientists (e.g. graduate students) believe they are actively discouraged from undertaking science communication by senior colleagues. This discouragement was described by Andrews *et al.* (2005) as 'Very few of the faculty members explicitly discussed the subject of outreach[3] with their students. Although they generally did not explicitly discourage their graduate students from participating in outreach programs, most never discussed the topic with their students at all.' Interestingly, this lack of dialogue led both the graduate students and faculty members to believe there was little interest in science communication activities from either side.

It seems an obvious conclusion that when scientists perceive there is little institutional support towards science communication activities, this will have a negative impact. A recent paper examined the emphasis given to science communication among 40 research institutes throughout Europe (Neresini and Bucchi, 2011). It revealed that very few of these institutes seemed to have

---

[2] The authors state that the completed questionnaires were from a representative population with a slight over representation of scientists at higher career levels.

[3] In the US, outreach is a term that has similar meaning to Public Engagement in the UK.

embedded science communication within their core activities. This lack of institutional support was conceptualised as a lack of:

- institutional funding;
- support for the design, organisation, delivery and evaluation of activities;
- training to support activities.

The work by Andrews *et al.* (2005) and McDaid (2008) recount scientists describing a lack of institutional incentives or recognition of communication activities. The consequences are that scientists feel obliged to conduct their outreach and science communication activities below the institutional radar, and out of sight of colleagues. In addition, a scientist often associates personal success with a positive reputation among peers and they are concerned that communicating with the public risks 'being deemed a popular thinker' (The Royal Society, 2006; McDaid, 2008). The 'Reward and Recognition of Public Engagement' report (Science for All Expert Group, 2009) also highlighted the idea that a scientist who is a 'science communicator' is probably a scientist who is an unsuccessful researcher. A junior scientist admitted he was worried that if he took part in science communication activities his peers would say 'he is doing that because he could not build a good enough research career'. Female scientists have an additional concern. They worry that science communication actually reinforces the stereotype of a ditsy female scientist being less serious or professional than their male colleagues, 'I have been gently warned by senior colleagues that "if you are female [in a certain topic] then you need to avoid light and fluffy topics," i.e. communication of science with the public.'

3.2.3 Discomfort, exposure and vulnerability as inhibiting factors

Scientists are also constrained from undertaking science communication activities at an emotional level. Poliakoff and Webb (2007) identified a 'minority of scientists' who fear the 'unknown-ness' of communicating with the public, although they found that most scientists did not share this concern. The findings of the ScoPE project (Burchell *et al.*, 2009), conveyed in the 'Public Culture as Professional Science; Final Report', were in sharp contrast to this. The report described barriers such as 'discomfort, exposure and vulnerability'. Scientists shared real concerns about their ability (or lack of) to learn the communication skills required for interacting with the public. Throughout their careers, scientists have honed a language to accurately communicate the complexity of their science and research to their peers. This attention to accuracy and complexity has led to the development of a scientific language and jargon that can be impenetrable to a lay audience, including scientists from other disciplines. To communicate effectively with a public audience, scientists must learn to translate the technical complexity of scientific writing into jargon-free, simple language, that can be understood by the public without losing the scientific truth or misrepresenting scientific uncertainty (see Section 4.7). The report demonstrates that scientists have recognised that they cannot take a 'one size fits all approach'. This is because there is more than one 'public' (see Chapter 6) and care must be taken to ensure the scientific language and content is appropriate for each public audience. When this doesn't happen, the result is that science becomes remote and unattainable, and our audience feels disengaged, disenfranchised and excluded. Using inappropriate language can also result in misinterpretation by the public and this lack of clarity can have ethical implications (Pace *et al.*, 2010).

The interviews with scientists, undertaken as part of the ScoPE project, also highlighted an additional concern, raised by non-clinical scientists working on clinical diseases. These scientists are anxious about approaches from members of the public who want to discuss their diseases or diseases of their loved ones. These conversations need to be handled with sensitivity and care to minimise any misinterpretation that can lead to misunderstandings, false hope or lack of hope and ultimately distressed individuals. This requires real skill and training. Finally, there is considerable nervousness expressed by some scientists who work in research areas such as animal research that attract activist groups with hostile agendas. Quite understandably these scientists are unwilling to engage in a public dialogue because they fear for their personal safety or the safety of their family and colleagues.

3.3 Removing barriers and providing incentives

A key question that recent reports have addressed is how can more scientists be encouraged to undertake science communication activities? Clearly policymakers, scientists and the public can identify positive reasons for taking part, but significant barriers still need to be removed so that:

- more scientists are encouraged to participate;
- the scientists that participate are encouraged to do so more regularly;
- the scientists that participate are encouraged to get involved in a wider range of activities with a wider variety of audiences.

3.3.1 The political environment

The first stage in the process of removing the barriers associated with science communication has been to recognise that they exist and to identify what they are. The second step has been the political[4] will to begin removing them (Chapter 2).

There have been key changes in the political arena to encourage and promote science communication in the workplace, and the workplace is beginning to respond to this (Chapter 2). The evidence for this is:

- the number of HEIs that submitted bids to obtain either Beacon Status or Catalyst funding;
- the swathe of research institutes and HEIs who have signed up to the recent Concordat;

---

[4] Political is used to describe any person, body, society or party that can influence decisions.

Table 3.1 Prizes and awards for science communication activities.

Award	Eligibility	Nomination Process	Website
Royal Society Kohn Award	Early career scientists of any discipline	You can be nominated or you can self nominate	http://royalsociety.org/awards/kohn-award/
RSE Beltane Prizes for Public Engagement Senior Award	The Senior Prize is awarded to someone who demonstrates a sustained high-quality track record of public engagement and who shows exemplary communication and innovation skills.	Nomination by Fellow of the Royal Society of Edinburgh	http://www.rse.org.uk/667_RSEBeltanePrizesforPublicEngagement.html
RSE Beltane Prizes for Public Engagement Innovator's Prize	The Innovator's Prize is awarded to someone who does not have a long history of engagement but who is recognised as an emerging talent and whose innovations will help them develop their career.	Nomination by Fellow of the Royal Society of Edinburgh	http://www.rse.org.uk/667_RSEBeltanePrizesforPublicEngagement.html
Fame Lab	Over 21 years old working in or studying science, technology, engineering, medicine or maths in the UK. This includes private and public sector employees.	Enter into the competition	http://famelab.org/
Royal Society Michael Faraday Prize	Practicing scientist or engineer	Must be nominated	http://royalsociety.org/awards/michael-faraday-prize/
Science Writing Prize	UK professional scientists at postgraduate level and above UK anyone else with a non-professional interest in science (this includes undergraduate students).	Enter into a Competition	http://www.mrc.ac.uk/Sciencesociety/Awards/index.htm
Margaret Mead Award	Presented to a younger scholar for a particular accomplishment, such as a book, film, monograph, or service, which interprets anthropological data and principles in ways that make them meaningful to a broadly concerned public.	Must be nominated with two letters of recommendation	http://www.aaas.org/aboutaaas/awards/public/
Joshua Phillips Award for Innovation in Science Engagement (Josh Award).	Anyone who is recognised as an up-and-coming talent in science communication, with innovative and new ideas	Must be nominated	http://www.eps.manchester.ac.uk/about-us/features/joshua-phillips-award-nominations/
Sagan Award for the Public Communication of science	Outstanding communication by an active planetary scientist	Division of Planetary Science Member	http://cssp.us/public-understanding/sagan-award-for-public-understanding-of-science.html

Prize	Criteria	Eligibility	URL
Kalinga Prize for the Popularisation of Science	Candidate shall have made a significant contribution to the popularisation of science	Governments of Member States	http://www.kalingafoundationtrust.com/website/home.htm
Society of Biology	New researcher and established scientists	Must be nominated	http://www.societyofbiology.org/newsandevents/scicomm
BBSRC Innovator of the Year	Open to all BBSRC-funded scientists	Must be nominated	http://www.bbsrc.ac.uk/innovator
The Kelvin Medal and Prize	Outstanding contribution to public engagement within physics	Must be nominated	http://www.iop.org/about/awards/education/kelvin/page_38636.html
The Bragg Medal	To physics education and to widening participation	Must be nominated	http://www.iop.org/about/awards/education/bragg/page_38627.html
The Rooke Medal	To individual, small team or organisation	Academy Fellows and external applicants.	http://www.raeng.org.uk/prizes/pdf/Awards_Medals_Brochure.pdf
HEPP Group Science in Society Prize	Physicists early in their careers for outreach in particle physics	Must be nominated	http://www.iop.org/activity/groups/subject/hepp/prize/society/page_40789.html
SGM Peter Wildy Prize	To a microbiologist	Must be nominated	http://www.sgm.ac.uk/about/prize_lectures.cfm
SGM Outreach Prize	To a microbiologist for outreach	Must be nominated	http://www.sgm.ac.uk/about/prize_lectures.cfm
The JBS Haldane Lecture 2012	To an individual for outstanding ability to communicate topical subjects in genetics research	Must be nominated	http://www.genetics.org.uk/page/4386/The-JBS-Haldane-Lecture-2012.html
Examples of Individual Prizes awarded by individual Universities	University of Edinburgh Tam Dalyell Prize For Excellence in Engaging the Public with Sciences	Must be nominated	http://www.ed.ac.uk/news/all-news/dalyell
	UEA CUE East Individual Awards for Public and Community Engagement	Must be nominated	http://www.uea.ac.uk/ssf/cue-east/awardsfunding/publiccommunity

- the growing number of prizes being offered by scientific institutions (including The Royal Society and the Royal Society of Edinburgh) and several science societies (as listed in Table 3.1);
- the growing number of research institutes and HEIs that recognise science communication success with monetary rewards such as prizes, bonus awards and promotion. Information about many of these prizes, including who is eligible and how you can be nominated, are listed in Table 3.1.

3.3.2   Rewards and recognition

If scientists consider that rewards and recognition are an incentive, then the question is: What are the culture changes in the workplace that can incentivise us to engage with the public?

Many universities in the UK have started to include engagement activities as a promotion criterion for academic staff across all academic disciplines including science. It is worth investigating whether your workplace includes public engagement alongside other promotion criteria such as teaching, administrative duties, enterprise and research. For example, the University of East Anglia (UEA) embedded public engagement into its promotions criteria as a direct result of being awarded Beacon Status for Public Engagement in 2008. The process that led to this change is presented in Case study 3.1 written by Julie Worrall, the Director of the CUE East Beacon.

Case Study 3.1

Creating an Institutional Culture That Rewards Public Engagement
Julie Worrall

Between 2008 and 2012, CUE East (Community University Engagement East) at UEA, one of six national 'Beacons for Public Engagement', helped to create an institutional culture whereby public engagement (PE) is recognised and rewarded as part of academic practice in HE. In 2008, CUE East commissioned a baseline survey exploring academics' attitudes at UEA towards PE and the factors affecting their involvement. It revealed compelling evidence about the impact of the lack of formal recognition of PE:

> *It's not going to be anything we can use on our CV for future job applications, so I suppose the brutal truth of it is it's got to be done as an act of social citizenship rather than anything else. Senior Researcher, UEA*

—(McDaid, 2008)

It also provided a benchmark against which the changes could be assessed, recognising PE activity already taking place and pointing to the incentives and barriers for individuals getting involved.

From the outset, CUE East saw their role as providing the 'carrots' as opposed to the 'sticks' that would be provided by research funders via awards and funding criteria. In early 2008, UEA's Head of Human Resources invited CUE East to draft new academic promotions criteria, and work began on providing a clear and convincing rationale for a significant change that would be acceptable to all. Working with the then named 'Knowledge Transfer' Executive, CUE East produced a simple but effective model to aid discussion on assessing individual performance and impact in relation to PE. The discussions considered:

- What is engagement?
- Why should engagement be included?
- How might you measure inputs, outputs and impacts?
- What might an assessment framework actually look like?
- What complications/issues are we likely to confront in implementing the assessment?

CUE East also devised a simple typology of PE which reflects one-, two- and three-way types of engagement activity.

Communicating knowledge and enriching cultural life	**Providing a service and being in dialogue with the public and communities**	**Being in dialogue with the public and policy-makers**
1 way	2 way	3 way
e.g. public lectures, media work, writing for the non-specialist, exhibitions, show casing academic know-how, pro-bono schemes, communicating research to the public, acting as the lead for major festival themes, contributing to the organisation and delivery of engagement activities.	e.g. volunteering, promoting and employing user involvement in research and the co-production of research, forums, focus groups, seminars and debates that involve the public, pro-bono schemes, drama outreach, museum education, continuing education and lifelong learning, contributing to the organisation and delivery of engagement activities.	e.g. governmental committees involving the academic as the 'expert', such as an expert panel, government led public consultation and task forces, and active membership of professional bodies.

The draft criteria were then incorporated within UEA's revised promotions 'Green Book' and piloted for 1 year, during which CUE East obtained feedback on issues encountered by both individuals, and the school promotion boards. CUE East had provided lots of examples of activities and outputs in the draft criteria in the light of early concerns about what activities should be 'counted' as engagement. It seemed, however, that this served to confuse rather than clarify things and the criteria was deemed 'diffuse'. Final amendments were then made and formally ratified by UEA's Executive Team in 2010.

CUE East also introduced individual PE Awards in the form of ex-gratia payments at £1000 and certificates presented at UEA's Annual Congregation, providing another, more public, way of recognising and rewarding UEA and Norwich Research Park[6] staff for their

---

[6]The Norwich Research Park is a collaboration between the University of East Anglia, the Norfolk and Norwich University Hospitals, and four independent research centres; the John Innes Centre, the Institute of Food Research, the Sainsbury Laboratory and, since July 2009, The Genome Analysis Centre. The Norwich Research Park is also home to over 30 science and IT based companies.

commitment to PE. Award holders have stated that the recognition is more important than the financial reward. The Awards specifically aimed to:
- identify exemplar individuals who can act as role models to others;
- encourage and reward two-way dialogue between the university and the community;
- raise the profile of public and community engagement.

The key lessons from this case study are:
- be realistic about the challenge of shifting organisational culture;
- start small and don't promise too much in order to manage expectations;
- be strategic: systematically identify barriers and motivations, and target resources to address these;
- do your baseline research as this provides a vital comparator against which to measure success.

A longer version of the this case study can be found on the NCCPE website: http://www.publicengagement.ac.uk/how/case-studies/recognition-promotion-uea
A 'mini' case study on the CUE East Individual Awards can be found on the NCCPE website: http://www.publicengagement.ac.uk/mini-case-study-cue-east-individual-awards

---

Gaining promotion as a direct result of your engagement activities with the public isn't confined to the UK. Erin Dolan (2008) writes that in the US 'a number of institutions offer promotion and tenure for outreach scholarship'. As an example, she provides the promotion criteria for the University of Wisconsin: faculty must demonstrate excellence in a primary area (i.e. research, teaching, or outreach/extension) and significant accomplishment in one of the two remaining areas as grounds for promotion or tenure (Dolan, 2008).

As scientists in general we are comfortable with building a case for research achievement based around the measures and metrics used to establish research success which can be measured by;
- grant awards and funding success;
- published research papers, both the quantity, and the quality as judged by impact factors;
- your H index (measure of citation rate);
- invitations to speak at conferences;
- prizes and rewards for research success;
- impact of research (e.g. enterprise activities).

Similarly, scientists that work in HEIs are also developing teaching portfolios that highlight their teaching success and achievements. With the inclusion of overt recognition of science communication through awards, tenure or promotion, it is clear that there is a growing requirement to develop your curriculum vitae or public engagement portfolio to document and highlight your achievements. Even if your workplace does not currently recognise or value your science communication activities, remember that other sources offer prizes and rewards (Table 3.1) that you can apply for. Take time to consider the raft of metrics that you can use to display your successes and achievements and allow them to be measured as highlighted in Box 3.1.

> **Box 3.1   Metrics used to measure success and achievement**
> - Obtaining funding is a direct result of a good idea and a well written grant application: it provides evidence of funding success.
> - Publishing case studies, evaluation reports, or sharing good practise and tips with colleagues: this provides evidence that can be clearly identified and assessed.
> - Receiving invitations to speak at conferences (both science communication and subject specific conferences): provide good indicators of success, respect and good practice.
> - Being nominated for and receiving prizes and rewards for your science communication activities: provide evidence and recognition of your good practice and achievement (Table 3.1).
> - Sharing good practice and drawing on your experience: can be evidenced by requests for consultancy and mentorship.
> - The number and range of activities and the types of audiences that you engage with can be documented within a science communication portfolio: evaluation of your activities that show a demonstrable societal impact is a key indicator of success and achievement in this area.

There is evidence that public engagement including science communication activity has been a key factor in gaining promotion. An appendix document to a Beacon report,[7] describes promotion on the basis of engagement activity occurring at the University of East Anglia, the University of Edinburgh, the University of Manchester, the Manchester Metropolitan University, Cardiff University, the University of Glamorgan, and University College London. In addition, the RCUK recently published a booklet *What's in it for Me?* (RCUK, 2010). This booklet contains a selection of case studies that emphasise the positive benefits of science communication experienced by researchers from across the UK. The first case study in this booklet was written about Professor Jim Al-Khalili, an Engineering and Physical Sciences Research Council (EPSRC) Senior Media Fellow at The University of Surrey. He is quoted as saying that 'my promotion to Professor of Physics came a few years early' as a result of growing success in science communication activities.[8]

3.3.3   Funding as an incentive

Another incentive that can encourage scientists to communicate with the public is the potential to attract funding, one of the treadmills that the majority of research scientists have to endure.

In a recent paper written by Pace *et al.* (2010), one of the potential rewards of communication activities is described; 'broadening public awareness of one's research (e.g. through news stories in the media) can also attract the attention

---

[7] The Independent Review of Beacons for Public Engagement Evaluation Findings Report Achievements – Appendix 3 'Create a Culture within HEIs, Research Institutes and Centres' published in – Achievements published in 2010 by People Science & Policy ltd. http://www.rcuk.ac.uk/documents/scisoc/BeaconsAppendices.pdf

[8] http://www.rcuk.ac.uk/documents/scisoc/RCUKBenefitsofPE.pdf

of potential funders, including private donors and foundations'. In addition, the merits of engagement and outreach activities can 'raise a scientist's profile as well as that of his or her research institution among the broader community'. The following Case study 3.2 outlines how collaboration between the scientist Dr Michael Wormstone, a Senior Lecturer at the UEA and the national charity Fight for Sight, managed to raise awareness of vision research among the local community. The outcomes of this event included:

- an increased awareness and understanding of vision research;
- additional funding for the Fight for Sight charity that funds vision research.

Case Study 3.2

World Sight Day 2009
Michael Wormstone

On the 8th of October 2009 – World Sight Day – my laboratory and the charity Fight for Sight organised a day of events to raise awareness of vision research. During discussions with representatives from Fight for Sight, the UK's largest funder of eye research, it became apparent that the knowledge and support base for the charity in Norfolk was weak, relative to other parts of the country, even though world-leading research was conducted at the UEA. It was mutually agreed that some initiative was needed to highlight the importance and value that eye research has to the people of Norfolk and the world at large. With this in mind my colleagues and I at UEA started to develop an approach to do this.

Two events were organised. In the daytime we set up a stand in the Jarrold department store in Norwich. This was manned by eye research scientists, members of the East Anglian eye bank and representatives of Fight for Sight. This gave us the opportunity to engage with members of the public unfamiliar with the work at UEA, or the research funded by Fight for Sight and to explain the value of what we have done and what we aim to achieve in the future. The second event was an evening presentation by Michele Acton, the CEO of Fight for Sight, followed by a presentation by myself as a keynote speaker. A reception to enable informal discussion and questions took place afterwards. This evening activity was a ticket only event. Considerable effort was expended in organising the events on World Sight Day, but for all those who took part it was a rewarding experience with tangible benefits.

The stand in Jarrold department store was extremely well received by the public. We were well positioned on the ground floor near an entrance; therefore it was hard to ignore us. Several people were always present at the stand and members of the public happily engaged in activities we had prepared and were keen to see presentations illustrating our work. A wide variety of questions were asked, and through activities like this you develop a feel for what really matters to people, which in turn helps determine your next avenue of research.

In the evening, we decided to restrict the number of speakers to two. Michele Acton gave an overview of Fight for Sight, which was followed by a single keynote lecture given by myself. After the presentations, members of the audience were able to mix with laboratory members, the Fight for Sight team and clinicians from the Norfolk and Norwich University Hospital (NNUH). This worked extremely well as the audience were able to ask questions which were fully answered and they headed home more informed and happy.

Impact

It is now a couple of years on from the World Sight Day events and we can look back and reflect upon that time. We set off with a blank canvas and through hard work and commitment from both my colleagues at UEA and Fight for Sight I believe we delivered a truly memorable day for those involved.

The evening event allowed the charity to invite people who had already supported Fight for Sight and their friends and family. This in itself began the process of raising the profile of eye research and the work of the charity, UEA and the NNUH.

The charity was overwhelmed by the response that followed. Prior to the event there had been very little activity in Norfolk on behalf of Fight for Sight, but the event encouraged Jane Ridley from Burnham Market to hold a lunch to raise its profile. Jane commented:

> Having got to know the charity and the wonderful research it funds in Norfolk and elsewhere, I wanted to galvanise support for the work being undertaken to help prevent sight loss and restore sight.

In the two years since, supporters have organised a further 10 events and more are planned. These events have included lunches, a quiz, a coffee morning, a wine tasting, a bridge drive and an evening with the Sheringham Shantymen organised by Rosemary Rushbrook of Norfolk. Rosemary said:

> I attended the World Sight Day evening at UEA and given the importance of sight to me, I decided to help raise money for eye research.

Each of the events organised was attended by between 25 and 150 people. A stall was held at Creake Abbey Market, which regularly has 400 attendees, and a Fight for Sight tree decorated in colourful spectacles was on display at the Fakenham Christmas Tree Festival in front of 18 000 people.

Sarah Caswell, an artist in Great Walsingham, Norfolk, took part in the London Marathon 2011 and raised money for Fight for Sight:

> I wanted to support Fight for Sight because I am a painter of really large, really bright paintings. Having had an extreme detached retina and with myopia and familial glaucoma, I absolutely recognise the importance of eye research.

Overall, Fight for Sight was extremely pleased with the outcome of the initial day. The profile of eye research and Fight for Sight has been increased by the events mentioned above. It is also pleasing that these events have raised £7500 for vital eye research.

Michele Acton of Fight for Sight said:

> Fight for Sight is very grateful for the support from UEA and the NNUH. Working together we were able to demonstrate to the original event attendees the importance of the work that we are all doing. People have taken this message and cascaded it further across their networks. This is vital in helping raise the profile of eye research and the need for funding.

3.3.4   Addressing perceptions

Genuine concerns have been voiced by scientists that any attempts to communicate with the public would be looked at askance by their colleagues, especially if the forum for communication is through the media as 'academics are prejudiced against it as it isn't seen as proper scholarship, mere journalism' (McDaid, 2008). However, many respected scientists have little patience with this view (Taylor, 1988), Sir Lawrence Bragg states:

> They are apt to regard colleagues who attempt to give popular talks as actors aiming
> at popular applause, who cheapen science by over-simplification and spoil the dignity
> of its aloofness. I am quite out of sympathy with my fellow scientists when they adopt
> this attitude.

If you are one of these concerned scientists, then hopefully you can find additional reassurance in the findings of the 'Scientists on Public Engagement: from Communication to Deliberation' report (Burchell *et al.*, 2009) that provides a different perspective. This project was an in-depth social science study that investigated scientists' views about science communication. As part of this project, 30 scientists at various stages in their career were interviewed about their views on science communication, between March 2007 and June 2008. The report clearly showed that many scientists have a positive perception of science communication. It highlighted the fact that scientists who are enthusiastic and successful science communicators are viewed positively by colleagues because they are representing the scientific profession in a positive manner.

If you are a young career scientist then you should take note that another view that emerged from these interviews was that, young scientists were felt to be more important than their senior colleagues for public facing activities, especially activities focused at children and young people. Senior colleagues recognised that young scientists can replace the view, perception and stereotype of the mad, old, grey scientist with a young, exciting, attractive individual with interesting prospects (Figure 3.2).

Figure 3.2 Mad scientist.

What you need are young people, you need the vigorous, exciting people who are really having an impact through their science but at the same time communicating that enthusiasm . . . to the public.

—*Burchell et al. (2009)*

Case study 3.3 was written by Dr Phil Smith MBE, director of the Teacher Scientist Network (TSN). It agrees with the findings of the ScoPE project as it highlights the benefits for scientists as well as pupils and teachers when it comes to addressing the negative stereotype of scientists. In addition it describes the advantage of offering scientists the opportunity to improve their communication skills.

Case Study 3.3

Benefits to Scientists, Pupils, Teachers
Phil Smith

The Teacher Scientist Network (TSN) is an organisation which was conceived as a partnership scheme between teachers and scientists to support science education in local schools. Teacher–scientist partnerships work with pupils and thereby help to challenge the (often negative) stereotypes of what scientists do and what scientists look like. The long-term nature of these partnerships means that the pupils as well as the teachers can:

- build a relationship with the scientist;
- get to know the scientist in a more informal context;
- show scientists to be similar to other adults, with outside interests and not tied to the ivory towers of their laboratory.

Much evidence suggests the 'geeky' image of a scientist to be one of the factors influencing poor uptake of the sciences by young people, so any attempt to counter this is a welcome benefit to the programme.

It is clear that teachers and their pupils benefit from an input from the science community, but scientists also gain. Scientists have said they find working with teachers and their pupils very rewarding and a refreshing break from the laboratory. They find satisfaction in seeing outcomes quickly: in school, children react immediately to the activity, whereas in the research laboratory it might take weeks or months before the result is seen.

Partnership working in the classroom provides an ideal forum for the scientist to improve their communication skills, and some have said they see their research in a new, broader perspective. Probably in the long-run, the scientific community are the ones who will benefit the most – if our children grow up with a more positive view of science, our future citizens will perhaps be in a better position to make rational judgements about science within a democratic society.

Further detail about the activities of the TSN can be found in Chapter 10, Case study 10.8. http://wwww.tsn.org.uk

3.4   Seeking advice and support

Moving scientists from being willing to take part in science communication activities but lacking the confidence and skills required, is a priority if we want to increase the number of 'scientists who are taking part in science communication activities' (Figure 3.1). The ScoPE report provides an insight into some of the common concerns shared by many scientists who are considering science communication but feel unprepared and unwilling to expose a lack of skills and abilities. If you are one of these scientists then one of the first things you should remember is that you can participate in science communication activities without having a front line, public facing role. Every activity or event requires an initial idea, as well as a planning and development stage that offers opportunities for useful contributions without a public presence. The public face of the activity can be provided by other colleagues. For example, consider appointing an external presenter, or contact your local outreach officer (see Box 3.2) to help with this aspect of your communication activity. This potential additional expense can be incorporated in your requests for funding. Many HEIs and research institutes employ outreach officers to support the science communication activities of scientists. The support provided will vary depending on the institute and type of department as well as the skills and remits of each individual outreach officer. A typical role of an outreach officer including the support they can provide, is highlighted in Box 3.2.

Box 3.2   Outreach officers

Carl Harrington is an Outreach Officer with a science specialisation at UEA whose position is funded as part of the university's commitment to widening participation. His main role is to work with local schools, especially secondary schools (Key Stage 3 and 4) to make their pupils aware of HE and its benefits, plus encourage them to consider remaining in education after they leave school. As an Outreach Officer, part of Carl's role is to liaise with schools, making them aware of and encouraging participation in outreach events and services. He has also developed expertise in designing, organising and running outreach activities and events. As part of his role Carl has opportunities to share his experience with undergraduate students through their taught programmes and the student ambassador scheme at UEA (http://www.uea.ac.uk/teachers/pre16/amb). He has also worked with Masters, PhD students and academics who are keen to start working with schools and has found himself not only offering the initial training and support, but also acting as an informal mentor.

I am happy to work with any scientist who is considering starting out in science communication. I can give scientists opportunities to come along and see what I do, take part in school events with me, or I can work with them to design their own events. It is really rewarding working with pupils from schools but people can find it hard to get the pitch of an event right, especially if you haven't worked with schools before.

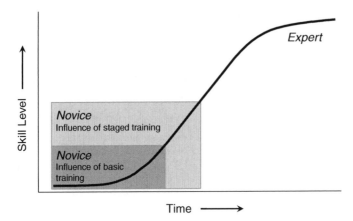

Figure 3.3 The graph represents the lifetime development of a full-time science communicator. The inset box shows the much smaller amount of time and development that may apply to science communication when practiced (and taught as a sideline) to the main career of a scientist with initial training. However, with support and planning throughout your career, further progress towards 'expert' can be achieved. Adapted from 'Towards a professional framework for scientists involved in public engagement work', a report prepared for the Wellcome Sanger Institute by A. Macleod (2010). Reproduced by permission of the Wellcome Trust Sanger Institute.

3.4.1 Training courses

In 2010, Alison MacLeod began to explore whether scientists working at the Sanger Centre are able to use training to develop the necessary skills required for science communication activities. She concludes there is enough scope for someone to make the journey from 'novice to seasoned communicator' although it may be difficult to give someone the right amount of practice and training necessary to become an expert. This journey is represented in Figure 3.3. MacLeod explains that the figure is a very simplified representation of the development of a novice science communicator into a seasoned communicator, and then into an expert. The time required to become an expert will differ according to:

- scientific discipline;
- time available to practice and train as a science communicator;
- the natural ability of the scientist involved.

Despite these differences, training plays a key role in developing the necessary skills for science communication (MacLeod, 2010). If you are a scientist who would like to start developing skills in this area but feel that you need training, there are several courses specifically designed with you in mind. These include short courses that last two or three days as well as credit-bearing, postgraduate courses. They can include training in the practical aspects of science communication, for example developing presentation skills and working with the media. There are also courses that include the theory, philosophy and history that underpins science communication. Details of several of these courses are provided in Box 3.3.

Box 3.3   Training courses

Institute of Physics

The Institute of Physics runs outreach workshops at various times and locations throughout the year.

www.iop.org/outreachworkshops

The University of the West of England (UWE)

UWE's offers a variety of training courses that include a Science Communication Masterclass, an intensive course created to provide professional development in science communication.

www.scu.uwe.ac.uk/index.php?q=node/199

SciConnect

Offers bespoke training courses in science communication and media skills to research scientists.

http://www.sciconnect.co.uk/courses.html

Science Communication Unit

Based at the University of the West of England, the Science Communication Unit aims to communicate science to various audiences using innovative means. It offers a series of short science communication workshops, including: 'Science Communication Masterclass' and 'Bespoke Training Portfolio'.

http://www.scu.uwe.ac.uk/index.php?q=node/201

Science made simple

Has over ten years' experience of presenting and developing science presentations. It runs a number of one and two-day training courses aimed at museum and science centre staff, science communication students, undergraduates, postgraduates and research scientists.

http://www.sciencemadesimple.co.uk/page24g.html

STEMNET

Offers training for STEM Ambassadors contact your local STEM Ambassador contract holder and see what they have available

http://www.stemnet.org.uk/content/ambassadors

The Royal Society—Science communication workshops

The UK's independent academy of science offers one-day communication skills courses and one day media training courses designed exclusively for scientists. These are tutored by journalists and communications professionals. In addition they also offer a two-day residential course which combines the communication and media skills courses below. Held at the Kavli Royal Society International Centre in Buckinghamshire.

http://royalsociety.org/training/communication-media/

The Walnut Bureau

A specialist issue-management consultancy that offers media training, presentation skills, crisis and issue management, public understanding of science, public dialogue, press information, public relations and communications.

http://www.walnutbureau.co.uk/page1/index.php

Your Own Institute

Some universities or research institutes offer public engagement or outreach training to post graduate students or as part of staff development programmes

MSc Courses in Science Communication

The University of the West of England (UWE)

The UWE offers an MSc that provides practical experience of the media as well as projects designed to bring science directly to the public

www.scu.uwe.ac.uk/index.php?q=node/199

Cardiff University

Offers an MSc in Science, Media and Communication: the course offers students a unique blend of practical and theoretical skills.

www.cardiff.ac.uk/socsi/degreeprogrammes/postgraduate/taughtmasters/degreeprogrammes/sciencemediaandcommunication/index.html

City University London

Offers an MA in Science Journalism that provides a thorough grounding in the best practices in health, science and environmental journalism, whilst teaching students to be a critical consumer of scientific information

www.city.ac.uk/courses/postgraduate/science-journalism

Dublin City University

Offers an MSc that explores social issues in science and technology, as well as the communications and controversies surrounding them.

www.dcu.ie/prospective/deginfo.php?classname=MSC&originating_school=60

University of Glasgow

Offers an MSc in Inter-Professional Science, Education & Communication that explores the different contexts in which science communication and education take place.

www.gla.ac.uk/postgraduate/taught/interprofessionalscienceeducationcommunication/

Imperial College London

Offers two postgraduate taught programmes, an MSc Science Communication and an MSc Science Media Production. These provide a balanced technical and theoretical training and are highly relevant to those seeking careers in public engagement or science policy.

www3.imperial.ac.uk/humanities/sciencecommunicationgroup

University of Kent

Offers an MSc in Science, Communication and Society which gives experienced, practical, professional and critical perspectives on science communication.

www.kent.ac.uk/bio/study/postgraduate/master/sc/index.html

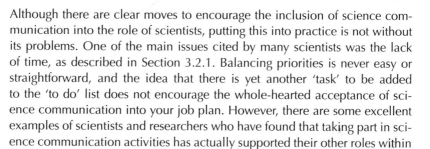

University of Manchester
Offers an MSc in Science Communication designed for students who want to focus on how scientists communicate with non-scientists, including not only the "general public" but also influential supporters, policymakers and experts in non-scientific fields.
    www.chstm.manchester.ac.uk/postgraduate/taught/courses/
routesciencecommunication/

The Open University
Offers an MSc in Science and Society for those who want to explore aspects of science and society at postgraduate level.
    www3.open.ac.uk/study/postgraduate/qualification/f48.htm

University of Sheffield
Has recently developed and now offers an MSc in Science Communication.
    http://shef.ac.uk/faculty/science/learning-and-teaching/msc-in-science-
communication

3.5   Embedding PE in your job

Although there are clear moves to encourage the inclusion of science communication into the role of scientists, putting this into practice is not without its problems. One of the main issues cited by many scientists was the lack of time, as described in Section 3.2.1. Balancing priorities is never easy or straightforward, and the idea that there is yet another 'task' to be added to the 'to do' list does not encourage the whole-hearted acceptance of science communication into your job plan. However, there are some excellent examples of scientists and researchers who have found that taking part in science communication activities has actually supported their other roles within the workplace.

3.5.1   Part of grant applications

One of the most obvious places to start has to be one of the most fundamental roles of the modern scientist – obtaining funding for research activities. The RCUK have changed the way that they fund science communication activities. They are keen to change the culture that surrounds science communication, so that it's no longer a 'bolt on activity' but is 'built in' from the start. This has manifested itself in the withdrawal of funding for specific science communication projects. Instead, researchers are being asked to apply for funding to resource their science communication activities *within* their grant proposals for research funding. Although the RCUK requires that science communication activities link to and highlight the research area supported by the grant, they also offer flexibility:

- in the type and size of the science communication activity that they will fund;
- in the delivery of the science communication activity itself; partnering with professional science communicators is acceptable.

This move towards more 'publically engaged research' requires scientists to embed science communication within their research practice. Although this doesn't summon up additional time to take on this new task, it does allow these activities to move from being under the radar to being worthwhile activities deserving of your time, energy and effort.

3.5.2   Embedding science communication within research

Research is tightly woven into the tapestry of science. But it is no longer the sole preserve of the professional scientist. There are new and exciting developments that allow science communication to be brought into all stages of the research process. The role of the citizen scientist and the benefits that they can offer your research is described in detail in Chapter 8.

3.5.3   Embedding science communication in teaching

In the US, science communication activities are commonly viewed as 'service'. In 1990, Ernest L. Boyer, the then President of the Carnegie Foundation, wrote a seminal report he named 'Scholarship Reconsidered: Priorities of the Professoriate'. He asked the question 'what activities of the professoriate are most highly prized?' He looked at the changing role of the professoriate over time. The report discusses the emergence of the different roles of the academy. It starts with the initial vision for HE brought across the Atlantic by the Pilgrim Fathers. The vision was of a student focused experience which advanced learning through teaching. In the late nineteenth century, the challenges of a growing nation with a new emerging economy based on advances in industry and agriculture, called for changes in the education requirement of students. In response to this, American HE in both private and public universities added service to their mission statements, where service is short for 'not only serving society but reshaping it'. At the same time, the advancement of knowledge through research had taken firm root in American HE, but it wasn't until after the Second World War that scientific progress through research was seen as a priority by the American government and adopted by the nation's colleges and universities. Boyer proposed that the work of an academic should 'encompass four general views of scholarship –discovery, integration, application and teaching'. Higher Education is to 'be of greater service to the nation and the world'. To Boyer:

- 'discovery' represents the academic perception of research;
- 'integration' represents the disciplined work that seeks to interpret, draw together and bring new insights to bear on original research;
- 'application' moves towards engagement as the scholar asks how can knowledge be responsibly applied to consequential problems? How can it be helpful to individuals as well as institutions?
- 'teaching' builds bridges between the teachers understanding and the students learning.

According to Boyer both 'teaching' and 'application' are central to the role of the academic and the academic scientist. Although it may not seem obvious, there are some clear examples and opportunities for science education and science communication to the public to be mutually supportive

(Boyer, 1990). Dr David Lewis has provided a case study that highlights this mutually supportive interaction. This case study describes how science communication has provided:

- biology students with opportunities to use experiential learning to develop transferable skills that enhance their career prospects;
- a solution to a teaching problem; namely the lack of suitable wet, laboratory-based research projects for final year students;
- opportunities to promote your institute, department and research to local communities and build links with local schools.

Case Study 3.4

Biological Science Undergraduates as *'Science and Society'* Ambassadors

Dave Lewis

Background

The Faculty of Biological Sciences (University of Leeds) was finding it increasingly difficult to provide enough wet, laboratory-based research projects for all final year students and needed to develop alternative types of projects. As most biological sciences graduates don't go into careers in scientific research, we wanted to offer students the opportunity to undertake a project which more closely matched their final career destinations. The *'Science and Society'* projects were developed to meet these needs.

Undergraduate students designed a 1- to 2-hour teaching session, suitable for delivery to either primary, years 3–6 (Key Stage 1 and 2), secondary, years 9–11 (Key Stage 3 and 4) or AS-Level students. The students were free to choose the format of their sessions but it had to be:

- interactive;
- support and enhance the National Curriculum;
- match their supervisors' research interests;
- capable of modification to suit different age groups and learning styles.

Topics selected by the students included: 'Spinal cord injuries', 'Science behind healthy lifestyle choices', 'Ethics and embryos' and 'Don't bug me!' and included interactive elements designed for school pupils.

Once in schools, students evaluated the effectiveness of their teaching sessions using pre- and post-session knowledge or opinion (for ethics sessions) based questionnaires, audience response systems within the session, and post-session pupil and staff feedback questionnaires. For the pupils, feedback included questions concerning what they had enjoyed the most and the least, what they were going to take away from the session and whether they would they like to take part in similar sessions in the future. For the teachers, feedback included questions on quality of the teaching, whether it enhanced the curriculum, any suggestions for improvements or ideas for sessions for the next academic year. Both pupils and their teachers really enjoyed these sessions; feedback from the latter included:

> 'I can't believe that's the first time he's done that – he's a natural.'
> 'A pleasure to host X and his workshop on Healthy Lifestyles. His delivery was excellent, enthusiastic and knowledgeable.'
> 'Very appreciative of the contribution of your Department and students.'
> 'Love to host again next year because of the great impression this year's cohort made.'

Students were also required to regularly reflect on their progress and post these reflections in their own individual project blogs, both during development of the activity and between teaching sessions, using these reflections to modify and improve the activity from one delivery to the next. Whilst they find the projects very challenging initially, they quickly come to realise and appreciate the employability skills they are developing and the transferability of these to their future careers.

> 'Challenging, enjoyable and rewarding.'
> 'Communicating and engaging with children, thinking diversely and reasoning-transferable into medical career.'
> 'Confidence and leadership, able to speak clearly and explain rather difficult concepts in simple terms.'

The projects also have received excellent feedback from External Examiners.

> 'A truly exceptional project; a delight to read.'

These projects have enabled us to develop close partnerships with local schools. All the schools we have worked with are highly appreciative of these activities and want us to deliver the activities again in future years. We have been working with some of these schools for three years.

'Science and Society' projects are:
- an excellent way to engage young people in science, both school pupils and undergraduate students;
- a valuable tool for promoting your faculty and it's research, as well as your institution to the local community;
- useful for developing close partnerships with local schools;
- an ideal vehicle for developing key employability skills in undergraduate students;
- an excellent source of projects that we are using to populate a database of resources that we can use in subsequent years and also in other Faculty outreach events;
- a viable alternative to wet lab projects.

3.6   Personal benefits and benefits to the wider society

There are other incentives for taking part in science communication activities that are described by scientists. These incentives can be separated into personal benefits and the benefits for the wider society. The personal reasons described by scientists are that they enjoy taking part in science communication activities and sharing their passion and love for their subject or science in general with a wider audience. The recent booklet *What is in it for Me?* published by the RCUK contains case studies that describe some of the benefits scientists have found.[9]

Scientists have discovered that they begin to appreciate their research and to understand it at a deeper level. Undertaking science communication activities with a wider audience can also allow transferable skills to be developed and honed such as teaching skills, time management and presentation skills (Andrews *et al.*, 2005). Finally, additional incentives are: the potential to attract funding; promotion; to influence or develop collaboration as an outcome.

The primary literature and recent reports have described some of the more altruistic reasons why scientists take part in science communication activities. They describe these as offering benefits to the wider society such as:

- combating the negative stereotypes and images of scientists;
- addressing the misinterpretation of science by the news media;
- sharing their personal enthusiasm and knowledge of science with a wider audience;
- the desire to ensure that scientific breakthroughs are able to benefit society;
- encouraging and attracting the next generation of new scientists by providing a positive role model and an enthusiasm for science.

In summary there are many incentives for scientists to become involved in science communication. Dudley Shallcross and Tim Harrison have provided a case study that illustrates this beautifully. Their case study (9.1) describes the development of a demonstration lecture for school pupils (see Chapter 9). However, Shallcross and Harrison outline the additional benefits that they encountered as direct outcomes of their science communication activities (Box 3.4).

Box 3.4   Gain from developing a Pollutant's Tale

A Pollutant's Tale (APT) was created by Dudley Shallcross, a professor of atmospheric chemistry and climate change and Tim Harrison, a school teacher fellow, both employed at the School of Chemistry, Bristol University. It is a demonstration lecture designed:

- to teach atmospheric chemistry and climate change to school pupils;
- to be adapted to school pupils of different ages, and at different stages of education both in the UK, Ireland, Singapore, and South Africa.

Shallcross and Harrison acknowledge that there have been several advantages to the demonstration lecture that were unforeseen at its conception. Five of them are noted.

1 The UK's funders of research grants brought in 'Impact Statements' that required, as part of a grant application, a methodology to impart a dissemination of research to the wider public. There have been variations of APT to include more recent climate chemistry research into appropriately targeted performances.

2 Aspects of APT have been the subject of research by experienced teachers and other researchers as part of their masters' degrees.

---

[9] (http://www.rcuk.ac.uk/documents/scisoc/RCUKBenefitsofPE.pdf)

3 Articles have been published in journals, aimed at school pupils and teachers such as Chemistry Review (UK) and Science in School (Europe-wide). They have been written on a number of aspects of atmospheric chemistry and climate change, partially as a response to teachers wanting to know in more detail how the atmospheric systems work and, how to perform the demonstrations and practicals.

4 There is of course impact on the young research chemists involved in delivering APT, as with any other outreach work. The students develop the 'soft skills' much in demand by future employers. Such skills involve communication to audiences with differing science knowledge, improved presentation skills, time-management and team work. Indeed the contact with young people in an educative role has even lead to some researchers taking up positions in school teaching and in lecturing.

5 The existing collaboration between Bristol (UK) and Rhodes (South Africa) has even seen a spin off as research collaboration in recent months, with a multi-million Rand grant in the measurement of natural halocarbon emissions being written. This is a nice example of outreach leading the way for research rather than the more usual outreach as an example of research dissemination.

Finally, the past few years has been challenging for the scientific community. Once again the spotlight has focused on science and scientists. Stories such as the 'Climategate affair', the recent Fukushima disaster and the Deepwater Horizon oil spill in the Gulf, have joined the steady stream of science stories that have grabbed the media headlines. These stories have one thing in common; they lead to a disgruntled public that seeks answers and explanations from the scientific community. We have no doubt that there will continue to be a call on scientists to engage with the public and we must respond to this need by learning to be better communicators. Nancy Baron points out that responding to this challenge can provide the pay-off 'that good communication can make you a better leader, and a better scientist' (Baron, 2010).

References

Andrews, E., Weaver, A., Hanley, D. *et al.* (2005). Scientists and public outreach: participation, motivations and impediments . Journal of Geoscience Education **53**, 281–293.

Baron, N. (2010) Stand up for science. Nature **468**, 1032–1033

Boyer, E.L. (1990) *Scholarship Reconsidered*. San Francisco: Jossey-Bass

Burchell, K., Franklin, S. and Holden, K. (2009) Public culture as professional science: final report of the ScoPE project – Scientists on Public Engagement: From Communication to Deliberation? BIOS, London School of Economics and Political Science

Burns, D. and Squires, H. (2011) Embedding Public Engagement in Higher Education; Final Report of the National Action Research Programme. https://www.publicengagement.ac.uk/sites/default/files/Action%20research%20report_0.pdf (accessed 5 May 2012)

Dolan, E. (2008). *Education Outreach and Public Engagement*. Blacksburg, VA, Springer

Mcdaid, L. (2008) A Qualitative Baseline Report on the Perceptions of Public Engagement in University of East Anglia Academic Staff. Report No. Rs7408. The Research Centre, CCN, Norwich. http://www.theresearchcentre.co.uk/files/docs/publications/rs7408.pdf

MacLeod, A. (2010) Towards a professional framework for scientists involved in public engagement work' report. The Wellcome Trust Sanger Institute. http://www.sanger.ac.uk/about/engagement/docs/professionaldevelopmentframework.pdf

Neresini, F. and Bucchi, M. (2011) Which indicators for the new public engagement activities? An exploratory study of European research institutions. Public Understanding of Science **20**(1), 64–79

Pace, M. L., Hampton, S. E., Limburg, K.E. *et al.* (2010) Communicating with the public: opportunities and rewards for individual ecologists. Frontiers in Ecology and the Environment **8**(6), 292–298

Poliakoff, E. and Webb, T. L. (2007) What factors predict scientists' intentions to participate in public engagement of science activities? Science Communication **29**, 242–263

The Royal Society (2006) The Science Communication Report. http://royalsociety.org/uploadedFiles/Royal_Society_Content/Influencing_Policy/Themes_and_Projects/Themes/Governance/Final_Report_-_on_website_-_and_amended_by_SK.pdf (accessed 5 May 2012)

Science for All Expert Group (2009) Reward and Recognition of Public Engagement; Report for the Science for all Expert Group – People Science &Policy Ltd. http://interactive.bis.gov.uk/scienceandsociety/site/all/files/2010/02/Reward-and-recognition-FINAL1.pdf (accessed 5 May 2012)

Taylor, C. (1988) *The Art and Science of Lecture Demonstration*. Institute of Physics (IOP)

CHAPTER FOUR

Communication, Learning and Writing

4.1 Communication theories

The roots of people's interest in communication go back to the fifth century BC and the philosophers Plato and Aristotle. Communication became an academic study in the twentieth century (Miller, 1959) and it can be defined as 'a systemic process in which people interact with and through symbols to create and interpret meaning' (Wood, 2003). The early models suggested that communication moves from a source to a receiver via a channel. This is a linear, one-way approach to communication that is akin to the deficit model of communication discussed in Chapter 1, where information is passed from scientists to the public. The linear model most often referred to is the Shannon–Weaver Model developed in 1949 (Figure 4.1) and while more sophisticated models have since been developed, it is credited for initiating many other models within communication theory. The model was developed by Claude Shannon and Warren Weaver, engineers at the Bell Telephone laboratories who tried to optimise the efficiency of telephone cables and radio waves. It was this model that allowed a mathematical theory of communication to be developed.

As illustrated in Figure 4.1, the Shannon–Weaver model has an information source producing a message, which the transmitter encodes into a signal. The signals are transmitted through a channel to the receiver, which decodes (reconstructs) the message. The noise is anything which can interfere with the intended message. The strength of this model lies in its simplicity, but its weakness is that it fails to take into account more complex communication, such as feedback from the receiver to the sender.

In 1960, David Berlo applied the Shannon–Weaver model to human communication (see Figure 4.2). Berlo thought that the most important aspect of successful communication lay in the relationship between the communicator (source) and the listener (receiver). The engagement model of science communication (outlined in Section 1.5.4), certainly values this communication relationship. Most communication scholars, however, do not regard these linear models as being relevant to communication because they are too simplistic. They prefer to view communication as a transactional model, such

Science Communication: A Practical Guide for Scientists, First Edition. Laura Bowater and Kay Yeoman.
© 2013 John Wiley & Sons, Ltd. Published 2013 by John Wiley & Sons, Ltd.

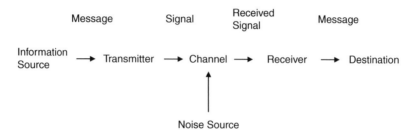

Figure 4.1 Shannon and Weaver model of communication, 1948.

as the example shown in Figure 4.3 which was developed by Wood (2003). This model recognises that communication is both interactive and two-way. People communicate with each other at the same time through verbal and non-verbal communication. Both participants are defined as 'communicators' (i.e. communicator A and B), rather than source or receiver. It also takes account of the environment of the communication process and it acknowledges the communicators' range of experiences. The model suggests that there is a shared range of experience and indicates that this can change over time. This transactional approach is certainly more akin to the dialogue rather than deficit model of science communication (Section 1.5.4).

4.2   Learning and learning theory

The communication models are important in determining the flow of information, but we can also consider science communication as a process of

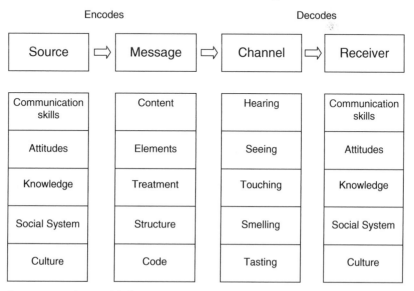

Figure 4.2 Berlo's model of communication.

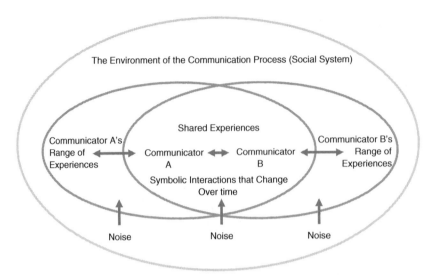

Figure 4.3 Transactional model of communication adapted from Wood, 2003.

learning, using either a deficit or a dialogue approach (see Chapter 1). The deficit approach is the learning of factual material and tends to be one-way (linear communication): from the information provider to the information receiver. On the other hand the dialogue approach is more reflective, and involves learning through the examination of experience via conversation. Within this approach, both the participant and the facilitator are involved in the learning process which is a two-way interaction (transactional communication). When you are designing a science communication activity, it is important to take into account the different ways that people can actually learn. Good communication can be seen as saying the same thing in as many different ways as possible so that as many different people as possible are able to learn from the experience.

Learning theory is an important part of educational research because it allows us to understand the different mechanisms of learning. Carlile and Jordan, (2005) state that 'educational theory may be considered as the distilled experiences of others' and they believe it matters because without it, education is just 'hit and miss'.

Learning is something we all do from birth and continue to do throughout our lives. Dierking et al. (2003) suggest that general learning, and science learning in particular, will not often happen from a single experience; it builds up over time through many different interactions. These interactive experiences include learning from:
- reading books and magazines;
- watching television;
- listening to the radio;
- visiting museums, science centres and zoos;
- through interaction with the internet;
- socially, through conversations with friends and family.

4.3 Learning theory frameworks

There are several philosophical frameworks for learning theory; the four covered here are behavioural, humanistic, cognitive and constructivism. Humanistic learning involves a desire on the part of the individual to learn and it is value driven. Many science communication activities with the public fall into this framework as the participant 'chooses' to go to the event and the event designers become the 'facilitators' for the learning experience.

Behavioural learning occurs in response to a stimulus and involves learning through repetition. The learner is an empty vessel which is then filled with 'the learning' after which behaviour occurs; you may recognise that this is similar to the 'deficit model' of science communication.

Cognitive learning is about understanding, processing thoughts and developing insight. It involves learning by experience – often called experiential learning. To illustrate experiential learning, the American educationalist David Kolb devised the Kolb's learning cycle outlined in Figure 4.4. This cycle draws together experience, perception, cognition, and behaviour (Kolb, 1984). The majority of science communication events will have some aspect of experiential learning where people have an experience at the event, reflect upon it and form ideas about it.

Finally, constructivism is a theory of learning which suggests that all new knowledge is built upon what the individual already knows and/or has experienced, and new ideas are accepted or rejected within this context (Stocklmayer, 2001).

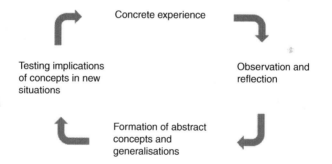

Concrete experience

Testing implications of concepts in new situations

Observation and reflection

Formation of abstract concepts and generalisations

Figure 4.4 Kolb's learning cycle, adapted from Kolb (1984).

4.4 Constructivism and how it applies to science communication events

Constructivism has been put forward as the most useful and relevant theory that can explain how people learn about science in both formal (e.g. in school) and informal situations (e.g. science centres and museums). Science learning does not occur in a vacuum; it's a cumulative process, involving input from a variety of sources including the media, formal learning at school and extracurricular (informal) learning (Falk and Dierking, 2000).

In order to apply the ideas behind constructivism to the design of a science communication event or activity, you need to think about including the following points.

Relevance – there can be no learning unless there is something to build on. It's quite good to try and insert local appeal to an event. For example, a public event that we designed was 'Norfolk Science Past, Present and Future' (see Case study 4.1 by Jaeger Hamilton). You can also use 'hooks' to encourage people to come to your event, and once there, to take part in an activity. The hook can be an existing piece of knowledge or experience, although designing this hook may be challenging if you don't know who your audience is going to be. A good example of a hook was the current interest in vampires sparked by the *'Twilight'* series of teen vampire novels written by Stephanie Meyer. This hook was used by Jo Verran to attract a young audience to attend a 'Twilight and the Science of Vampires' event described in Case study 6.3.

Participant learning and interaction – consider using hands-on activities, ask questions, or ask for opinions and encourage reflection and self-analysis. An example of this might be a public event where people are asked to record their views about full genome analysis from birth, or the use of genetically modified food, or stem cell technology or other alternative controversial issues. This can be done directly using face-to-face discussion, indirectly through a polling mechanism or by simply encouraging people to write on a message board. There are good examples of hands-on activities in Case studies 6.4 and 10.2.

We also really like the assertion by Stocklmayer (2001) that 'you don't have to be special to fully participate in science'. This is a view that we have certainly used to underpin the design and running of all our science communication activities.

4.5 Learning styles

Honey and Mumford (1992), based on the work by Kolb, identified four different learning styles outlined in Box 4.1. The styles are activist, pragmatist, theorist and reflector, and they can all be linked to Kolb's learning cycle. Most people tend to have parts of all these styles, but one usually dominates.

Box 4.1   Learning style definitions based on Kolb's learning cycle

Activists: An activist learns by doing – using concrete experiences and active experimentation

Reflectors: Reflectors look back over past experiences and are often imaginative and good at seeing things from different perspectives

Theorists: Theorists are adept at analysing observations and creating theoretical models. They can use inductive reasoning to justify decisions.

Pragmatists: Pragmatists learn by testing ideas to see if they work in practice. They are good at applying ideas practically and solving problems

Howard Gardner, an American education psychologist, extended the idea of different learning styles and he identified the following seven intelligences:

- **Linguistic**-mastery of language, using language to express, but also to store information;
- **Logical**-to detect patterns and to reason, often associated with scientific and mathematical thinking;
- **Spatial**-ability to create mental images-not limited to vision, blind children can have spatial intelligence;
- **Musical**-pitches tones and rhythms. Auditory function is required for pitches and tones, but not for rhythm;
- **Bodily Kinaesthetic-**to use ones mental abilities to control body movement;
- **Interpersonal**-understanding the feelings and intentions of others;
- **Intrapersonal**-understanding ones own feelings and motivations.

These have been distilled into the VARK system (Fleming and Bonwell, 2009), which stands for visual, aural, read/write and kinaesthetic. Most people will find that they use a mix of these different learning styles, but some people find they have strong tendencies towards one style (see Table 4.1).

If you are interested in finding out about your own learning style, there are several learning style websites (some are free and others are not), where you can explore this aspect of yourself. The Peter Honey Publication website has either a 40-question or an 80-question survey, both based on the original Honey and Mumford learning style questionnaire (www.peterhoney.com); both require payment. However, both questionnaires can be completed online and the data analysed immediately. The VARK website consists of 16 questions, can be completed online and an immediate analysis of the data sent to your email address (http://www.vark-learn.com/english/index.asp). VARK is inexpensive and it also gives useful advice about how to tailor information for each learning style. The learning styles website (www.learning-styles-online.com) is a free site, which asks you 70 questions, so it does take a bit of time to complete. At the end you get a graphical output of the results, but no further information about how to use the knowledge about your learning style(s).

Thinking about your own learning style preferences can be useful when you are designing a science communication activity to appeal to a variety of learners (Bultitude, 2010). Table 4.2 shows how you can then incorporate

Table 4.1 Learning preferences (adapted from Fleming and Bonwell, 2009).

Learning preferences	Definition
Visual	Preference for information presented e.g. as charts, graphs, flow charts and other devices to represent what might have been presented as words
Aural	Preference for information which is spoken or heard
Read/write	Preference for information displayed as words either read or written
Kinaesthetic	Preference related to the use of experience and practice (learning by doing)
Multimodal	A mix of the above learning styles, can be a combination of two, three or all preferences

Table 4.2 Linking learning preference to the design and delivery of a science communication activity.

Learning preference	Intake of information	Design of activity
Visual	○ Charts ○ Maps ○ Graphs ○ Diagrams ○ Colours ○ Patterns ○ Word pictures ○ Highlighters	○ Use a colourful poster to convey information ○ Make good use of images in the activity ○ Think about use of different fonts ○ Highlight or underline key words ○ Use flow charts to explain concepts ○ Use gestures when explaining ○ Whitespace-blank areas around images and text ○ Replace words with pictures or symbols ○ Think about spatial arrangement of information
Aural	○ Explaining ○ Discussion of topics with other learners or the teacher/facilitator ○ Describe pictures and other visual material	○ Think about your verbal explanation of the activity ○ Allow opportunity for individual or group discussion ○ Use sound recordings ○ Ask the child to explain the activity to the parent/carer or if in a school setting to the teacher or peer group ○ Point out and describe visual information ○ Try using a debate ○ Get them to read out the instructions for an activity ○ Ask questions about prior knowledge ○ Describe what has been learnt
Read/write	○ Lists ○ Definitions ○ Quotations ○ Printed handouts ○ Essays ○ Reports ○ Webpages ○ Taking notes ○ Instructions	○ Use PowerPoint for presentations ○ Give out printed information ○ Use quotes in posters or other written material ○ Use definitions in printed information, maybe as a glossary ○ Give out printed instructions ○ Use online material ○ If doing a school activity get the pupils to write down what they did and what they learnt
Kinaesthetic	○ Video ○ Field Trip ○ Applied opportunities ○ Experimentation ○ Hands-on activity ○ Use of the senses ○ Collections of materials ○ Real life examples ○ Trial and error ○ Samples of things, e.g. collections of rocks ○ Examples of principles	○ Design a hands-on activity or experiment to demonstrate concepts ○ Use video material to help explain concepts ○ Get people to make or draw things from observations ○ Make their own video or recording ○ Get plenty of examples of material which people can touch or smell ○ Go outside and collect or make observations ○ Measure things ○ Use games to demonstrate principles ○ Allow people to make mistakes ○ Use real life examples of how concepts are applied in verbal or written material

the idea of learning styles, first when you are designing the activity and then when you are actually running it. Remember that the aim is not to 'pigeonhole' learners, but to allow an event to be designed that offers multiple learning methods. This will ensure that the same thing is communicated in as many different ways as possible.

Building different learning preferences into the design of your science communication activities will ensure that each participant will have activities that relate to their own specific learning style or combination of styles. Table 4.3 shows how to incorporate learning styles into a school lesson plan. In addition, ideas surrounding learning styles can also be used to help design public

Table 4.3 Relating learning styles to an after-school science club session on blood for pupils in Key Stage 2 (aged 10–11).

Concept	Activity	Learning preference
Prior knowledge	Ask questions (you could use a personal response system).	Audio and kinaesthetic
Circulation	A pupil lies on a long length of paper and a body outline is drawn. The circulation system is mapped onto the body outline, this activity also involves discussion of what to include and why.	Visual, aural, kinaesthetic Read/write
	Measure pulse rate before and after activity.	Visual and Kinaesthetic
	Videos of human/dog and hummingbird heart beat. Electrocardiogram using BioPac. The printed output can be labelled.	Visual, Kinaesthetic, read/write
Components of blood	Bloodmobile video and images of different blood cells.	Visual and Kinaesthetic
	Discussion on the function of different blood cells.	Aural
	Blood component ratio worksheet.	Read/write
	Give cuddly blood cells (giantmicrobes.com).	Kinaesthetic
Function of platelets	Game: Two lengths of ribbon represent the artery. Some of the children are different blood components, running up and down the artery. Some of the children are 'invading bacteria'. The ribbon is cut at a specific point, and the platelets need to rush to the cut site before the bacteria can enter.	Kinaesthetic and aural
Diseases of blood	Discussion on blood diseases, use a real life example of haemophilia.	Aural and Kinaesthetic
Veins and Arteries	Vein and artery models showing normal and diseased states.	Visual and Kinaesthetic
	Build your own artery with a toilet paper inner tube, and the different blood components made from coloured clay.	
Learning summary	Question and answer match cards game, where the answer to one question is next to a different question.	Aural

events. Case study 4.1, written by Jaeger Hamilton, shows that by thinking about how different people may want to interact with your material, you can provide different opportunities and experiences within a single activity.

Case Study 4.1

Embedding Learning Styles in a Public Engagement Event

Jaeger Hamilton

Background

This event, 'Norfolk Science, Past, Present and Future', was part of the National Science and Engineering Week and it was designed to have local appeal and relevance. The water mould *Saprolegnia* spp. was first identified as a pathogen of fish by the eighteenth century Norfolk philosopher, Royal Society Fellow and naturalist William Arderon. We decided to work with *Saprolegnia* not only to commemorate him and highlight the science done by local people, but also because *Saprolegnia* is interesting in its own right. The event took place at Norwich Castle Museum, one of Norwich's main visitor attractions. It has extensive natural history galleries and is an ideal location for a family science event.

Planning the activity

The most important point in planning the activity was that it should be accessible to every type of learner and that it was a dynamic, creative and open learning experience. It needed to be, in a sense, 'self-evident' and immediate. The science underlying the activity should emerge from taking part in the activity. My activity would be centred on the learner and I would adopt the role of facilitator. I thought the process of the *Saprolegnia* zoosporangia hatching, releasing its swimming zoospores into the outside world, quite dramatic, and would appeal to most people. The advantage of using *Saprolegnia* is that it does not exist in isolation, the sugars and exudates released by *Saprolegnia* into its external medium make it a brilliant 'home' for other microbes, such as *Vorticella*, rotifers and *Paramecia*. Under the microscope, *Saprolegnia* hyphae appear as branching reticulum with a lively microcosmos of creatures dancing in and out, and encounters with these microbes can transform 'pond-life' from something dull and inanimate into a vivid appreciation of the interconnectedness of all life.

I planned my activity by imagining what the actual event might entail. I realised the event could be very busy with large groups of people and few opportunities to talk to individuals but everyone would still need something to do. My aim was to give learners a mix of my attention but also their own opportunities of action and discovery. I designed the activity to be laid out across two tables. There was a laptop displaying the *Saprolegnia* on a slide under the microscope and displayed using a video camera. Learners had the opportunity to search the slides after I started them using the microscope. The right-hand table also contained handouts and feedback forms. Behind the tables were two posters, containing the scientific information that might appeal to different learners. I planned to refer to the posters to illustrate particular points that would be relevant to the science activity.

I chose to consider Kolb's theory of experiential learning when designing the activity. Kolb's learning cycle involves a cyclical four stage process of experience, perception, cognition and behaviour. There are four learning styles (pragmatist, activist, reflector, theorist), which correspond to each stage of the learning cycle; each learning style denotes a certain preference by which the learner will approach a new learning situation.

- For an activist, I recognised that the event should involve some form of concrete experience; something would have to be *done*. Thus the learning activity offered plenty of 'hands-on' experience: the learner could take a slide of *Saprolegnia* and place it on the microscope. I explained how to use the microscope and how to explore the slide, with an image of what's beneath the microscope being displayed on the laptop screen. This provided the activist learner with the opportunity to share their experience with their family, friends or myself.
- Reflective learners need space to think, watch and ponder the learning activity and would benefit from not getting 'caught-up' in it; they need space to observe. For reflective learners plenty of material was on display for them to consider. Two posters were prepared, one describing oomycetes (*Saprolegnia* is an oomycete) and their phylogenetic relation to other major kingdoms (animals, plants, fungi), and also some descriptions of *Saprolegnia* and some of the organisms that associate with it (such as *Paramecia*). The reflective learner also had plenty of activity to see: whatever was under the microscope was displayed on the laptop screens through the video camera. The reflective learner would learn without having to become directly involved in the activity.
- Theorists prefer to incorporate their experiences into logically sound theories. The learning activity catered for theoretical learners; it provided plenty of information to be gathered from posters and the demonstrator/facilitator. My role was to enable the theoretical learner to assimilate the new knowledge and put it into context with other scientific knowledge (with the help of posters).
- Pragmatists learn best from situations that have a clear practical application to 'real-life'. In order to gear the activity toward a pragmatic learner, I dedicated a whole poster to explaining that we were studying an oomycete and that oomycetes have great consequences for agriculture. This way, the pragmatist learner will be able to appreciate that studying such organisms as *Saprolegnia* can have real and important implications for the way we live.

Personal Gain

My activity attempted 'multi-way' dialogue for informal science learning events with the emphasis on 'learning on all sides', and not providing a deficient lay audience with 'scientific facts'. The value of the activity was created by a sharing experience in which all parties learnt. It is a paradoxical notion that in order to teach others, we should think of how we might learn with others. But it seems to work. I'm sure that many of the learners will remember *Saprolegnia* as a result of this 'multi-way' process; whereas if I had only tried to inform them of what it is I doubt they will have remembered anything. For my part, I have learnt much about how people learn in informal settings, how to communicate with people better and I have a deeper understanding of the science as a result of trying to explain it to others.

4.6 Model of family centred learning

Five years ago I, Kay Yeoman, obtained funding from the Wellcome Trust for the Mobile Family Science Laboratory (MFSL). I wanted to create an exciting way for children and adults of all ages and abilities to interact with biomedically related science as well as our specially developed art-in-science activities. The mobile laboratory travels around Norfolk and it is used to run school and public events. The MFSL enables people to discuss and debate

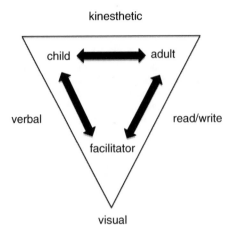

Figure 4.5 Model of family centred learning (Yeoman and Hamilton, 2012).

topical issues related to science. The MFSL employed two undergraduate students, one of whom is now a science teacher, and the other is undertaking a PhD. Many other undergraduate and postgraduate students also help in designing and running events. During the first three years we ran 10 public events held in both city and rural locations attracting approximately 13 600 visitors. While running these public events, participant observation was used to observe how families interact with each other and with the hands-on activities within a public setting. This allowed us to establish the model of family centred learning shown in Figure 4.5. In this model, the hands-on activity (represented by the inverted triangle) caters for different learning styles: verbal, visual and kinaesthetic. Whilst taking part in the activity, there is a three-way conversation between the facilitator, the child and the adult, represented by the arrows within the triangle.

We found that children tend to have a strong kinaesthetic approach to learning, although some will watch a friend or sibling do a hands-on activity before getting involved themselves. Some children immediately draw in the adult by saying 'come and look at this!' but conversely a nervous child can be encouraged to participate by an adult using the same phrase. Most of the children like to communicate verbally with the facilitator and another adult (usually a parent) about what they are seeing and doing. Children tend not to read or discuss written information, but their attention can be directed to it by either the facilitator or adult. Older children ($\geq$12 years old) can discuss what they have read, but the majority of younger children lose interest and return to the kinaesthetic part of the activity. Generally we found it was the adult learners who benefited from written information; they *read* the information while their children *do* the activity. The adults talk to the facilitator about the information and the activity, and then talk to and watch the child learner *doing* the activity, most often without getting involved directly themselves.

We observed that only a few children spontaneously refer to prior knowledge in relation to the new knowledge gained during the activity, but most can do this when prompted by the facilitator. The adult learners tend to ask more pragmatic questions, for example 'Why is this important to society?' and 'How does this help?' In turn, the adult learners verbalise the information from the facilitator back to the child learner, for example 'that's a ... you're seeing, we can find that at home!' In contrast, few children ask any obvious questions trying to link how the new knowledge could be applied, but many of them are practically minded and wanted to 'get on' with the kinaesthetic learning. Generally, this 'triangle of communication' continues for quite long periods of time (~15–60 minutes) and some family groups return for further interaction with the activity.

Clear differences were found between the generations. We learnt that the adult learners tended to show theoretical, reflective learning tendencies, as well as a pragmatic concern for how the knowledge was applicable, and the child learners tended to be very much the 'activists'– learning by doing.

4.7 Successful scientific writing for the public

A book should be written in the language of its readers, but a very considerable number of scientific writers fail to realise this. A few write boldly in the dialect of their science, and there is certainly a considerable pleasure in a skilful and compact handling of technicalities; but such writers do not appreciate the fact that this is an acquired taste, and that the public has not acquired it.

—H.G Wells, in 'Popularising Science' (Nature, 1894)

When you are designing a science communication event, as well as thinking about communication models and learning styles, you must also make sure that you are communicating effectively to allow learning and engagement to take place. To allow this to happen one of the crucial communication skills that you need to develop is a clear writing style. As a scientist you have different opportunities and different mediums to write for the public. These include:
- books;
- magazine articles;
- newspaper articles;
- blogs;
- web pages;
- material to support school activities;
- posters and other material for science communication events;
- promotional material, including press releases.

There are several excellent books that provide advice about specific types of writing, such as press releases, and newspaper articles. They have been listed in a recommended reading list at the end of this Chapter. This section is intended to give you some general information to help you get started with writing for the public. Most scientists have spent many years being trained to write in the formal 'scientific' style associated with the scientific papers of peer-reviewed journals. As scientists we have been taught to use the past tense and passive voice, for example 'the protein sample obtained after the final purification step, was mixed with equal volumes of 10% (v/v) glycerol prior to flash freezing in liquid nitrogen to ensure the safe storage of the sample at −80 °C'.

Using this tense and a passive voice can make scientific writing seem more objective (Gregory *et al.*, 1998), but it doesn't make it particularly readable from the public's perspective. Readability is a measure that assesses how easily a piece of text can be read and understood. When communicating to the public in any form, including writing, your goal is to make your message understood by them in the context of why it's important (Baron, 2010). You must take the time to consider the following points, issues and concerns.

Consider your audience: the most important thing to consider before you begin to write is your audience. Who are you writing the material for? How old are they? What are their backgrounds? Answering these questions will influence the complexity of the language that you use. Weitkamp (2010) suggests that you build a mental picture of your audience, perhaps by imagining the members of your audience as your children, your parents, grandparents, siblings, aunts, uncles or neighbours.

Look at different examples of writing: a good piece of straightforward and sensible advice is provided by Cornelia Dean, author of *Am I Making Myself Clear?* She suggests that you find previous examples of writing that echo or relate to what you want to write about. Next, ask yourself whether you think they are good examples of clear writing or whether they represent examples of writing that could be improved. This is followed by thinking about what the particular factors are that make these pieces of writing clear to read and easy to understand or hard to read and difficult to understand. Once you have defined the qualities associated with a clear writing style you can begin to build these qualities into your own work.

Define scientific terms: there is no reason why you should have to avoid the use of scientific terminology, but you do need to ensure that all scientific terms are explained well and the language surrounding them is kept as simple as possible. There is no getting away from the fact that science can be difficult to understand, especially for a non-expert audience. In addition, as scientists, we often use scientific terms that have different meanings to different people and different audiences. A good example is the word 'proof' which is a commonly used and well understood principle in the scientific community, but 'proof' means something completely different to a publisher (Shortland and Gregory, 1991). These differences are not confined to a scientific and a non-scientific community. There are certain key words, such as 'nucleus' that can mean different things to different types of scientists; to a particle physicist the nucleus is the centre of an atom, while a biologist considers the nucleus as the cellular compartment that contains the DNA.

Do use images: images can be especially useful in posters or other similar material that you wish to provide for public events. One tip is to make

Table 4.4 Organisations which supply images.

Organisation	Access	Web Address
Science Photo Library	Paid	http://www.sciencephoto.com/
Wellcome Trust Images	Paid	http://images.wellcome.ac.uk/
Biosciences Image Bank	Free	http://bio.ltsn.ac.uk/Imagebank/
ARKive	Free	http://www.arkive.org/

sure that you are allowed to use the image and that you are not breaching copyright laws. There are organisations that provide images on websites, some of which are free providing an appropriate reference is used. Others allow you to use the image on payment of a fee. Table 4.4 provides examples of useful organisations that offer images that can either be downloaded from their websites, or ordered, and high-quality images files sent to you via email. When you start to write and produce posters for a public audience, consider consulting a professional for the actual layout and design, or consider taking advantage of your own institution's design departments. The cost of image use and professional design can be included in your funding application (Section 6.8).

Reduce the amount of text: 'When in doubt, leave it out' (Dean, 2009). In contrast to previous posters that you may have produced in the past, for scientific conferences, posters intended for public display need fewer words. When we write the content for pull-up display banners measuring 2 m × 1 m, we aim to use no more than 250–300 words. Examples of our banner displays can be found on the book website.

Don't use jargon: it can really confuse people, even other scientists. In our own field of molecular biology, research papers can be full of impenetrable jargon and acronyms, including peculiar gene names, names of enzymes and the names of different molecular techniques.

Avoid too many abbreviations: if you use abbreviations or acronyms make sure that you clearly define them. Don't place the definition in brackets, as people generally skip over this information. Instead consider providing the abbreviations using the following format.

Example: Bovine Spongiform Encephalopathy can be shortened to BSE.

Use the active voice in your writing: verbs have two voices, passive or active. The active voice is when the subject of the sentence performs the action of the verb. The passive voice is where the subject of the sentence receives the action of the verb. Examples are provided in Table 4.5.

Use the present tense: information and stories tend to feel more immediate and relevant when written in the present tense. The audience can connect better with the information you are providing.

Use numbers with care: numbers can be confusing, so try and make them easy to understand, for example 50% is half, 25% is a quarter, or perhaps one in four, 1×10^6 is a million. Avoid equations if you can.

Keep it short and simple (KISS): this acronym can apply to many different aspects of your communication, including verbal communication and

Table 4.5 Examples of sentences in the active and passive voice.

Active Voice	Passive Voice
The dog chased the cat.	The cat was chased by the dog.
Scientists have discovered a new species of bat.	A new species of bat was discovered by scientists.
James Watson and Francis Crick discovered the structure of DNA.	The structure of DNA was discovered by James Watson and Francis Crick.

the design of hands-on experiments. In relation to writing, your sentence length should be less than 20 words. However, don't use lots of short sentences as this can give a 'choppy' feel. Use only one idea per paragraph and have at least three sentences in each paragraph. Finally, to keep it short and simple, aim to write for an audience with a reading age of an average 12-year-old.

Replace complex words and phrases: understanding science can be challenging, so don't make it any harder by using complex words or long phrases when simple words will do. Table 4.6 provides examples of some common words and phrases which can be replaced by a single word.

Use metaphors: metaphors can be very useful, as they allow difficult concepts to be understood by comparing them to something that is more easily visualised or understood by your audience. Plenty of metaphors are used in science, for example DNA can be compared to a 'blueprint', 'ladder' or a 'recipe', and an enzyme substrate specificity is often compared to a 'lock and key mechanism'. However, avoid using mixed metaphors, and don't overuse them.

Table 4.6 Ideas of how to replace words of long syllables, or phrases with single words.

Word or phrase	Replacement word
Subsequently	then
Demonstrate	show
Frequently	often
As well as	and
In the event of	if
As a consequence of	because
Due to the fact that	because
Similar to	like
The majority of	most
In the interim	meanwhile
Makes an attempt	tries
A considerable amount of	much
Optimal	best
Approximately	about

Use pronouns with care: Pronouns such as he, she, it, they, you and we are useful in sentence construction, but they can cause confusion. Shortland and Gregory (1991), give this good example;

> 'The committee ordered the fish to be removed from the laboratory because it smelt dreadful. They were incinerated.'

The 'it' used in this example could refer to the fish, but 'it' could also refer to the laboratory. Similarly the 'they' could by the fish, but 'they' could also be the committee. Even with this example of ambiguous pronouns the chances are that common sense would allow you to correctly surmise what 'it' and 'they' refer to. However if this sentence was also peppered with unfamiliar scientific terminology, guessing, what 'it' and 'they' refer to may not have been so straightforward especially for a non-expert audience such as the public.

Readability formula – SMOG: there are several formulas that you can use to assess the readability of your text; these include the Fog Index and the Flesch formula. We like to use the SMOG index. In 1969, Professor Harry McLaughlin developed a readability formula, which he called the Simple Measure of Gobbledegook (SMOG index). This is a useful formula that provides an indication of the years of education needed to understand a piece of writing. Table 4.7 shows how this relates to the UK education system. However, it is important to remember that calculating the readability index does not assess whether your writing makes sense. You can produce a piece of writing that has a SMOG index of 7 (seven years in education) but your writing may not make sense. So while it is a useful tool, it should not be the only method you use to assess the clarity of your writing. Box 4.2 provides the method used for calculating a SMOG index. In addition, there are several websites that allow you to copy and paste text to calculate the SMOG Index.[1] Box 4.3 provides examples of sentences that have been put through this website to have their SMOG index calculated. The sentences have then been re-written to reduce the SMOG index.

Table 4.7 Linking SMOG index to approximate education level the UK system.

SMOG result	AGE	UK
7	9–13	Years 5–8
8	12–14	Year 8–9
9	14–15	Years 9–10
10–11	15–16	Years 10–11 (GCSE)
12	17–18	Years 12 and 13 (A-level)
13–16	18+	College or University
17–19	21+	Postgraduate Education

---

[1] http://www.wordscount.info/wc/jsp/clear/analyze_smog.jsp

Box 4.2 How to calculate the SMOG index

SMOG Grading

1. *Count 10 consecutive sentences near the beginning of the text to be assessed, 10 in the middle and 10 near the end.* Count as a sentence any string of words ending with a period, question mark or exclamation point.

2. *In the 30 selected sentences count every word of three or more syllables.* Any string of letters or numerals beginning and ending with a space or punctuation mark should be counted if you can distinguish at least three syllables when you read it aloud in context. If a polysyllabic word is repeated, count each repetition.

3. *Estimate the square root of the number of polysyllabic words counted.* This is done by taking the square root of the nearest perfect square. For example, if the count is 95, the nearest perfect square is 100, which yields a square root of 10. If the count lies roughly between two perfect squares, choose the lower number. For instance, if the count is 110, take the square root of 100 rather than that of 121.

4. *Add 3 to the approximate square root. This gives the SMOG Grade, which is the reading grade that a person must have reached if he is to understand fully the text assessed.*

Reproduced from McLaughlin, G.H. (1969) SMOG Grading – a new readability formula. Journal of Reading, with permission from John Wiley & Sons, Inc.

Box 4.3

Rhizobium leguminosarum is a bacterium capable of forming a mutualistic symbiotic relationship with leguminous plants. (SMOG of 19)

Is re-written to:

Some bacteria can form a partnership with plants such as peas, beans and clover. (SMOG of 11)

■　■　■

Iron is essential for nearly all organisms, but in its most frequently encountered ferric, Fe^{3+}, form, it is virtually insoluble. To counteract this, microbes use various systems to acquire Fe^{3+} from their environment. (SMOG 16)

Is re-written to:

Most organisms need iron, but iron is not very soluble in water. To solve this problem bacteria use different ways to get iron from where they live. (SMOG of 11)

Always edit: after you have finished your piece of writing, leave it for a few days and then return to it with 'fresh eyes'. This process allows you to:

- detect simple typos;
- correct any ambiguity or misunderstanding in the text;
- add additional material for clarity and additional understanding;
- remove material for clarity and additional understanding – 'when in doubt leave it out'.

4.8   Concluding remarks

As a scientist reading this book, you will have no problem with the 'science' part of science communication. We hope that after reading this chapter, you have gained more insight into how your own scientific understanding can be communicated more effectively to a lay audience. Attempting to distil two major areas of research, that of communication and learning into bite-size chunks was no easy task, and it will be up to you, the reader, to decide if we managed it. Our goal, however, to quote Dr J.E. Taylor;

> . . . is not to be esteemed learned, so much as to be deemed useful.
>
> —J.E. Taylor (*Science Gossip Magazine,* 1872)

Recommended additional reading for writing for the public

Baron, N. (2010) Escape from the Ivory Tower. A Guide to Making your Science Matter. Island Press

Christensen, L.L. (2007) The hands-on guide for science communicators. Springer

Dean, C. (2009) Am I Making Myself Clear? A scientist's guide to talking to the public. Harvard University Press

Shortland, M. and Gregory, J. (1991) Communication Science. A handbook. Longman Scientific and Technical

Weitkamp, E. (2010) Writing Science, In: Introducing Science Communication, Eds Brake, M.L. and Weitkamp, E. Palgrave Macmillan

References

Berlo, D.K. (1960) *The Process of Communication: An Introduction to the Theory and Practice.* Holt, Rinehart and Winston

Bultitude, K. (2010). Presenting science. In: *Introducing Science Communication* (eds Brake, M.L. and Weitkamp, E.). Palgrave Macmillan

Carlile, O. and Jordan, A. (2005) It works in practice but will it work in theory? The theoretical underpinnings of pedagogy. In: *Emerging issues in the practice of University Learning and Teaching* (eds O'Neill, G., Moore, S., McMullin, B.). Dublin: AISHE

Dierking, L.D., Falk, J.H., Rennie, L.J., *et al.* (2003) Policy statement of the 'Informal Science Education' Ad Hoc Committee. Journal of Research in Science Teaching **40**, 108–111

Falk, J.H. and Dierking, L.D. (2000) *Learning from Museums: Visitor Experiences and the Making of Meaning.* AltaMira Press

Fleming, N. and Bonwell, C. (2009) How do I Learn Best? A Students' Guide to improved Learning. http://www.vark-learn.com/english/index.asp (accessed 5 May 2012)

Gardner, H. (2011) *Frames of Mind. The Theory of Multiple Intelligences,* 10th edn. Basic Books

Gregory, J., Miller, S. and Earl, A. (1998) *Handbook of Science Communication.* Institute of Physics Publishing, Bristol and Philadelphia

Honey, P. and Mumford, A. (1992) *The Manual of Learning Styles.* Peter Honey Publications, Maidenhead

Kolb, D. (1984) *Experiential Learning: Experience as the Source of Learning and Development.* Prentice-Hall, Englewood Cliffs, NJ

McLaughlin, G.H. (1969) SMOG Grading- a new Readability Formula. Journal of Reading, 639–646

Miller, K. (1959) Communication Theories, Perspectives, Processes and Contexts. 2nd Edition. McGraw-Hill

Stocklmayer, S.M. (2001) The background to effective science communication with the public. In: *Science Communication in Theory and Practice* (eds Stocklmayer, S.M., Gore, M.M. and Bryant, C.). Kluwer Academic Publishers

Weitkamp, E. (2010) Writing science, In: *Introducing Science Communication* (eds Brake, M.L. and Weitkamp, E.). Palgrave Macmillan

Wood, J.T. (2003) *Communication in our Lives*, 3rd edn. Thomson

CHAPTER FIVE

Monitoring and Evaluating your Event or Activity

Everything that can be counted does not necessarily count; everything that counts cannot necessarily be counted.

—Albert Einstein

5.1 Introduction

Monitoring and evaluation are the processes used to discover whether your science communication project has achieved what you aimed to do. Or if it didn't, it can help you understand the changes required to achieve your aims in the future.

Before we go any further, now is a good time to define three important terms that are used within this book:

- **Activity**: an activity describes something that is designed to provide a direct experience of something. It stands alone as a single entity.
- **Event**: an event is an organised occasion that is often composed of several individual activities.
- **Project**: the project is the overall task that requires time and effort to complete and involves the planning, organisation and delivery of a body of work such as an activity or an event.

Having gone through the entire process of designing, planning and running your project, which could be either a science communication event or an activity, you may be thinking that it is now a good idea to take stock and retrospectively monitor and evaluate what you have achieved. You are both right and wrong; it is a good idea to undertake monitoring and evaluation but leaving these processes until after you have completed your project will not allow you to monitor or evaluate it successfully. Instead these processes should be considered as soon as you start to design your activity or event. To be undertaken effectively and to provide meaningful information, evaluation needs to be 'built into' and not 'bolted-on to' your project, and these processes should occur at each and every stage. Evaluation is especially important for an activity or event that you have undertaken for the first time as it will allow

Science Communication: A Practical Guide for Scientists, First Edition. Laura Bowater and Kay Yeoman.
© 2013 John Wiley & Sons, Ltd. Published 2013 by John Wiley & Sons, Ltd.

you to run it more successfully in the future. Additionally, evaluation is often a key requirement for other stakeholders, for example:

- the audience or participants who may have invested time and effort in the design and delivery of the project;
- the funder of your project may be keen to ensure their money is well spent;
- your institution may have a clear agenda towards public engagement;
- governing bodies, for example the RCUK (Research Councils UK) and HEFCE (Higher Education Funding Council for England) are seeking to establish the impact of research in academia as emphasised in the new Research Excellence Framework (REF) (http://www.hefce.ac.uk/research/ref/) and your project could have the potential to be included in new reports and studies.

5.2   Key Stages in undertaking an engagement project

There are at least three different stages to undertaking a public engagement project and we need to consider each of these stages in turn (Figure 5.1).

Stage one – The Starting Point

The initial idea: it may be yours or it may be someone else's. Perhaps the idea is in response to a request or maybe it is a requirement of your research funding. It may be topic led or event focused. It is at this stage that you need to consider the **aim** of your project. You need to ask 'what are you hoping to **achieve**?' There are many aims that lead to initial ideas for engagement, and some may be more altruistic than others. The British Science Association produced the 'Science for All Final Report 2010' that outlines some of these reasons and they include (in our suggested order of altruism):[1]

- to make the world a better place;
- to enhance social cohesion and democratic participation;
- to be ethical accountable and transparent;
- to create a more efficient, dynamic and sustainable economy;
- to develop skills and inspire learning (in others but could include yourself);
- to win support for science;
- to increase the quality and impact of my research;
- to enhance my career.
  In addition there may be other reasons that are closer to home:
- your funder or institution may require you to communicate your research to the wider public;
- you may be keen to demonstrate how your research has an impact out with the academic community and within a wider society.

---

[1] http://www.britishscienceassociation.org/NR/rdonlyres/D6B1ACFC-2F42-4F07-A5D1-938E1D83F3ED/0/ScienceforAllFinalReport.pdf

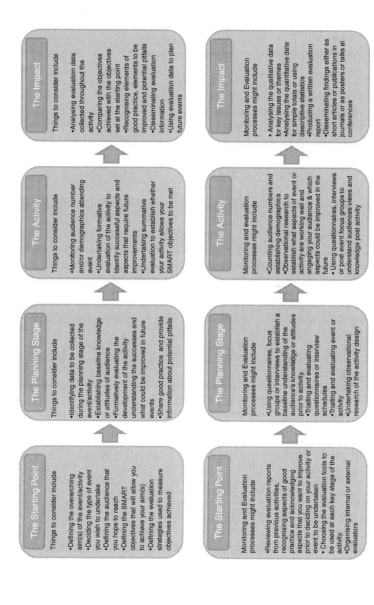

Figure 5.1 The steps involved in an evaluation process of an event or activity. Information that outlines what should be considered at each stage is included. The book website also provides a proforma that can be used when you evaluate your project.

The Starting Point

Things to consider include

• Defining the overarching aim(s) of the event/activity
• Deciding the type of event you wish to undertake
• Defining the audience that you hope to reach
• Defining the SMART objectives that will allow you to achieve your aim(s)
• Defining the evaluation strategies used to measure objectives achieved

The Planning Stage

Things to consider include

• Identifying data to be collected during the planning stage of the event/activity
• Establishing baseline knowledge or attitudes of audience
• Formatively evaluating the development of the activity
• understanding the successes and what could be improved in future events
• Share good practice and provide information about potential pitfalls

The Activity

Things to consider include

• Monitoring audience number and/or demographics attending event
• Undertaking formative evaluation of the activity to identify successful aspects and aspects that require future improvements
• Undertaking summative evaluation to establish whether your activity allows your SMART objectives to be met

The Impact

Things to consider include

• Analysing evaluation data collected throughout the activity
• Comparing the objectives achieved with the objectives set at the starting point
• Recognising elements of good practice, elements to be improved and potential pitfalls
• Disseminating evaluation information
• Using evaluation data to plan future events

The Starting Point

Monitoring and Evaluation processes might include

• Reviewing evaluation reports from previous activities, recognising aspects of good practice and acknowledging aspects that you want to improve prior to deciding on your activity or event to be undertaken
• Choosing the evaluation tools to be used at each key stage of the activity.
• Organising internal or external evaluators

The Planning Stage

Monitoring and Evaluation processes might include

• Using questionnaires, focus groups or interviews to establish a baseline understanding of the audience's knowledge or attitudes prior to activity.
• Trialling and evaluating questionnaires or interview schedules
• Trialling and evaluating event or activity
• Undertaking observational research of the activity design

The Activity

Monitoring and evaluation processes might include

• Counting audience numbers and establishing demographics
• Observational research to establish what aspects of event or activity are working well and engaging your audience & which aspects could be improved in the future
• Using questionnaires, interviews or post event focus groups to understand audiences views and knowledge post activity

The Impact

Monitoring and Evaluation processes might include

• Analysing the qualitative data for key issues or themes
• Analysing the quantitative date for simple totals or using descriptive statistics
• Producing a written evaluation report
• Disseminating findings either as short articles or publications in journals or as posters or talks at conferences

95

Once you have understood the key aim(s) for your project, the next consideration is to define the objectives required to achieve your aim(s). This is the stage where you will begin to decide the type of event or activity and the audience(s) that you are hoping to reach. Box 5.1 offers you some guidance on defining your key objectives. This process will have at least two useful outcomes; it will allow you to:

- develop a viable evaluation strategy;
- move forward to the next stages of undertaking the activity/event – the planning and the delivery.

Box 5.1   What makes a good objective?

A good objective is a **SMART** objective.

S Your objective should be **S**pecific. Ideally it should also be **S**imple and **S**traightforward to achieve.

M Your objective should be **M**easurable. You should be able to **M**easure if you have met your objective. It should also be **M**anageable.

A Your objective should be **A**chievable within your timeframe, resources and budget. It should also be **A**ppropriate for achieving your overall aim(s).

R. Your objective should be **R**elevant enabling you to meet the overall aim. It should also be **R**ealistic. It should not be too ambitious: it should be achievable.

T Your objective should have a **T**imescale that will allow you to achieve your objective. It should also be **T**angible or able to be evaluated.

Stage two – The Planning

The planning stage: when an initial project idea is turned into an event or activity it will include the design, the planning as well as the organising and resourcing that occurs prior to undertaking your activity or event. The planning stage is easier if a clear set of objectives has been defined. This stage can be time intensive as you may need to:

- secure funding;
- organise a venue;
- organise the resources, such as the people and the money, that you will require to run the activity or event (see Chapter 6);
- publicise the event;
- ensure you have the relevant paperwork such as CRB checks, or insurance (see Chapter 6);
- trial the activity or event.

Stage three – The Delivery

The delivery includes running the actual event or activity. It is the culmination of and conclusion to the initial idea and the planning stages. Examples of successful projects have been detailed through the case studies and these can be found throughout this book.

5.3   Monitoring and evaluating

5.3.1   Monitoring

The process of monitoring your project includes collecting and recording data and information in a systematic, informative and methodical manner. The data collected are usually numerical and will provide useful information about your project. They could include;

- numbers of lectures delivered in a lecture series;
- audience numbers and audience demographic at each lecture;
- the number of hits on a website;
- the number of requests for reading material.

Monitoring is an important method of gathering useful information. Counting the number of people attending your event will let you know whether you have reached your desired audience numbers and may even provide you with information about the audience demographics. It can inform you about the success of the scheduling and whether the location or venue of the event has worked well or whether these elements should be reconsidered. Simply counting audience numbers will not provide you with information about whether your event was enjoyed by your audience, or if they found it informative or even life changing, but clearly if you are not attracting your anticipated audience then you are less likely to meet the objectives defined prior to undertaking your project.

5.3.2   Evaluation

Evaluation can be formative or summative. Either the *process* at each of the three key stages outlined in Section 5.2 (formative) or the *delivery* of the outcomes (summative) of the event can be evaluated; both strategies are informative and important. Formatively evaluating the process will allow you to:

- improve your activity/event through feedback and discussion, throughout the planning stages;
- understand not only what worked well in the planning and the delivery stages of your activity/event, but also acknowledge what could be improved in the future;
- disseminate the information to others to share good practice and provide information about potential pitfalls and how they can be avoided during the planning stages.

Formative evaluation is best undertaken at each of the three key stages (Section 5.2) and it is important to ensure that you provide sufficient time and resources to allow this to take place. Considering your evaluation strategy as soon as you start to develop your project will allow you to undertake this successfully, as highlighted in Figure 5.1.

On the other hand, summative evaluation reveals whether the overall aims and outcomes of your project have been achieved. This information allows you to establish the impact of your project. A case study that highlights the differences between formative and summative evaluation is written by James Piercy. This case study details the design of the interactive science show built around the Bloodhound SSC, a unique British engineering project to build a car capable of reaching 1000 m.p.h. The case study indicates where summative and formative evaluation can be included at different stages of the project and demonstrates the value of evaluation being included at the initial stages of the project planning.

Case Study 5.1

Bloodhound SSC
James Piercy; science made simple

Background

Bloodhound SSC is a unique British engineering project to build a car capable of reaching 1000 m.p.h. The design and engineering team behind the project were keen to exploit its educational potential. The Bloodhound Education team (BET) funded *science made simple,* an award winning science communication company, to develop an interactive science show called *Bloodhound SSC on the Road*.

A formative evaluation process was undertaken to ensure that the show met the needs of the BET, school audiences and the funders. The formative process involved a focus group held with 20 year 6 school pupils, a review of the national curriculum at Key Stage 2 and 3, as well as consultation with BET and the engineering team. The process identified key learning objectives for the show and provided ideas for the core structure. The formative evaluation process resulted in the content of the interactive show: demonstrations, video clips, images and opportunities for audience participation were all included to increase effective audience engagement. Part of the planning stage was that further evaluation was conducted through the delivery of pilot shows with the target audience, and following a review, was offered to science festivals during summer 2009.

The show lasts 45 minutes and is targeted at years 5–8, it was launched at the Cheltenham Science Festival in June 2009 and has since been delivered at schools and festivals across England. The Bloodhound team used a professional and experienced company to produce and deliver the project. Since *science made simple* has many years experience working with schools the time spent on insurance and safety issues was minimized. All presenters are CRB checked and a comprehensive risk assessment of the show material was carried out in line with company policy. Funding from the BET meant that equipment could be built or purchased for the presentations. Three sets of show equipment were produced to enable delivery in a large number of venues. Marketing was done through STEMNET as well as existing databases and BET's other activity.

What went well in the event?

The show was well received and had an impact on the audiences. Ongoing iterative evaluation meant that developments with the design project and feedback from students and teachers continually informed the content and delivery of the show. Using an established company such as *science made simple* allowed the BET to reach a large number of pupils in schools across England. During the 2009–10 academic year the show had been seen by over 22 000 people in over 60 schools and five science festivals.

What didn't go quite so well as expected?

The results of our evaluation strategies showed some evidence of a bias towards male pupils. Boys were more likely to have enjoyed and been affected by the show than girls. Although we had initially tackled this by the use of social context and female presenters we had to consider different possibilities to adapt the show to make it more attractive to a female audience.

Summative Evaluation

We evaluated the show using a questionnaire given to pupils. These were distributed to pupils immediately after the show to achieve a large number of responses. The pupils' questionnaire was produced in line with the Generic Learning Outcomes framework developed by the Museum, Libraries and Archives Council to measure informal learning (http://www.inspiringlearningforall.gov.uk/toolstemplates/genericlearning/). The strategy looks for attitudinal and behavioural change in audiences as well as judging their level of satisfaction with an intervention. The questionnaire included ranking of attitudes on a Likert scale as well as opportunities for open ended responses and comments.

Evaluation has shown high levels of pupil enjoyment and increase in positive attitude to science and engineering. There is also a demand for more shows from staff and pupils. During December 2009, the show underwent a further review based on formative evaluation supplied from presenter experience and feedback from teachers and pupils. The visuals were also updated at this time to reflect changes in the car design and to allow the inclusion of new information. The show continues to undergo periodic review as new audiences receive the show and iterative evaluation is undertaken. As the Bloodhound project develops further it will be useful for the *science made simple* team to meet with the engineering team to update the show content and highlight the current challenges.

It has proved difficult to get feedback from large numbers of pupils. We received 586 responses, which only represents 2.57% of the show audience. The need for schools to move pupils to their next lessons and demand on staff time makes it difficult to distribute and gain responses from large numbers of pupils.

Of these responses, 53.9% were from males and 46.1% from females. The overall results showed high levels of satisfaction with the show, and that the pupils felt they had learnt something new by watching the performance. They were also given the opportunity to tell us *what* they had learnt during the show by allowing space for open-ended comments. Their answers covered a range of topics from the very basic:

- 'I learnt that there was a Bloodhound project' (year 8 girl);
- 'cars can go quite fast' (year 9 boy);
- 'How technology and science is heading in the future and how the features of Bloodhound work' (year 8 girl);
- to the detailed:
- 'How the jet and rocket work, and a new compound H_2O_2' (year 8 boy);
- ' I learnt how rockets work, how difficult it is to make sure that bloodhound stays on the ground and how much science goes into it' (year 8 girl);
- 'How aerodynamics affects movement' (year 7 boy).

One aim of the project was to promote positive attitudes towards science and engineering and to encourage pupils to consider careers in STEM. They were asked questions to allow us to explore this aim. Once collated, the data was analysed and the responses were split by gender to identify any imbalance in the impact of the show. More than half of respondents reported that they had more positive attitudes towards STEM subjects following the performance. However, it was found that boys were more likely to show increased engagement than girls, 62.5 % vs 51%. Without a pre-visit baseline assessment it is impossible to know what the girls' previous attitudes were. The large number of pupils and schools involved meant undertaking a full baseline assessment was not possible; however, we used previous research undertaken on similar issues to help us interpret the results.

Teachers were also given questionnaires to allow them the opportunity to rate the show and to comment on its suitability for their pupils. All the teachers surveyed rated the show, 'very good' or 'excellent'. Comments received included:
- 'pitched perfectly;
- 'engaging and friendly;
- 'enthusiastic and knowledgeable
- 'very enthusiastic and well organised.

The full results of the evaluation are available at:
http://www.sciencemadesimple.co.uk/page202g.html

References
Analysis of the evaluation data was done by comparison with previous projects:

Brosnan, Mark (2007) Factors Predicating Attitudes and Success upon a Science/ Engineering Project. http://tinyurl.com/yasx9uf

DCSF (2008) After School Science and Engineering Clubs Evaluation Interim Report. http://tinyurl.com/ye3ymz2

NOISE (2006) You and Work. www.noisemakers.org.uk/modules/articles/show_press.cfm? id=23

Information on *science made simple*'s other projects and copies of final reports are available in the 'Research and evaluation" section of www.sciencemadesimple.co.uk

http://www.bloodhoundssc.com/

5.4   Undertaking evaluation

5.4.1   Measuring your SMART OBJECTIVES

Clearly the process of evaluation relies on upstream planning and the SMART objectives can enable you to do this, see Box 5.1. To establish whether you have achieved your objectives it is important that you consider the types of measures that may be useful. Box 5.2 provides an example of a project where the objectives have been defined, and measures have been identified to establish whether they have been achieved.

Box 5.2   Aims and objectives and methods to measure them

Aim: To inform my local community about Charles Darwin and his theory of evolution by natural selection using microbes as illustrative examples.

Objective 1: To develop a series of resources that will provide information about Darwin and his theory of evolution by natural selection that will appeal to a variety of audiences.

Measures: The number of resources developed: which resources attract which audiences: which resources were not appealing to audiences.

Evaluation tools: Observational research, interviews with participants and colleagues.

Objective 2: To provide opportunities for students to participate in science communication events and develop skills in science communication activities.

Measures: The number of students that took part in the activity, number of students that felt they had developed new skills, understanding the different skills developed by the students.

Evaluation tools: Focus group with students, interview with students, self reflective diaries completed by students.

Objective 3: To increase knowledge about Darwin and the theory of evolution in different audiences.

Measures: Views of audiences that took part in the activity, views and observations of event organiser, percentage of audience that learnt something new.

Evaluation tools: Questionnaires before event perhaps with ticket issue, focus group with students, interview with students, self reflective diaries completed by students, questionnaire after event, exit interviews with audience members, focus groups.

Setting SMART objectives that involve changing attitudes or increasing knowledge about specific scientific issues are worthwhile. However, if you are intending to measure a change whether it is a change in attitude or a change in knowledge, then it is really important to establish a baseline or a benchmark prior to your project in order to determine whether a subsequent change has occurred.

5.4.2 Evaluation tools

Once you have established the aim and defined the objectives of your project and considered the measures that can be used to determine whether they have been achieved, the next step is to think about the evaluation tools you can use to collect the data. There are at least four different methods for collecting data and each can provide valuable information. Designing an effective evaluation strategy may well use a combination of some or all these methods and may be used at the different stages of your project. The different methods used include:

- questionnaires;
- interviews;
- focus groups;
- observational research.

Each method is dependent on collecting data by using an effective sampling rationale. In general you should aim to avoid bias within your evaluation data by capturing responses from a representative selection of your audience and/or your participants. Unfortunately, this is not always synonymous with achieving a substantial amount of responses. For example, asking people to complete a questionnaire as they leave a public lecture may give you a large number of responses, but they may not be representative of your audience, as shown in Box 5.3. There are other caveats to data collection. Firstly, you can receive a more positive evaluation than your project justifies (see Box 5.3). Secondly, it can be the other way round; people may have really enjoyed your event or activity but may not have filled in a questionnaire or provided you with this data. Therefore it is really important to remember that you can only work with the data that you receive and this may not be an accurate representation of what your audience actually thinks or feels.

Box 5.3   An evaluation scenario

Let us imagine that you have decided that an excellent approach to meet your aim (as stated in Box 5.2) 'to increase knowledge about Darwin and evolution in different audiences using microbes as illustrative examples' is to deliver a public lecture on the subject.

You monitor the number of people in the audience and you are delighted to see that you have filled the 400-seat lecture theatre. Questionnaires are handed to each audience member as they arrive. After your talk (which you thought went well) questionnaires are collected and you are pleased to see that you have 200 responses; a response rate of 50%. Now let's imagine that 100 people in your audience 'really enjoyed your talk' and are happy to impart this information to you. They all complete and return the questionnaire.

On the other hand there are 200 people in the audience who thought your talk was 'okay' but you spoke a bit fast; they didn't really follow all of it and they still can't understand what evolution has to do with microbes. However they felt that you 'tried your best', 150 audience members from this 'okay' group decide they don't really want to fill in your questionnaire just to let you know that your talk was okay: so we have 150 audience members in the 'okay' group who do not complete the questionnaire. On the other hand, 50 audience members from the 'okay' group decide they do want to give you feedback on your talk so that you can improve it when you give it again and they complete and return the questionnaire.

Lastly there are the 100 people in the audience who thought your talk 'was terrible'. In fact 50 of them left before the end without completing a questionnaire. The remaining 50 are so annoyed by how terrible your talk was that they feel obliged to share this with you by completing and returning the questionnaire.

As you can see, although your response rate is quite impressive, the views that you have solicited do not accurately represent the views of your audience. The feedback you have gathered indicates that 50% of your audience 'really enjoyed your talk', 25% thought you 'tried your best' but spoke a bit fast and 25% of your audience thought your talk was 'terrible'. Let's compare this to the actual views held by your audience: 25% 'really enjoyed your talk', 50% thought you 'tried your best' but spoke a bit fast and 25% of your audience thought your talk was 'terrible'.

This shows that your evaluation strategy has over-represented the positive aspects of your lecture and under-represented the negatives.

There are different methods of sampling that can be undertaken to try and overcome some of the issues highlighted in Box 5.3 to gather a more representative view.

- You can ensure that you collect information from every participant that takes part in your activity.
- You devise a strategy to ensure you reach a representative sample of your audience based on your audience demographic. This demographic can be based on age, the school they attend, their sex or their ethnic origin.
- Your strategy involves systematically asking every 'nth' person to fill in a questionnaire or provide feedback.

These methods require considerable forethought and a prior knowledge of your audience. There are more pragmatic approaches to sampling that are often the most convenient, realistic and straightforward to use. However, it is important to recognise that these approaches will not provide a random sample. They may not be representative of your audience and are open to bias. These approaches include:

- collecting feedback from a convenient audience. If we think back to our lecture described in Box 5.3, this may be the first 50 people to leave the lecture theatre after you have finished your talk (not the ones that left half way through!);
- collecting feedback from a purposive sample. In other words you deliberately collect feedback from a group that you feel represents the wider community or audience.

Clearly not everyone who is asked, whatever sampling method you choose, will agree to take part; however, as shown in Box 5.3, the audience members that enjoyed the activity are more likely to respond to requests for feedback than those that did not. Ensuring that you get as high a response rate as possible will mean that you increase your likelihood of achieving a representative sample. Therefore developing strategies for achieving a high response rate is important. These strategies can include:

- using interviews to complete questionnaires; people are more likely to take part as they don't want to be rude;
- offering a small prize draw for all completed questionnaires;
- handwritten post-it notes on each questionnaire may increase your response rate;
- supplying a stamped, addressed envelope with a postal questionnaire can encourage greater returns;
- if you enjoy using new technology, consider supplying your event or activity with a Quick Response Code that can be scanned using smart phone technology. This code will direct participants to your evaluation website.

5.4.3 Questionnaires

Questionnaires are an extremely convenient method for obtaining a large quantity of both quantitative and qualitative data. Quantitative data can include responses to questions such as 'How many participants attending the lecture felt that it was pitched appropriately for the age range of the audience?' However, questionnaires can also be used to gather more qualitative data such as 'Why did you decide to go to this lecture?', or 'What was the most interesting thing you learnt from this talk?' There are several methods of distributing questionnaires and each has their own pros and cons as shown in Table 5.1.

5.4.4 Designing an effective questionnaire

It is really important that you introduce your questionnaire by explaining:
- who you are;
- why you are asking the respondent to answer the questions;

Table 5.1 Comparing types of questionnaires.

Type of questionnaire	Pros	Cons
Self completion: usually paper, often distributed to people as they arrive or leave an activity	Inexpensive	Low response rate
	Can generate large quantities of data	May not be representative
	Easy to distribute	Cannot guarantee that the questions have been understood
	Instant feedback from activity	
Postal: usually when you have a mailing list, most often will require postal reminders	Can get better response rate if reminders are sent	Need to send several reminders
	Allows time to reflect on activity	May not be representative
	Allows time for action to occur as a result of the activity	Cannot guarantee that the questions have been understood
		Postal cost
Email: using survey tools such as Survey Monkey	Easy to distribute with an email list	Requires internet access
	Offers systematic sampling potential	Cannot guarantee that the questions have been understood
	Inexpensive	
Web survey: pops up when someone logs onto a website	Offers systematic sampling potential; every fifth person that visits website	Low response rate so may not be genuinely representative
	Inexpensive	Cannot guarantee that the questions have been understood
		Requires internet access
Interviews	Can reach a representative sample	Resource intensive
	Opportunities to ensure questions are understood	Expensive
	Provides more detailed information	Interviewer may unintentionally bias responses
	Encourages good response rate as respondents are less likely to say no	

- to the respondent what you intend to do with the information that that they are providing.

You should then consider 'What do I really want to know?' Returning to the objectives that you developed at the beginning of your project is always a useful starting point when planning your questionnaire. Understanding the key information you want to find out will make the process of designing the questionnaire much more straightforward. It is also important to consider who is going to be filling in the questionnaire and then adjust it accordingly; a questionnaire designed for schoolchildren will be different to one designed for adults. You should take into account the length of time the respondent will *have* to complete the questionnaire and ensure this is *similar* to the time *needed* to complete it. A general rule of thumb is to keep it short and simple (KISS) and avoiding jargon and scientific or technical terminology is also desirable (see Section 4.7). You need to make sure that the design of your questionnaire answers the question 'What do I really want to know?' and avoids answering 'Wouldn't it be interesting to know?' Hopefully this will help to keep the size manageable. Finally, an important step is to trial your questionnaire ahead of time. This will allow you:

- to find any ambiguous questions, and to check that questions are in the correct order;
- to ensure that the questionnaire is straightforward to complete and can be done in the allotted time;
- to check that the questionnaire provides the answer to 'What do I really want to know?'

5.4.5 Types of questions

There are two main types of questions that can be included:

- **pre-coded** – the participant selects an answer from a list that you have supplied often by selecting a tick box; this provides quantitative data;
- **open ended** – the participant is able to provide their own thoughts, ideas, views and suggestions; this can provide opportunities for qualitative data.

A pre-coded question can be a simple, straightforward multiple choice question such as Yes, No, with a Don't Know as an option **or** a True, False, Not Sure as illustrated in Box 5.4.

Box 5.4 Examples of precoded questions

Would you recommend this lecture on Evolution and Microbes to your friend?

Yes ☐   No ☐   Don't know ☐

However if you decide to use a simple multiple choice you need to make sure that your question is simple and straightforward. Avoid questions that are actually asking two different things, for example:

Did you enjoy this lecture on Evolution and Microbes and will you recommend it to your friends?

Yes ☐   No ☐   Don't know ☐

The respondent will not know whether to respond to:

Did you enjoy this lecture on Evolution and Microbes?

or

Will you recommend this lecture to your friends?

Other multiple choice question formats commonly used to understand more abstract concepts such as the audiences attitudes and experiences include Likert scales. This format offers 'scaled' responses to questions but it is important that the response options are balanced; there should be equal positive and negative responses. Responses that are skewed with more options for positive than negative, or vice versa should be avoided. Avoid leading questions that force the respondent towards a specific answer. Questions should be neutrally worded and avoid stating 'truths' that respondents will be reluctant to disagree with (Box 5.5).

Box 5.5   Providing scaled responses to questions

Did you find the lecture on Evolution and Microbes

Very interesting	☐	Very Interesting	☐
Quite interesting	☐	Interesting	☐
Not very interesting	☐	Quite Interesting	☐
Not at all interesting	☐	Not very Interesting	☐
Not sure	☐	Not sure	☐

Good Design: Equal positive and negative responses

Bad Design: Skewed options more positive responses

Neutral Question

We are interested to know your views about microbes. Do you believe that they are evolving?

Yes ☐   No ☐   Not Sure ☐

Leading Question

Microbes clearly demonstrate evolution in action. Do you agree?

Yes ☐   No ☐   Not Sure ☐

Another approach you can use is to provide a group of different statements and you ask your respondent to choose one as illustrated in Box 5.6.

Box 5.6   Supplying different statements

Going back to our Evolution and Microbes lecture a group of statements might be:

Please tick the statement that most closely represented your views towards evolution prior to attending the lecture this evening

1. There is evidence that shows the theory of evolution is not the explanation for the variety of life on Earth   ☐

2. There is evidence to underpin several different theories that explain the variety of life on Earth   ☐
3. The evidence may or may not support the theory of evolution as an explanation for the variety of life on Earth   ☐
4. The evidence only supports the theory of evolution as an explanation for the variety of life on Earth   ☐

An alternative approach to glean the views of respondents toward evolution would be to include an open ended question and provide your audience with a text box for their answer.

> Can you tell us/me your views on evolution prior to attending the lecture this evening?

Your questionnaire should look attractive and, although you may feel an urge to squeeze as many questions as you can into a small space, the layout should be well spaced and easy to read, so avoid a small font size. Consider the most effective way to order your questions. Ideally:

1 the first couple of questions should grab the respondent's attention;
2 you should lead the respondent through the questions in a logical order;
3 if you want to ask sensitive questions, for example asking details about age, ethnicity and income, it may be best to leave these to the end as they may put people off from answering the questions you really need the answers to.

When asking for details about age or income, ensure that the ranges do not overlap; age ranges should be 0–9, 10–19, 20–29 and **not** 0–10, 10–20, 20–30. The questionnaire should include boxes that are easy to tick or complete and if you have used open questions, provide sufficient space for respondents to write their views. Finally, just to re-emphasise, always take time to pilot your questionnaire prior to using it for your activity or event. This will check that your questionnaire provokes the anticipated responses from your respondents. This can be part of the formative evaluation process during the design and planning of your project.

5.5   Interviews

Conducting interviews can be an effective method for obtaining both qualitative and quantitative feedback on your project. Interviews can be conducted with stakeholders and with colleagues who are helping to deliver the event as well as with the participants. They can be undertaken either:

- face to face;
- using the internet, including social media networking sites such as Facebook and Google+;
- via the telephone or Skype.

Interviews can provide opportunities to gather more in-depth information compared to questionnaires, but the downside is that they are much more time and resource intensive. A robust interview process will always use an interview schedule, which is a series of questions (very like a questionnaire) that the interviewer uses to direct the interview. The interviewer can stick tightly to these scheduled questions, which assures that there is parity between all the interviews that are undertaken. Alternatively, the interviewer can use the schedule as a guide, but is free to adjust the interview in response to information provided by the respondent. In both cases, preparation is key; prior to the interviews taking place, the schedule should be trialled and the interviewee should be well briefed. Before beginning an interview it is important that the interviewer:

- identifies themselves;
- asks permission to undertake the interview;
- explains why they are undertaking the interview;
- informs the interviewee how long the interview is likely to take.

If you intend to interview children it is necessary to ask permission from an accompanying adult. Once the interview is underway, record the answers accurately as they are given and do not rely on your memory to provide the information after the interview is completed. Alternatively consider using recording equipment (dictaphones can be purchased quite cheaply) as this will allow you to transcribe the interview at a later date. You must seek permission to use this equipment before you begin the interview process.

5.6   Focus groups

Focus groups are an extremely useful evaluation tool. As with interviews, they can be undertaken with stakeholders and with colleagues who are helping to deliver the event as well as with participants. This evaluation tool can enable views and experiences to be explored in depth and they can provide detailed qualitative data. A successful focus group will:

- usually have less than 10 participants;
- clearly state the reason for the focus group;
- be well facilitated to allow all members of the group to participate and to avoid domination of the discussion by one or two vocal characters;
- have a clear idea of the topics and the subjects to be covered;
- be well recorded either through clear and comprehensive note taking or using recording equipment that will allow the discussion to be transcribed later.

Focus groups can take time to organise. They can also be time and resource intensive. Although they can provide useful, in-depth qualitative data they are not particularly useful for providing quantitative data. It is important to consider the make-up of your group to ensure that you have access to the views of all relevant parties; to facilitate this, the focus group must meet at a time and a place that is convenient for all the group participants. It is also worth considering organising focus groups using online discussion fora. There are free tools that you can use to organise focus groups in this way. These include creating groups in Facebook and undertaking a whole group chat. Alternatively, Google + allows you to create 'huddles' where people can come together and chat using webcams. Skype also has the potential to be used for online discussion groups through their conference call application which is free if you and all participants are signed up to Skype. In addition, for a subscription fee, Skype will allow you to undertake group video calls. Undertaking focus groups in this way has the advantage that it may be more convenient to the group participants. However, the caveat is that you are assuming that the participants have access to computers (and webcams); this isn't always the case.

5.7   Observational research

Observational research can provide a wealth of evaluative data. There are different methods and approaches for undertaking this type of research. For example:
- keep a reflective diary throughout the formative evaluation process and as the project is designed and developed;
- record the ebb and flow of different participants attending the event or activity;
- record observations during the event/activity itself, such as the participants' reactions to certain exhibits or which exhibits attract the most interest;
- have a visitors' book or a post-it-note board where participants can leave comments; collect voxpops or ask participants to draw pictures that represent their experiences;
- take part yourself and reflect on your experiences.

5.8   Deciding which evaluation tools to use for your project

Clearly, there are several tools you can use to evaluate your project. The choices you make will be dictated by many different factors that can include:
- the type of event or activity;
- the outcomes and outcome measures associated with activity or event;
- the available resources and expertise.

As outlined above, each evaluation tool has its advantages and disadvantages and it may be the case that your evaluation strategy needs a variety of different tools to measure the outcomes and allow impact to be established. Case study 4.1 written by Jaeger Hamilton details how to design an activity that appeals to different learning styles and different audience types. The activity described was part of a science communication event funded by the Wellcome Trust as part of the Mobile Family Science Laboratory (Chapter 4) delivered during a National Science and Engineering Week. In addition Case study 5.2 has used this activity to provide an example of the aims, objectives and evaluation tools used to evaluate this activity. Case study 5.2 also demonstrates the type of data that can be gathered using these evaluation tools. We have used this data presented within the case study to provide examples of how data can be analysed. In addition, an observational tool for evaluation has been applied to a wide range of science communication events funded by the Wellcome Trust as part of the Mobile Family Science Laboratory. This has allowed us to develop a model of family centred learning which is detailed in Section 4.6.

Case Study 5.2

Evaluating an Activity for the 'Norfolk Science Past, Present and Future' Event

Jaeger Hamilton

Introducing the activity

My activity was one of several designed by students from the school of Biological sciences at the University of East Anglia as part of our Science Communication Module. The activities formed the event, 'Norfolk Science Past, Present and Future', which took place at Norwich Castle Museum, one of Norwich's main visitor attractions. My activity involved using microscopes to learn about the water mould *Saprolegnia* which was first identified as a fish pathogen by the local eighteenth century philosopher and scientist William Arderon. It releases sugars and exudates into its external medium which make a brilliant 'home' for other microbes, such as *Vorticella,* rotifers and *Paramecia*. Under the microscope, *Saprolegnia* filaments appear as a branching reticulum with a lively microcosmos of creatures dancing in and out, which I hoped would be enjoyed by the visitors. This event was planned to take place during National Science and Engineering Week (NSEW) organised by the British Science Association. It was important that my activity was *accessible* to every visitor, reflecting their learning styles (pragmatist, activist, reflector, theorist), and that it was a dynamic, creative and open learning experience

The aim and objectives

The overall aim of my activity and the event that it was part of, was to raise awareness and stimulate interest in science/microbiology, with a public audience. The primary objectives were that:

- my activity design should appeal to a mix of ages and to a broad range of learning styles including, activists, pragmatists, reflectors and theorists;
- my activity should allow the scientific facts of the interconnectedness of different life forms as highlighted by the microscopic world of *Saproneglia* and its associated microbes to unfold as the participants took part;
- I provide learners with a mix of my attention as well as their own process of action and discovery;
- I was to design an activity that would enable me to develop my knowledge of and skills in, science communication as part of my undergraduate degree programme.

Setting the objectives, and defining the measures and the evaluation tools that I chose to use are shown in Case study table 5.2.1.

Case study table 5.2.1 Planning the evaluation tools and defining the measures.

Objective: The design should appeal to a mix of ages and to a broad range of learning styles.
Measures: Which resources appealed to which visitors: could all visitors engage in the activity: did my activity attract visitors with a variety of ages and learning styles?
Evaluation tools: Observational research of my activity: feedback book for my activity: post event questionnaire/emails.
Objective: The activity should allow the scientific facts of the interconnectedness of different life forms to emerge.
Measures: Did visitors comment on/ describe/acknowledge/discover information about the interconnectedness of different life forms?
Evaluation tools: Observational research: feedback book for my activity: post event questionnaire/emails.
Objective: The design allows participants to participate in a process of action and discovery in addition to receiving my attention.
Measures: Could visitors participating in the activity without my support make their own discoveries: was I able to support the visitors learning as and when required
Evaluation tools: Observational research: feedback book provided for my activity: my personal reflections of the dynamics of the activity
Objective: The activity should enable me to actively develop my knowledge of and skills in science communication.
Measures: Did I develop any science communication knowledge or skills as a result of designing and participating in this activity?
Evaluation tools: personal reflections of the process of designing and delivering my activity: observational research; feedback book provided for my activity.

When designing my evaluation strategy the pragmatic decisions that I made included:
• recognising that I did not have the resources to allow a pre-activity questionnaire to assess visitor's understanding of the interconnectedness of different life forms prior to the event;
• awareness of my limitations: I knew that I wouldn't be able to undertake interviews with the attendees as I was actively running my activity;
• that I would have a pragmatic approach, and distribution and completion of the questionnaires (sampling) would be post-activity.
Some of the free text comments written by participants on the exit questionnaire that related to my activity are shown in Case study table 5.2.2.

Case study table 5.2.2 Amazing things learnt today: free text comments left on exit questionnaire.

'bout smaller life forms'	'all about microscopes'
'microbes can be found in different places'	'how to use microscopes properly'
'there are living microbial things'	'that tiny things can move'
'protozoa can eat bacteria'	'there are tiny water creatures'
'you can see bacteria and fungi under the microscopes'	'there are tiny things that move'
'using the microscopes'	'about *Paramecium*'
'it was fascinating to see *Paramecium* up close'	'*vorticella*'

Evaluation data gathered from observational research

Monitoring the activity highlighted that it was constantly busy, but not unmanageablely so. On average, the group size was ~three people at a time (not including myself): often one/two children with one/two adults, there was usually one to three groups present at any given time.

The children showed a strong interest in using the microscope and asked questions about what they were viewing – displaying an 'activist' approach to learning. Occasionally a child showed 'reflective' preference by watching a friend/ sibling do the technical bits. None of the children showed a strong 'theorist' tendency, although some children were more enthusiastic about 'information' than others. The adults tended to ask more practical 'pragmatic' questions such as 'Why is this important to society?' and 'How does this help?' I would refer to the posters that explained the important effects that water moulds have had on human society. In turn, the adult learners would reflect this information from me back to the child. The children did not ask any obviously 'pragmatic' questions and none of the children explicitly asked about the posters. Occasionally their attention would be directed to something that was mentioned in the posters, and some children (usually the ones ≥ 12 years old) would discuss that information, but most of the children lost interest and returned to the excitement going on beneath the microscopes. It was mostly the adult learners who seemed to benefit from the posters: they read the posters while their children *did* the activity. So there was a clear disparity between the generations: the adult learners tended to show theoretical, reflective learning tendencies, as well as a pragmatic concern for how the knowledge was applicable, and the child learners tended to be very much the 'activists', learning by doing. I felt that my activity had met the objective of appealing to a broad range of learners with several groups staying for approximately an hour; some of them even came back for more.

Questionnaires completed by visitors

A questionnaire was also designed to provide feedback about the entire event covering all activities including my own. A pragmatic approach to sampling was taken: everyone that left the event was provided with a questionnaire and asked to complete it before they left, but a low return rate was achieved (10% when compared to ticket sales to the event).

Information obtained from the questionnaires included:

- the average distance travelled to attend the event was 17.4 miles and on average visitors spent over an hour travelling to and from the event (however whether they came specifically for the event was unknown);
- socioeconomic information about the visitors based on post codes provided and using the acorn category; showed that of the 89% of visitors from Norfolk, ~67% were high-income families, but only 14.5% were low-income families. The levels of education, however, were more evenly dispersed: 38% came from a postcode where a high proportion of people are educated to degree level, 34% medium proportion, and 27% of visitors were from postcodes where occupants are estimated to have a low number of degree level qualifications;
- information about how enjoyable adult visitors found the event using a Likert scale of 1 for poor, to 5 for excellent, to ask 'Did you enjoy the event?', 62 adult visitors responded and the average was 4.56.
- children's feedback showed that 60 children thought that the event was 'fantastic', 19 thought it was 'good', and 4 thought the event was 'okay'. None thought the event was boring or terrible;
- a description of what three amazing things did you learn today?

Visitor feedback left in the facilitator's booklet and e-mails sent after the event

The learners were also invited to leave comments/feedback in a booklet designed for my activity. In addition feedback was received from visitors who wrote to the Castle Museum after the event. Some of the comments are included in Case study Box 5.1.

Case Study Box 5.1   Comments left in facilitator's booklet

'Our boys-listening intently, were excited and enthusiastic, looking entranced in microscopes for an hour, asking sensible questions, captivated in the Museum until the end and wanting to come back-it's a miracle !! Not a Games console in sight and they weren't bored for a second !!'

'We really appreciated the Science weekend thank you.'

'My older son spent the rest of the weekend finding things to look at under his decent microscope and from his conversations it was clear that your presentation really sparked something in his thinking and interest. If you can have that kind of influence on even one person, that's great'

'I think the Paramecia is amazing. I couldn't believe they were alive.'

'I never thought analysing the processes of decay could be so interesting. The microorganisms were quite beautiful too, which surprised me. I would like a few more contextual examples of how oomycetes affect human (and other) life. The potato famine really helped me to understand the concept.'

'My son enjoyed looking through the microscope. Excellent instructions from the volunteer, made it very accessible and interesting for the children. Volunteers are extremely knowledgeable.'

'Very interesting! Fascinating to see how Fungi is the playground for bacteria and stuff.'

'Good idea giving people the chance to use the microscope which I wouldn't have had the chance to use!'

'I think it is very clever but a bit weird. It reminds me of the bubbles in a Jacuzzi.'

'They are really fascinating and look really cool under the microscope. They move very fast.'

'It's amazing how many things are going on in a tiny space! It's really cool!'

'Brilliant – science made fun and interesting for all ages.'

'Presentation could be better, a bit messy'

'I enjoyed myself!!! 'I think it looks like a germ, it makes me feel sick.'

'Thanks, learnt so much.'

'It's strange how they move about and we found a big fat one!'

'I want to become a scientist now!'

Personal reflections from running the event

One of my objectives was to actively develop my knowledge of and skills in science communication . For my part, my personal reflections after the event allowed me to realise that I have learnt much more about how people learn in informal settings such as the 'Norfolk Science Past, Present and Future' event at the Castle Museum, and how to communicate science more effectively to others. In addition I feel that I have a deeper understanding of the science underpinning my activity as a result of trying to explain it to others.

5.9   Analysing the results

Having gathered your data, you need to consider the approaches you can use to analyse them. You should consider whether you can analyse the data manually or whether you require additional technical support or training to analyse the data using spreadsheets (e.g. Excel) or using a package such as the Statistical Programme for Social Sciences (SPSS) for quantitative data or INVIVO for qualitative data. If you feel that your data analysis requires this additional support then it is worth exploring whether your institution or place of work:

- has a license that will enable you to freely access these programmes;
- provides training for these programmes.

5.9.1 Quantitative data

Quantitative data obtained during evaluation and monitoring of your project can be analysed quite simply if the information you require is totals, frequencies or descriptive statistics. This can be done using data inserted onto a spreadsheet and then using Microsoft Excel to calculate descriptive statistics. This data can be displayed using different graphical forms, including bar and pie charts. Using data from Case study 5.2, Figure 5.2 illustrates how the same data can be displayed in different ways. If you are keen to analyse quantitative data to detect information about any similarities or differences between different groups, then you should consider using non-parametric tests such as the Wilcoxon rank-sum test and the Mann–Whitney test (approaching a statistically able friend or colleague may be useful at this stage). This technique can be useful for:

- comparing the views of a group expressed numerically before an activity or an event and the views of a group after;
- comparing the views of participants expressed numerically from different age groups.

If you have obtained data from open-ended questions or free text responses on questionnaires, it is possible to quantitatively analyse this data using a programme such as Wordle which is freely available on the internet

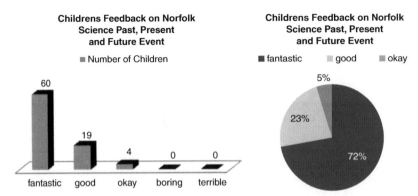

Figure 5.2 Graphical representations of quantitative evaluation data. There are different ways to represent quantitative data: the same data has been represented numerically and as percentages using different styles of charts. Note that the pie chart does not indicate that 'boring' and 'terrible' were options that children could have chosen!

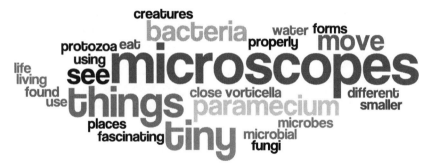

Figure 5.3 A Wordle diagram from the free text feedback from the questionnaire used in Case Study table 5.2.2 in response to the question 'Which three amazing things did you learn today?'

(www.wordle.net/). 'Wordle generates "word clouds" from text that you provide. The clouds give greater prominence to words that appear more frequently in the source text'.

An example of a Wordle undertaken using the free text feedback from the questionnaire used in Case study table 5.2.2 in response to the question 'Which three amazing things did you learn today?' is shown in Figure 5.3.

Clearly, the generated Wordle shows that at least one of the three most amazing things gleaned by the visitors after attending the activity involved the microscopes, other words with prominence include tiny and things, closely followed by paramecium, bacteria and moves. This feedback indicated that the visitors who left comments had actively engaged with the activity, enjoying the microscope and the microscopic life that they revealed.

5.9.2 Qualitative data

If you have obtained qualitative data either from open-ended questions on questionnaires, interviews or focus groups, then the next step is to analyse the data to begin to identify information that:

- may give you insight into the views and opinions of the participants or respondents, and you may be able to identify changes as a result of the activity or event;
- will ultimately allow you to measure whether the objectives that you set as part of your evaluation strategy, have been met.

There are several steps that will help you to analyse your qualitative data as highlighted in Figure 5.4. It is important to study the data that you have obtained to begin to identify the main issues or themes that begin to emerge. The analysis undertaken and the issues and themes identified can often be supported by quotes that illustrate your findings. An example of this type of analysis is shown in Box 5.7. This analysis has been undertaken using the qualitative data left by visitors to the 'Norfolk Science Past, Present and Future' event highlighted in Case Study 5.2.

Analysing Qualitative Data

Norfolk Science, Past, Present and Future

Gather and organise any relevant qualitative data. Remember that you may have received qualitative data from interviews, free text comments left on questionnaires, focus groups, post event feedback.

Qualitative data received from
• free text comments left on feedback forms
• comments left in facilitator's book
• emails sent from visitors after the event

Take time to careful scrutinise and read the data until you become familiar with it. It is during this stage that you will begin to code your data and identify ideas or themes.

Compiled data from all three sources is scrutinised until it becomes familiar and positive, and negative themes begin to emerge.

The evaluation data can be coded in different ways but it is most likely that you want to know what's working and what isn't working. Consider categorising the data into positive comments and negative comments. This can be followed by descriptive coding where you note the different topics or themes.

Positive	Negative
• demonstrator shows knowledge of activity	• limited contextual information
• use of microscopes	• disquiet at seeing microbes present in water droplets
• chance to see microscopic life form interactions	• untidy activity
• activities were enjoyable and fun	
• the activity was inspiring	

Next consider
• looking for any changes in participants views, skills, attitudes or knowledge
• distinguishing between those whose views have changed and those whose views have not
• looking at the **actual** outcomes of the project and the **desired** outcomes of the project

Changes in attitudes or knowledge: outcomes of activity

• positive influence on children '*sparked something in his thinking and interest*'
• using microscopes in new ways
• positive view of science and scientists
• understanding and wonder of the interconnectedness of life viewed under a microscope
• activity appealed to different members of the audience (positive comments were left by children and adults)

Reflect on the analyses and possible changes to the activity/event. However Remember that the data you have maybe limited and the ideas or themes that emerge through your analysis may not represent the views of all participants especially if your sampling technique has been pragmatic and you have received a limited number of responses.

Reflect on analyses and possible changes to the activity

• successfully achieved outcomes for the activity: it appealed to different learning styles and audience members: it showed interconnectedness of different life forms: learners had a mix of my attention as well as their own process of action and discovery
• consider changing format of activity to provide more contextual data and a more appealing presentation

Figure 5.4 The steps involved in analysing qualitative data. Information from the qualitative data provided in Case Study 5.2, is provided as an illustrative example.

> **Box 5.7   A qualitative analysis of the 'Comments left in the facilitator's booklet' shown in Case study 5.2**
>
> On the positive side, a key theme expressed by visitors was an appreciation of the opportunity to use a microscope. In addition the view was expressed that learning the technical aspects of how to use a microscope enabled this learning to be applied in new ways when they returned home: 'my older son spent the rest of the weekend finding things to look at under his decent microscope and . . . it was clear that your presentation really sparked something in his thinking and interest'. The design of the activity clearly appealed to children who were 'listening intently, being excited and enthusiastic, looking entranced in microscopes for an hour, asking sensible questions'. Adult visitors also appreciated the activity 'I never thought the process of decay could be so interesting'. Another comment left by a visitor emphasised this, stating that this was 'science made fun and interesting for all ages '. An objective for the activity was to 'allow the interconnectedness of different life forms to unfold'.
>
> Feedback from participants included that it was 'fascinating to see how the Fungi is the playground for bacteria and stuff'. The activity clearly allowed several visitors an intriguing insight into microscopic life found in water 'the *Paramecia* is amazing. I couldn't believe they were alive' and 'it's amazing how many things are going on in a tiny space! Another objective of the activity was to provide an opportunity to develop 'skills in science communication', feedback from visitors included 'excellent instructions from the volunteer, made it very accessible and interesting for the children'. Feedback from children also included the views that the activity was 'cool and inspirational' and 'I want to be a scientist now' implying that the event may have influenced their future ambitions towards studying science; one of the overall aims of the event.
>
> Negative feedback included the presentation of the activity which 'could be better' as it was 'a bit messy'. There was also the view that although the microscopic world was fascinating to view under the microscope, the activity could have been improved by relating it to our everyday lives: 'I would like a few more contextual examples of how oomycetes affect human (and other) life'. In addition, viewing the life forms in water that become apparent when viewed under a microscope can be unsettling for some children as illustrated in the comment 'it looks like a germ, it makes me feel sick'.

5.10   Reporting the results

Hopefully you are beginning to realise that any evaluation data that you gather can provide you with a wealth of useful information and insight, not only about the project you have undertaken but also for future projects that you or your colleagues deliver. In most cases your evaluation will take the form of a written report. Although it may seem that the obvious thing is to provide a summary of the responses to the questionnaires, the report should be succinct but it should also seek to;

- formulate key learning points;
- highlight successes but also identify weaknesses;
- provide recommendations.

Often if you are writing a report for a stakeholder such as a funder, they will ask for a report to be written in a specific format and it will ask for specific information. Indeed some funding bodies may have a proforma that they will ask you to complete. It is worth establishing the format and information required before you begin to compile your data and write your report. There is a lot to be said for communicating your project to a wider audience to share good practice and/or prevent others from avoiding the pitfalls that you have identified. You can consider publishing your evaluation as a case study (such as those provided in this book), or a short article in one of the specific science society's publications or newsletters such as Microbiology Today or The Biochemist. Alternatively, there are often conferences organised by science societies that provide opportunities to report on science communication events and activities either as poster presentations or short talks. In addition, reports can be published on your personal website or your institutions' web pages. If you can write up your project as a case study, it can be shared to allow others to learn from your experiences. Consider posting it onto:

- the database called collective memory, http://collectivememory.british scienceassociation.org/;
- the collection of case studies housed on the NCCPE website, http://www .publicengagement.ac.uk/how/case-studies.

Finally consider presenting the evaluation of your event to your colleagues in established forums such as departmental meetings or newsletters.

5.11 Assessing impact

Undertaking evaluation is also part of the process that is required to begin to measure impact. Public engagement is at a stage where there is a real will to establish the relationship between impact and engagement as described by the RCUK Pathways to Impact Statement and used in the Higher Education For England and Wales (HEFCE) to embed impact into the Research Excellence Framework. Assessing impact is proving to be challenging. However, a useful model that can be used to assess impact is the Kirkpatrick model which has proposed four levels of potential impact that occur over a linear timeframe. These levels are:

- **reaction**: the initial response to your activity or event;
- **learning**: has your audience changed or developed a more in depth understanding or knowledge base as a result of your activity or event?;
- **behaviour**: has your audience changed its behaviour as a result of the knowledge or understanding they have developed because of your activity or event?;
- **result**: has your activity or event resulted in any long-term measurable outcomes such as a long-term behaviour change?

More information about the Kirkpatrick model can be found on the Kirkpatrick website that contains free to use resources after you register your contact details (http://www.kirkpatrickpartners.com/Home/tabid/38/Default.aspx).

Lastly and unfortunately, it is difficult to design and deliver an evaluation strategy for your project that can provide information about its long-term impact. This is because:

- maintaining ties with the transient audiences that often attend these events is difficult and therefore obtaining information about the long-term impact is equally difficult;
- even if you have an audience who you regularly keep in touch with, encouraging this audience to supply you with the long-term feedback required to assess long-term impact can be problematic and challenging.

Let's assume that you have tracked your audience and even better, you managed to achieve a healthy, longitudinal feedback rate; there is still another issue – the lack of a control group or a group that didn't experience the intervention.

5.12   Ethical issues associated with evaluation projects

As you will have gathered from this chapter, undertaking evaluation is a key component of running a successful event or activity. However, it is important that no harm or excessive inconvenience is caused when you obtain your data. It is important that you undertake any evaluation in an ethical manner. Important steps that should be taken are:

- to keep the respondent or participant fully informed about the reasons for undertaking the evaluation, the demand on their time, ensuring they understand what will happen to their data and what the collected data will be used for and ensuring that they give fully informed consent before they are interviewed, take part in focus groups, complete questionnaires etc;
- to ensure that their data is kept securely, confidentially and anonymously unless they have clearly indicated that they are happy to be named;
- offer contact details that can be used to provide more information if required;
- to ensure that you have had a CRB check if you are obtaining information from vulnerable adults or children;
- to seek guidance from others if you are unsure about any ethical aspects of your evaluation. University ethics committees can offer help, advice and support if required.

Other useful resources and references

More information on writing SMART objectives

Doran, G. T. (1981). There's a S.M.A.R.T. way to write management's goals and objectives. Management Review, Volume 70, Issue 11(AMA FORUM), pp. 35–36

Further information on ethical issues associated with undertaking evaluation

The British Educational Research Association's revised ethical guidelines for educational research (2004) http://www.bera.ac.uk/publications/guidelines/
The information Commissioners Office website for the data protection act www.dataprotection.gov.uk
The Market Research society website www.mrs.org.uk
The Criminal Records Bureau www.direct.gov.uk/en/Employment/startinganewjob/index.htm?CID=EMP&PLA=url_mon&CRE=crb

More information on evaluation

Guide to teaching evaluation (National HE STEM Programme Other Useful Resources for Information on Evaluation http://www.hestem.ac.uk/evaluationtools
Resources for Evaluating Science Engagement Activities (NCCPE). http://www.publicengagement.ac.uk/how/guides/introduction-evaluation
The Arts and Humanities Research Council Understanding Your Project: A Guide to Self Evaluation http://www.ahrc.ac.uk/FundedResearch/Documents/Understanding%20Your%20Project.pdf
Evaluation: Practical Guidelines (RCUK). http://www.rcuk.ac.uk/Publications/policy/Pages/Evaluation.aspx
Project Planning—Evaluation Plan (JISC) http://www.jisc.ac.uk/fundingopportunities/projectmanagement/planning/evaluation.aspx
The NCCPE website provides useful information about evaluation http://www.publicengagement.ac.uk/how/guides/introduction-evaluation
A website that provides lots of detail about evaluating Public Dialogue Activities Including the Evidence Counts – Understanding the Value of Public Dialogue – Report is www.sciencewise-erc.org.uk

More Information on assessing impact

The RCUK toolkit for impact provides detailed guidance and tips. The site also provides links to each of the Research Council's own interpretations and guidance on impact.
The ESRC pathways to impact toolkit includes information on developing an impact strategy, promoting knowledge exchange, public engagement and communicating effectively with your key stakeholders. It is aimed particularly at Social Scientists.

CHAPTER SIX

Getting Started with Public Science Communication

Audiences, Locations, Planning, Hooks, Funding, Marketing and Safety!

Science knows no country, because knowledge belongs to humanity, and is the torch which illuminates the world.

—Louis Pasteur

6.1 Introduction

Chapter 1 introduced you to the background of science communication and the development of models to explain how the public interact with science and scientists. Chapters 2 and 3 described the science communication environment and provide reasons why you might or might not want to get involved. This chapter details the steps you can take to get started with science communication projects.

We have chosen to split science communication into:

1 events and activities with an adult or family audience (Chapters 7 and 8) and;

2 events and activities with schools (Chapters 9 and 10).

We made this decision because science communication with a school audience and with a public audience are different and 'feel' remarkably different upon delivery. However, note that the **processes** involved in designing school events and activities are equally applicable to a public setting and vice versa.

A school science communication event/activity tends to be more structured. For example, if it's within school time, it needs to fit within a timetabled lesson. Not all pupils will be interested in or want to do what you have planned. In effect you take on the role of 'teacher'. A public event/activity, which could also involve children, is a more open experience. People 'choose' to attend the event (humanistic learning – Section 4.3), and will be interested in what you have to say. There tends to be less structure and you take on the role of 'facilitator' during the event. However, there are certain elements

Science Communication: A Practical Guide for Scientists, First Edition. Laura Bowater and Kay Yeoman.
© 2013 John Wiley & Sons, Ltd. Published 2013 by John Wiley & Sons, Ltd.

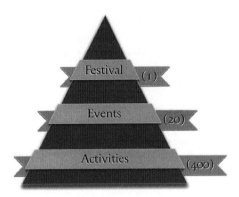

Figure 6.1 How activities build up into events which in turn can build up into a festival. The numbers show a hypothetical example of how many activities and events are required within a festival: one festival could need 20 events, and each event could have 20 separate activities.

associated with all science communication projects undertaken with any audience, such as the planning steps, which include information about health and safety, marketing and funding, which we will describe in this chapter. We have also made the distinction between events and activities. Activities are the discrete building blocks of an event, and in turn, events can build up into a science festival (Figure 6.1). Activities can be defined as a process or procedure intended to stimulate learning through actual experiences. Within science communication this is more often a hands-on activity, but it can also be discursive, for example a debate.

There are a multitude of different ways in which scientists can reach the public, through both direct and indirect means using both information and conversation as shown in Table 6.1. Examples of many of these methods

Table 6.1 Types of science communication activity.[a]

	Direct communication	Indirect communication
Information	Lectures	Television programmes
	Festivals	Radio programmes
	Plays	Science writing, e.g. popular
	Science 'stand-up' comedy	books and magazine articles
	Public events and hands-on	Open Access
	activities (e.g. at a museum or	Websites
	science centre)	Information leaflets
Conversation	Cafe scientifique	Social Networking, e.g.
	Festivals	blogging and tweeting
	Book clubs	Online opinion fora
	Science 'stand-up' comedy	Citizen Science projects
	Citizen's panels	
	Citizen Science projects	
	Policy formation	

[a]Examples provided are not limited. Some examples feature in more than one section, as they can have both information and conversation.

are described in the different case studies found throughout this book. These have been written by scientists who have designed and delivered science communication activities and events.

6.2 Understanding your audience

As identified in Chapter 1 (Table 1.3), the 'public' can be defined as every person in society (Burns *et al.*, 2003). We recognise that the concept of public is a complex mix of individuals of varying age, gender, race, socioeconomic status, educational and cultural background.

Before designing and running a science communication event or activity, it is very important to consider who is receiving the scientific message, i.e. who is your public and who is your target audience? If you concentrate on the content of the message alone and don't consider for whom it is intended, then your communication attempt will fail. Bultitude (2010) identifies three key reasons for considering audience targeting:

1 **effectiveness** – if audience needs are taken into consideration the message will be more effective;
2 **connection** – the audience will be more connected to the event as it will have been designed with them in mind;
3 **receptiveness** – the content of the event will be tailored to suit the needs of the audience and thus the audience will be more receptive.

The public can be grouped in different ways, and when you design or evaluate your science communication project you can use demographic data, public attitude survey data and public interest groups to shape your decisions and choices.

6.2.1 Demographics

The public can be split into separate groups according to demographic data consisting of age, gender, race, socioeconomic status and education. Information about different regions of the UK can be found at the Office for National Statistics[1] which is free to use. Another useful site is the Acorn Classification System, which segments small neighbourhoods, postcodes, or consumer households into 5 categories, 17 groups and 56 types. This can reveal information on socioeconomic status and likely level of education[2]. There is a charge associated with using the site, but you could include the cost of using this website into your grant application (Section 6.8) as part of the development and subsequent evaluation of your project. By collecting postcode data from the people attending your event, you will be able to determine if you were effective at reaching your target audience. This postcode information can be gathered at the same time as you obtain feedback data on your project. Section 5.4.3 offers suggestions on how to increase your feedback from participants. In the UK, social grades are determined by the occupation of the chief income earner (CIE) (see Table 6.2). Social grade is not the same

---

[1] http://www.statistics.gov.uk/hub/
[2] http://www.caci.co.uk/acorn-classification.aspx

Table 6.2 Social grade of chief income earner.[a]

Social grade	Description	% of UK population
A	High managerial, administrative or professional	4
B	Intermediate managerial, administrative or professional	23
C1	Supervisory, clerical and junior managerial, administrative or professional	29
C2	Skilled manual workers	21
D	Semi and unskilled manual workers	15
E	State pensioners, casual or lowest grade worker, unemployed with state benefits only	8

[a]Reproduced with permission from Ipsos Media CT (2009).

as social class, the latter is known as National Statistics Socio-economic Classification (NS-SEC) and is derived from the census; occupations are grouped by employment conditions and relations (Ipsos MORI, 2009).

6.2.2 Public attitudes to science

Over the past 40 years there have been different public attitudinal surveys carried out or sponsored by many different organisations across the world. These have been conducted to provide evidence about what the public think about science, scientists and science policy. Examples of these surveys are given in Table 6.3. By accessing these surveys you should be able to get specific information about your own country. These organisations also conduct public opinion surveys on more specific areas, for example stem cell research and nanotechnology. If you are interested in finding out further information about public opinion towards these more specific topics, the Eurobarometer surveys are very useful.

The most recent survey in the UK on public attitudes to science was done by an Ipsos MORI poll for the Department for Business Innovations and Skills (PAS, 2011) and in collaboration with the British Science Association. There have been a number of such surveys over the past 10 years in the UK and Table 6.4 provides details of where you can access the survey information. The PAS studies are the method used by the UK government to measure trends in opinion. The first study was conducted in 2000 and it used questionnaires, workshops and discussion groups to obtain information. For the 2011 survey, 103 adults aged 16+ took part.

The results of the 2011 survey seem to offer encouragement to the scientific community as they indicate that the public are very supportive of science; 86% are 'amazed by the achievements of science', 82% agree that 'science is such a big part of our lives we should all take an interest' and 88% think that 'scientists make a valuable contribution to society'. However, people are worried about the speed of scientific advances, as well as being concerned that these advances are going against nature. Specific areas such as GM crops,

Table 6.3 National and international organisations which have sponsored or conducted public understanding of science surveys.[a]

Acronym	Organisation
BAS-IS	Bulgarian Academy of Sciences, Institute of Sociology, Sofia
CAST	China Association for Science and Technology
CEVIPOF	Centre d'Etude de la Vie PolitiqueFrancaise, Sciences Po, Paris
CNPq	Brazilian National Council for Scientific and Technological Development
EB	Eurobarometer
ESRC	Economic and Social Research Council, UK
FAPESP	State of Sao Paulo Research Foundation, Brazil
ISS	International Social Survey Consortium
MORI	British Public Opinion Research Company
NCAER	National Centre for Applied Economic Research, Delhi, India
NISTEP	National Institute for Science and Technology Policy, Japan
NISTAD	National Institute for Science, Technology and Developmental Studies, India
NSF	US National Science Foundation
MST	Ministry of Science and Technology (Canada, China, Brazil)
Observa	Science in Society, Italian Not-profit Centre for Science in Society Research
OST	Office of Science and Technology, UK
PISA	Programme for International Student Assessment, Organisation for Economic Co-operation and Development, Paris
RICYT	Ibero-American Network of Science and Technology Indicators
STIC	Strategic Thrust Implementation Committee, Malaysia
Wellcome Trust	Research Foundation, London, UK

[a]Reproduced from Bauer (2008) by permission of Cengage Learning Services.

nuclear power and animal experimentation are still areas of concern. The report splits citizens into six attitudinal clusters depending on their responses to the questions in the survey. These have been compiled into Table 6.5 and illustrated in Figure 6.2.

After dividing citizens into attitudinal groups, you need to consider the ways in which these groups can be reached. However, one issue you should

Table 6.4 UK Public attitude to science surveys.

Date	Survey title	Organisations
2011	Public Attitudes to Science*	Ipsos MORI poll for the Department for Business Innovations and Skills
2008	Public Attitudes to Science	Department for Innovation, Universities and Skills and the RCUK
2005	Science in Society	Ipsos MORI poll for the Office of Science and Technology
2000	Science and the Public:	Office of Science and Technology and the Wellcome Trust

*See Public Attitudes to Science (2011) for links to all these reports.

Table 6.5 Description of the six clusters of public attitudinal groups.[a]

Cluster	Description	% of population	Media	Demographic
The concerned	The largest cluster; religion plays an important role in their lives. Strong views on the limitations of science, least convinced by the economic benefits of investment. More likely to have reservations about the intentions of scientists.	23	More likely to read tabloid newspapers, less likely to read specific science and technology websites.	More likely to be female from a younger age group (16–34). Lower social grades (C2DE). High percentage of people from ethnic minority backgrounds.
The indifferent	Less likely to feel informed about science, but are not especially negative or worried about it.	19	Television and newspapers, tend not to use the internet	Tend to be older (quarter being over 75) and half retired. More likely to come from social grades C2DE.
Late adopters	Did not enjoy science at school, but now take a strong interest in science and interested in public consultation. Appear to engage more strongly with science when placed in a wider context. Have strong environmental and ethical concerns. Have reservations about specific areas of science, e.g. GM crops. They want to hear scientists talk about social and ethical implications of their work.	18	More likely to have internet access and more likely to use social networking sites.	More likely to be female from a younger age group (16–34). Relatively few older people in this cluster. Come from all social grades.

Confident engagers	Strong positive attitude to science, relatively few concerns. Confident that science is well regulated and are likely to trust scientists to follow the regulations. Want to get more involved in decisions.	14	More likely to come from higher social grades (ABC1s). Higher level of education. More likely to come from a middle age group (35–54). Very few from older age group.
Distrustful engagers	Many have backgrounds in science or engineering. Very interested in science and think it beneficial and feel well informed. Less trusting of those who work in science and less confident in the Government's ability to regulate them. Tend to think that the public should have a wider role in decision making and are interested in being involved in this.	13	Predominantly male from high social grades (ABC1s). Higher level of education, more likely to be aged 55 or over. Relatively few people from ethnic minority backgrounds.
Distrustful engagers			More likely than average to respond to online engagement through specialist sites and blogs. Not especially likely to use social networking sites.
Disengaged sceptics	Feel less informed about science; tend to be less well educated and put off science at school. Find science overwhelming and do not see it as useful in everyday life. Have concerns about scientific developments and the ability of Government to control them. Do not trust scientists to self regulate. Not keen to get involved in public consultation, but do feel that the public should be listened to.	13	More likely to be female from less affluent social grades (C2DEs). Less well educated. Come from a wide spectrum of age groups. Relatively few people from ethnic minority backgrounds.
Confident engagers			More likely to read broadsheet newspapers, but also use a wide variety of media, including blogs and other websites for science and technology. Tend to use social networking sites.
Disengaged sceptics			Television is an important source of information, more likely than average to read tabloid newspapers.

[a]Information compiled from PAS (2011). Reproduced with permission of the UK Department for Business, Innovation & Skills.

Figure 6.2 Attitudinal groups. Figure illustrating the PAS (2011) data. Reproduced with permission of the UK Department for Business, Innovation & Skills.

recognise is that it can be difficult to reach all these audiences within a single event or activity. Deciding on the audience to be reached should be part of the initial design of the project (Figure 6.2), as this will then determine the potential mechanism in reaching the desired attitudinal group. Often, the location of the event can help you target different attitudinal groups (Table 6.6). A good example where the audience was key, was the Darwin Radio show (Case study 8.1).

6.2.3 Public interest groups

There are many societies, clubs and organisations which bring people together who have similar interests. There are bee keeping societies, natural history groups, local history societies and sports clubs, to name just a few, and all of

Table 6.6 Ideas for event venues with advantages and disadvantages (including socioeconomic grades).

Venue	Potential audience	Advantages	Disadvantages
Museum	Family audience and adults Can be free, but other Museums may charge Depending on the entry costs, all socioeconomic grades	Can link event to other museum exhibits Good facilities You may get help with marketing the event	People may need to pay an entrance fee Not always easy to get to Tied into set opening hours Physical environment is fixed
Science centres	Good for families with young children Depending on the entry costs, all socioeconomic grades	Lots of other hands-on exhibits specifically focusing on science Good facilities You may get help with marketing the event	People may need to pay an entrance fee Not always easy to get to Tied into set opening hours Physical environment is fixed
Town/village hall	Good for the older audience Good for community groups Wide variety of socioeconomic grades	Good for smaller groups and discussion events Easy access for the local community	You may need to hire this space You will have to do your own marketing
Public house	Good for reaching a wide ranging adult audience Wide variety of socioecconomic grades Potential for gaining access to hard-to-reach groups	Good place for discussion Relaxed atmosphere Easy access for the local community Open during the evening when people have finished work	You will have to do your own marketing Limited space for hands-on activities

(continued)

Table 6.6 (*Continued*)

Venue	Potential audience	Advantages	Disadvantages
Café	Good for reaching a wide ranging adult audience and older teenagers Wide variety of socioeconomic grades	Good for discussion Relaxed atmosphere Easy access for the local community	Not good for a large audience – limited space May be restricted opening hours Might have more problems with health and safety being a food outlet You will need to attract your audience with good marketing and a good 'hook'
School hall (primary or secondary)	Family audience. Depending on the school location may attract a wide variety of socioeconomic grades	Good open space for activities and you can decide on where activities should go within the space You may be able to get the venue free of charge, or at low cost Good facilities	Unless the event has been organised by the school, you will have to do your own marketing Restricted audience won't attract passers-by
School grounds (e.g. fete)	Family audience Depending on the school location may attract a wide variety of socioeconomic grades	Good open space for activities and you can decide on where activities should go within the space You may be able to get the venue free of charge, or at low cost Good facilities	Equipment needing a power supply might be a problem Weather maybe a problem

Venue	Advantages	Advantages	Disadvantages
University laboratories	Depending on the correct marketing strategy you can attract a wide audience including families More likely to attract higher socioeconomic grades	Can easily do wet laboratory hands-on activities Can support large audiences Good facilities	Can feel intimidating Audience may have to travel Good marketing is essential to get audience numbers
University lecture theatre	Depending on the correct marketing strategy you can attract a wide audience including families More likely to attract higher socioeconomic grades	Good facilities, e.g. internet access Good for large audiences Parking can be organised	Can feel intimidating Audience may have to travel Good marketing is essential to get audience numbers
Zoo	Very good for a young audience Good for families Can attract a wide socioeconomic groups depending on cost of entry and location	Potentially have a good open space Good for capturing the 'passing audience' Can link your activity to the other exhibits Good for biology and conservation issues	Equipment needing a power supply might be a problem Weather maybe a problem Your activity maybe lost within the scope of the venue Restricted opening hours
Theme park	Very good for a young audience Good for families	Good for demonstrating laws of physics	Equipment needing a power supply might be a problem Weather maybe a problem

(continued)

Table 6.6 (*Continued*)

Venue	Potential audience	Advantages	Disadvantages
Music festivals	Good for a young audience 16–35	Good audience numbers Good open spaces for science shows Good opportunity for discussion	Equipment needing a power supply might be a problem Weather maybe a problem Your activity may be lost within the scope of the venue
County show	Good for a family audience	Good audience numbers Good marketing of the event	Equipment needing a power supply might be a problem Weather maybe a problem Smaller size audience
Library	Good for mothers and fathers with young children	Good opportunity for discussion Good for capturing the 'passing audience'	
Public park	Good for families with young children	Good open spaces Good for 'passing people'	Equipment needing a power supply might be a problem Weather may be a problem
Leisure centre	Good for young men and women as well as families	Good facilities Good for capturing the 'passing audience'	People are probably not specifically coming for your event, unless it has been well marketed

these can be ideal groups to target for science communication events. A good example of a targeted audience with similar interests, i.e. people who knit (knitters can be young, old, female or male) is shown in Box 6.1. Other good examples of events or activities designed with public interest groups include the series of demonstration lectures developed by Dr Stephen Ashworth for the Women's Institute (Case study 7.2) and also the Bad Bugs Book Club run by Professor Jo Verran (Case study 7.8).

Box 6.1   The Big Knit

The British Society for Immunology ran a public event called 'Multiple Sclerosis: the big knit' which explored science of the disease through the craft of knitting. Groups of knitters came together to produce a woolly-art installation for the 2011 Cheltenham Science Festival. Over 90 knitters produced 300 individual pieces of knitted art from brain cells to DNA helices.

6.2.4   Location and timing of your science communication event or activity

The PAS 2011 study showed that people viewed visiting a zoo, museum or science centre as a family activity. This indicates that where the event is held is closely related to who attends. In Case study 4.1 by Jaeger Hamilton, the target audience was families, thus it was held at the Norwich Castle Museum which attracts a family audience. There is really no limit as to where you can hold a public science communication event or activity – it is anywhere you will find the public. Table 6.6 gives you some ideas for different venues and their relative advantages and disadvantages, including which types of visitors are most likely to frequent these venues. Wherever you decide to hold your event or activity, it is essential that you visit it beforehand to see if it's suitable for your expected numbers of visitors, has adequate access (including for people with disability) and power sources available for any pieces of equipment, should you need them. You also need to know how much the venue costs to hire.

Another important consideration is the timing of your event.

- National Science Weeks (in the UK held during March) offer good opportunities to showcase your activity or event, and there are quite a few funding sources that you can apply for, specifically to run an activity during Science Week. During this week there tends to be many events happening at schools and in public venues, so you might find that people suffer from 'event overload'.
- Look out for the United Nations 'Year of . . .', for example 2010 was the Year of Biodiversity and 2011 the Year of Chemistry. In 2012 it will be the Year of Cooperatives and the Year for Sustainable Energy for All[3].
- Events during half term and other school holidays can be good, particularly if you wish to target a family audience.

---

[3] http://www.un.org/en/events/observances/years.shtml

- Try not to clash with other big local events, such as the Olympics 2012!
  Examples of science communication events described in the case studies which have been held on specific dates include;
- Case study 7.3 'In your Element' by Elizabeth Stevenson – this event was held at the National Museum of Scotland during the Year of Chemistry in 2011;
- Case study 6.1 'World Sight Day' by Dr Michael Wormstone – this event was organised with the charity 'Fight for Sight', and was held on World Sight Day. It also used an interesting venue, a department store located in Norwich city centre.

Case Study 6.1

World Sight Day 2009
Michael Wormstone

Background
On the 8th of October 2009 – World Sight Day – my laboratory and the charity Fight for Sight organised a day of events to raise awareness of vision research. Two events were organised. In the daytime we set up a stand in the Jarrold department store in Norwich. This was attended by eye research scientists, members of the East Anglian eye bank, and representatives of Fight for Sight. This gave us the opportunity to engage with members of the public unfamiliar with the work at UEA, or the research funded by Fight for Sight and to explain the value of what we have done and what we aim to achieve in the future. The second event was an evening presentation by Michele Acton, the CEO of Fight for Sight followed by a presentation by myself as a keynote speaker. A reception to enable informal discussion and questions took place afterwards. This evening activity was a ticket only event. Considerable effort was expended in organising the events on World Sight Day, but for all those who participated it was a rewarding experience with tangible benefits.

Why did we organise these events?
During discussions with representatives from Fight for Sight, the UK's largest funder of eye research, it became apparent that the knowledge and support base for the charity in Norfolk was weak, relative to other parts of the country, even though world-leading research was conducted at UEA. It was mutually agreed that some initiative was needed to highlight the importance and value that eye research has to the people of Norfolk and the world at large. With this in mind, my colleagues and I at UEA started to develop an approach to do this.

Setting a date
The choice of date is often important for the success of an engagement exercise and we considered our options very carefully to obtain maximum impact. World Sight Day (8 October) was chosen. This obviously had global relevance, but also provided a vehicle that enabled us to connect with the local community. However, this date had drawbacks. The UEA is

an academic institute and being a member of faculty, I had teaching duties and other responsibilities. Therefore the additional workload resulting from organising an event of this nature at an extremely busy time was challenging. In addition we chose to organise the evening event at the university; this would have been considerably more straightforward out of semester time because room booking, parking and catering would have been easier to coordinate. Despite these logistical drawbacks, it was unanimously felt that organising events on World Sight Day offered advantages which were far greater than the disadvantages.

Organising the event timetable

Having established the date, we set about planning the day's events. A meeting was held in Norwich that comprised academics and administrators from UEA and Michele Acton. At this point neither my laboratory nor Fight for Sight had organised an engagement activity of this nature; consequently we had no template on which to work. However, all parties were determined to make the events on World Sight Day a success. Discussions at this first meeting were open, with many ideas presented and their merits evaluated. While refinement would take place over the coming weeks the general format was in place and individuals were tasked with achieving specific goals. Essentially, UEA coordinated contact with Jarrold department store, designed the contents of the activities, along with room booking, parking and catering for the evening events at UEA. Fight for Sight was responsible for advertising the event and distribution of tickets. This arrangement was logical and through good communication everyone could see the progress that was being made.

Coordination of an event is not a trivial exercise, and if taken lightly will be a failure. It is not simply a case of booking a room and turning up. One must constantly keep on top of the situation and deal with problems that will inevitably crop up. One issue, for example, is parking. If more that 100 people are expected to attend a talk, you need to arrange parking and more importantly to consider individuals with disabilities, ensuring that the parking and the lecture theatre are adequate for their needs. A request for information relating to disability was made when distributing tickets to identify specific needs; people appreciate this level of consideration. Also we had to consider University policy with regard to parking, which could have meant that drivers would have to pay for parking and be spread across the University campus. Negotiations with the University Traffic Officer were productive as they agreed to waive the parking fee and allowed sections of the university car park adjacent to the lecture theatre to be restricted to event ticket holders. UEA provided signage to help guide attendees to the event, while members of my laboratory, sporting Fight for Sight T-shirts, were there to greet people and guide them to a central atrium for pre-lecture tea/coffee and biscuits. Again this took a lot of effort to coordinate, but the result left those attending knowing that they were valued and their support was appreciated.

The stand in Jarrold department store was extremely well received by the public. We were well positioned on the ground floor near an entrance; therefore it was hard to ignore us. Several people were always present at the stand and members of the public happily engaged in the activities we had prepared and were keen to see presentations illustrating our work. A wide variety of questions were asked and we did all we could to answer them. Through activities like this you develop a feel for what really matters to people, which in turn helps determine your next avenue of research. A number of people who visited the stand also attended the evening lectures. This was agreeable as it established a degree of familiarity and it was encouraging to see how enthusiastic people were.

The idea to make the evening event ticket-only had a couple of key benefits. It allowed the number of people attending the lectures to be monitored and ensured that an appropriate room had been booked and adequate catering provided. Ticket request for our event was good, so we knew our choice of a lecture theatre with a capacity of 130 was appropriate. In addition to this practical purpose, the idea of a ticket (even though it is free) seemed to provide a feeling of selectivity and exclusivity. I believe this helped further impress upon the audience what we were saying.

In the evening, we decided to restrict the number of speakers to two. Michele Acton gave an overview of Fight for Sight, which was followed by a single keynote lecture given by myself. The rationale is that the audience will get to see the person running the charity as well as one of the world's leading eye researchers, all within a relatively short distance from their homes. Moreover, with two experienced speakers it is easier to pitch the level of the lecture to the primary members of the audience, the public. Another decision we made was to deliver the lectures, but to leave questions for the reception. Asking questions in a large lecture theatre can be daunting to some people and often answers are limited due to time restaints. This approach allowed members of the audience to ask questions and to mix with lab members, the Fight for Sight team and clinicians from the Norfolk and Norwich University Hospital (NNUH). This worked extremely well as questions were fully answered and members of the audience headed home more informed and happy.

For the impact of this event see Case study 3.1.

6.3   Taking your first steps

If you are doing this for the first time, it's always good to start off small – perhaps volunteer to give a public lecture where the event is organised for you, and then your commitment is to write and deliver the talk. It is also a really good idea to get your activity involved in an event organised by a third party, for example a school fete, county show or festival. This can really reduce your costs, especially those associated with venue hire and marketing. You will still need to do your own risk assessment and make sure that you give it to your event organiser in plenty of time (Section 6.10).

I think that it's also important to really consider what you are good at, as we all have individual strengths and weaknesses. For example, if you like talking to small groups, where there is more opportunity for discussion and conversation, then try a Café scientifique (Section 7.3.2). If you prefer a more distant approach, but still want to engage in dialogue and you enjoy writing, try setting up a blog (Section 8.9.1 and Case study 8.6). Figure 6.3 gives you more ideas on taking your first steps towards science communication with the public.

I have always done science communication activities with schools, getting involved when my children were young. I started running public events for the first time when the British Science Association Festival came to Norwich. Festivals of this kind provide an excellent way to get started and can be an ideal first stepping stone into organising your own bigger events (Section 7.2.4).

6.4   Planning your own event or activity

If you have been involved in events or activities organised by others, you may reach a point where you feel the need to have more control of the overall proceedings, for example the theme and organisation of the event, or the design of an activity. Of course you do not need to take this next step; even if you never feel the need to direct an event, you can still build a very effective engagement portfolio.

Whether you are planning a single activity or a whole event, there are several steps you need to consider. Figure 6.4 provides you with a generic

IDEAS TO GET STARTED WITH PUBLIC ENGAGEMENT

Become *involved* in a science festival.

Join **Sense about Science**
(www.senseaboutscience.org/)

Set up a **BLOG** about your views in different aspects of science.

Apply for money for an engagement event when you apply for research funding.

VOLUNTEER to help with someone else's public engagement event.

Start **TWEETING** on scientific issues.

Find a *mentor* within your organisation to help get you started.

Respond to a scientific story in an online newspaper.

Think about joining **Bright Club** for stand up science comedy. (www.brightclub.org)

If you belong to a leisure group, club or other society, ask if you can **DO AN ACTIVITY** with them.

Take part in a public lecture series.

Volunteer to **help out** at a public open day at your organisation.

Go and observe a **café scientifique** and then volunteer for a session.

If you belong to a book club, **pick a book to discuss** with a scientific theme.

Consider going on a **training course** either one run by your organisation or an external one.

Write a piece about science for your local newspaper or magazine published by your learned society.

If you have a webpage about you and your research, **ADD INFORMATION** about how your research could effect the public.

Figure 6.3 Ideas for getting started in public engagement.

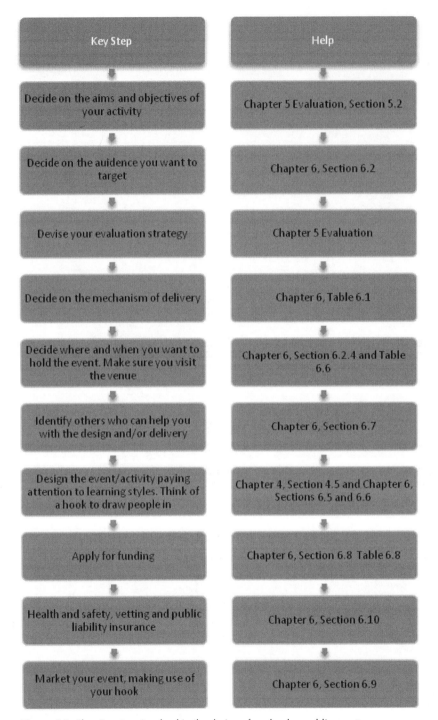

Figure 6.4 Planning steps involved in the design of a school or public event.

flow diagram of the different planning steps that are involved. It includes the section, figures and tables of this book that will provide you with assistance for each step.

Case study 6.2, 'Magnificent Microbes' by Dr Nicola Stanley-Wall at Dundee University provides an excellent example of a meticulously planned event. The timeline for this event is illustrated in Box 6.2 and is an example of how the planning steps of an event actually work in reality.

Box 6.2   Timeline for 'Magnificent Microbes' engagement event

✓ Molecular Microbiology Division formed at the University of Dundee (June 2008)

✓ Decision taken to hold a Division school and public engagement event (May 2009)

✓ Met with Dundee Science Centre and the University Public Engagement Officer (June 2009)

✓ Spoke with colleagues to gain support and commitment (June 2009)

✓ Organised dates for training, development day and event (July 2009)

✓ Training day, activities and objectives determined (October 2009)

✓ Activity development work sheets completed to allow the grant applications to be written (November 2009)

✓ Evaluation strategy planned

✓ Grant written and submitted to BBSRC (January 2010) – grant failed!

✓ Money secured from various sources internally (February 2010)

✓ Grants written to various academic societies (February 2010)

✓ Participants reminded to start thinking about their activities (February 2010)

✓ Advertising leaflets prepared and schools approached (February 2010)

✓ Risk assessments written and equipment lists prepared (March 2010)

✓ Adverts to all dentists and schools in area and some local cafes etc (April 2010)

✓ Development day (mid April 2010) and feedback

✓ Activities modified, floor plans drawn up, schedule for lunches arranged, etc. (April and May 2010)

✓ Event day (21–22 May 2010)

✓ Follow-up steps need to be done: evaluation of event assessed, reports to societies regarding how money was spent and final budget determined.

Magnificent Microbes
Nicola Stanley-Wall

Background

The Division of Molecular Microbiology was established in the College of Life Sciences at the University of Dundee in June 2008. I decided that it would be a valuable experience for all if we could hold an event for schoolchildren and members of the public to highlight microbiology. We had the capability to deliver a large-scale event at the Dundee Science Centre – Sensation, but for this to be successful I needed the commitment and support from other members of the Division. Before I approached the other academic members of staff in the Division, I wanted to establish what would be involved. Therefore I arranged a meeting with the public engagement officer for the University of Dundee and a senior member of the education staff at the Dundee Science Centre. At this first meeting we discussed the format of the potential event and put in place an outline of what needed to be organised i.e. people, money, dates, training, advertising and activities. One of the key requirements for holding an event at the Dundee Science Centre is that all of the facilitators are well trained and the activities are tested. The months of effort by team 'Magnificent Microbes' culminated in a two-day event that took place on Friday, 21st and Saturday, 22nd May 2010 at the Dundee Science Centre – Sensation.

Project planning, training and objectives

The event would involve a minimum time commitment of four days, the training day, the development day and the two event days (one for schools and the other for the public). With the help of the Head of Division, I raised the suggestion at one of our regular meetings. I obtained the full support and commitment of the other seven lecturers and professors in the division. We arranged dates that could be set aside for the fixed commitments. Once the dates were set and fixed in people's diaries, I emailed the Division to discuss what we wanted to achieve. I wanted to ensure that everyone invested in the event and that it was truly a joint effort where my role was a facilitator and participant, not sole organiser.

To ensure that the training day was successful, it was facilitated by external personnel who teach communication skills. The day was funded by Generic Skills, Staff development and the College of Life Science public engagement fund and was attended by 25 people including PhD students, postdocs and academic staff.

The training consisted of the following activities.

1 A discussion of different presentation styles and ways of communicating.
2 Staff from the Dundee Science Centre took us through their regular activities so that we could see how they got the concepts across to the audience. We then could ask questions (as if we were visitors) to see how they dealt with difficult people! Having witnessed this we were split into groups of five. We were then randomly given a piece of paper on which our role was named – parent, small child, teenage child or facilitator. We then had to recreate the activity that was demonstrated to us 15 minutes earlier and/or pretend to be the sulky teenager or annoying 5-year-old! This highlighted to us how hard it could be to get your message across even if you had just witnessed someone doing the same.
3 A visit to the Dundee Science Centre so that we could see the venue where we would have the event. This was followed by a discussion session to decide on the topics that could be

covered in the event and to identify people who wanted to 'be in charge' of a particular activity. We came away with an outline of the event that morphed into 'Magnificent Microbes'; which was funded by several sources including:

- the Society for General Microbiology;
- the British Society for Plant Pathology;
- the British Mycological Society;
- the College of Life Sciences;
- the University of Dundee.

After the training day our main objectives for the event were:

1 to use fun and interesting activities to make children and adults alike aware of how fascinating microbes, such as bacteria and fungi, really are;
2 to train PhD students, postdoctoral scientists, lecturers and professors in the art of communicating science to members of the general public;
3 to develop a bank of resources for future events.

Resources and help

- I contacted the University of Dundee's public engagement officer Dr Jonathan Urch. He helped me to contact the staff at the Dundee Science Centre and was a source of support throughout the preparation and delivery of the event.
- We used the University of Dundee's graphic design department to help prepare eye-catching advertising posters and templates for the posters used for illustrating the sciences behind each activity.
- The Society for General Microbiology, the British Society for Plant Pathology and the Scottish Crop Research Institute were all valuable sources for prepared resources that we distributed to the children, teachers and staff. This included pens and other giveaway items that helped to capture the children's attention, as well as information and follow-up booklets that could be used by the teachers in future activities.

The development day

The development day was a test run day, with the activities in as final a form as possible. The Dundee Science Centre has schools that they use to help them develop novel activities and one of them was asked to be our 'guinea pigs'. The day was held at the Science Centre and really helped many of the PhD students and staff to overcome first-night nerves. What was apparent was that while lots of thought had gone into developing the activities, deciding *how* they were going to be presented was lagging behind. There were in fact a lot of giggles at first coming from some of the younger PhD students who were a little overwhelmed by the schoolchildren and did not really know what to say. Therefore by including this test day, the facilitators could ensure that they knew how they were going to introduce their activity to the public. To make the most of the development day, feedback was obtained from the schoolchildren, teachers and from the Dundee Science Centre staff, that was then used to help alter the activities where they needed improvement.

The event

There were ten activity stalls for the visitors to enjoy. The stands were 'manned' by eight academic members of staff, five postdocs and nine PhD students. The first day we had ~180 schoolchildren (primary years 6 to 7), who were split into groups with different time slots throughout the day. The invitations to the schools were sent out though the Dundee Science Centre as they have very close contacts with the local council and the schools. The Science Centre obtained funding for the price of admission to the centre for the schoolchildren and the costs associated with transportation. This was very important since funds are very limited in individual schools. The schoolchildren had 45 minutes in Magnificent Microbes and 45 minutes to enjoy the other exhibits in the Dundee Science Centre. At any one time this meant that there were around 60 children in the building and 30 were at the Magnificent Microbe stalls. This felt like a good number as the children were distributed amongst the ten stands. It meant that the facilitators did not feel like they were just hanging around. There were two sessions like this in the morning and one in the afternoon.

On the public day, things felt much slower and to be honest the facilitators enjoyed the day with the schoolchildren much more. We had a total of ~150 people coming through the door.

Specific safety issues (for general safety advice see Section 6.10)

1 As 'Magnificent Microbes' included experiments with live bacteria, we had to ensure that all were safe and suitable for taking to an external event. The criteria that we were asked to adhere to were that the strains needed to be commercially available and that they were not genetically modified. We also had to ensure that we had plenty of antibacterial hand lotions to allow the children and families to use. This was something that was highlighted by preparing the risk assessments.

2 As we had several stands that used pieces of equipment and/or computers we had to ensure that we did not overload the Dundee Science Centre electricity supply. To do this I provided in advance a list of the number of electrical sockets that each stall needed. This allowed the staff at the Dundee Science Centre to ensure that the stands were placed appropriately throughout the venue (whilst maintaining the logical flow in the activity stands as far as possible).

Event highlights

1 The training day allowed everyone who was participating to be involved in controlling the direction of the event. Small groups of two to three people were designated as responsible for developing a specific activity related to the theme of 'Magnificent Microbes'. The activities developed during the training day and then through subsequent discussions were incorporated into the grant applications.

2 The development day was attended by a local journalist and an article was run in the local paper in the week before the actual event.

3 We wanted to allow visitors to see what bacteria were growing on their hands or within soil samples. As this does not provide an instantaneous result we needed a way of allowing members of the public/schools to get their images. For the schools this was easy since we simply had the children write their name on the plates and then the images where sent on a CD to the teacher along with a few additional resources (i.e. about hand washing from the SGM). For members of the public this was harder since we had to consider data protection and also ease of recording whose plate was whose. To overcome this, I contacted the College of Life Sciences website development team. I imagined a website where a member of the public could go and use a unique user name and password to allow them access the images of the bacteria that had grown from their hands. This appeared to be possible and actually served two further purposes:
   1. it was an ideal way of obtaining feedback about the event (i.e. they had to answer the questions set before getting the images!);
   2. it brought people to the College of Life Sciences impact website where they could find out about other events (http://www.lifesci.dundee.ac.uk/other/impact/index.html).

So at the practical level this involved the generation of a series of random usernames and passwords. These were printed out onto two stickers – one of which was placed on a piece of paper for the visitor and the other was stuck on the petri dish. The plates were grown, imaged and then the data provided to the web developers. I was then able to go into the site a couple of weeks later and see how many people had visited to get their images and what they thought of the event.

Problems

We were hit hard by the weather. The day that was open to the public happened to fall on the hottest day of the year and we lost out on visitors who went to the beach instead!

It was harder to obtain funding for the event than initially thought. At first we approached the BBSRC for funding, to be matched by a generous contribution from the College of Life

Sciences public engagement fund; however, we were unsuccessful. This generated a significant problem for us since we had already designed the event in the training day and both PhD students and staff were significantly invested in the science communication event. With the generous offer of underwriting from the College of Life Sciences and the University Public engagement fund, we set about obtaining funding from other sources – but knew that we had the funds to go ahead. We were successfully supported by the Society for General Microbiology, the British Society for Plant Pathology and the British Mycological Society. We also approached various companies (e.g. VWR, Fisher Scientific, etc.) for donations of resources that would be used in the event (e.g. petri dishes, media, sterile loops). Using this combined approach we were able to reach our target and the event was able to go ahead predominantly as planned.

Personal gain

I have benefited in many ways, including gaining a feeling of support and satisfaction by obtaining the interest and help of my colleagues in making the event possible and a far greater understanding of which types of activities are enjoyed by children (the messy hands-on ones!). Particular skills I have honed during the event are organisation and delegation. These are two skills needed in the university environment. Also I have very much realised that you cannot achieve perfection in such an event as it would be too time consuming – aim for 90% of perfection!

Top tips
1 Make a very comprehensive list and include a time line for each requirement.
2 Be prepared for a large amount of work and having to nag people to get things done.
3 Get a separate email address for everyone to use for event-related details as the number of emails involved in organising the event was very large.
4 Keep attempting to get funding, even if you are turned down at first. Try other sources.
5 Don't be put off by the amount of work that organising an event looks like – if you ensure that you have the right support, you will be able to succeed.

6.5   How to design hooks for your event or activity

Chapter 4 introduced the idea of constructivism, a learning theory which suggests that all new knowledge is built upon what the individual already knows and/or has experienced (Section 4.4). To apply the ideas behind constructivism to the design of a science communication event or activity you need to think about relevance. People will want to attend your event or activity because it has meaning for them and holds relevance to their past experiences. This means that you need to think very carefully about the 'hooks' that will draw people to your event or activity. Hooks are also an important feature of your marketing strategy (Section 6.9). In order to come up with hooks consider the following:
1 What is a current newsworthy issue? e.g., GM food and stem cells;
2 What is affecting your audience? e.g. personalised medicine and climate change;
3 What do your target audience enjoy? For example, local history and sport;
4 Is there a local angle? e.g. coastal erosion or TB and badgers.

   One excellent example is Case study 6.3 by Professor Jo Verran on the 'Science of Vampires'. This event drew on the interest and enjoyment that many young people have in vampire literature, TV programmes and films. She then cleverly linked this to the science of disease transmission.

Twilight and the Science of Vampires

Jo Verran

Background

In 2009 and 2010, the *Twilight* vampire novels and movies seemed to be an obsession for many young people. I began to wonder: why are vampires so fascinating to such wide-ranging audiences? There are plenty of theses on this topic, but as a microbiologist, I then began to wonder if this fascination with vampires could be used to stimulate interest in infectious disease. After all, once you are bitten by a vampire, you die, became a vampire or become an ongoing source of nutrient – certainly the first two scenarios are consistent with transmission of disease. If you want to avoid vampires, you eat garlic, wear a cross and don't invite them into your home – surely the same principles as modifying behaviour or immunisation as forms of disease prevention? The iconic image of the pale, thin vampire, with blood running down their chin is reminiscent of the consumptive, and the significant sensuality of vampires themselves could be used as a warning against sexually transmitted disease. The more I thought about it, the more common features I found.

I had the opportunity to explore the idea at the 2010 Manchester Children's Book Festival, held at Manchester Metropolitan University (MMU). This significant festival ran across an entire week, culminating in a family day event on the final Saturday. So in amongst the Dr Who costumes, storytellers and other activities, I ran my 'Twilight and the Science of Vampires' activity.

Activity design

We set up the activity room as a 16-seater *classroom*, and gave everyone laboratory coats to wear. In the Stephanie Meyer novels *Twilight*, Bella and Edward first meet in a biology class, whilst examining microscope slides of meiosis. After a general but vague introduction about how we were going to link microbiology and *Twilight*, I asked the audience how Edward first became a vampire (he had influenza in 1918). This led to a discussion on influenza generally, and its transmission, which we demonstrated by the use of a simulated sneeze, using an aerosol spray containing a gel that glowed when viewed under UV light (Wash and Glo, Glotech: info@glotech.co.uk). I had identified two readings from the novel that described Edward and Bella's interactions in the classroom, and a female drama student from MMU read them aloud. Meanwhile, the 'sneeze' had conveniently contaminated some laminated cards that helped the class to identify the specimens under their microscopes, and which were passed around during the reading. After the microscopy session, we used UV light to demonstrate how the germs transmitted by the sneeze had moved around the class. Then we could talk about disease transmission – initially by inhalation and contact, and then by biting – finally identifying ingestion as the other key route of transmission.

A final reading described how Bella was bitten by a vampire and Edward sucked out the poison. Then we could talk about treatment of disease, and contrast it with prevention. We finished the session providing soap (to wash off any remaining gel and to reinforce the hand hygiene message) and information sheets on influenza and hand hygiene.

The activity lasted for around one hour. We ran it three times during the day, and each time the age profile of the audience was very varied: grandparents, teenage *Twilight* experts, young budding scientists and families. Discussion varied accordingly, but feedback was very good.

I ran the activity with the help of the Society for General Microbiology, who provided the Wash and Glow kit, information sheets and other goodies (soap, brooches), and moral support.

Funding and resources

All resources were either provided by the venue (tables, chairs), my laboratory (microscopes, slides, lab coats), or the Society for General Microbiology (Wash and Glow kit, gifts, information sheets). The organisers of the festival were helpful, but their prime objective was for the running of the overall event rather than specifically my activity. I developed and ran the activity myself, and was very grateful for the assistance of my colleague from the Society for General Microbiology.

Reflection on the activity

The main attraction was the microscopy – everyone loved playing scientist.

> *'When you sneeze it goes everywhere and you don't know it is on you' (Sam aged 5).*
>
> *'I have loved this workshop. I have learnt loads about the spread of infection. I love Twilight'.*

The audience varied for each session, and it took a few minutes to assess their character, and hence the content of the discussion. Teenage girls interested in the *Twilight* novels were less interested in the science, and also less keen to participate in discussion, unless it related directly to the story.

The future

As a result of the promotional material for the festival, I was asked to repeat my activity at the annual conference of the National Association for Gifted and Talented Children (NAGTC). In this case, I ran two sessions, with 10- to 12-year-old and 13- to 16-year-old children. I had help from one of my postgraduate students, and my daughter did the readings. I received some payment from NAGTC, which I shared with these helpers.

The younger students were enthusiastic and enthralled

> *('I really loved it! I loved the microscopes and the readings were read fantastically!' 'This was very enjoyable and I've learned loads about biology'). The older students were really hard to engage, although their feedback was positive, albeit brief ('Thank you crazy vampire people', 'Thankz').*

Again it was the microscope activity which significantly improved participation.

I really enjoyed making the links between vampires and microbiology, and am now planning activities incorporating zombies and werewolves as well! Which diseases are they most like? Which of the three would cause the worst epidemics? The enthusiasm and awe of the 'science' amongst the younger children was really lovely.

6.6   Designing a science communication activity

Big public events need separate activities to fill them. When designing your science communication activity, use the following as a guide.

Aims and Objectives – your activity should have one clear aim. Objectives then help you meet your aim.

Interaction – make your activity interactive, something participants can do, see, or discuss. Good communication says the same thing in many different ways. When designing an activity for family audiences, incorporate something that can be made to take home. These objects can often then be used for future family discussions. Think about different ways in which people learn (see Section 4.5), use both verbal and visual explanations of your material. Try and build discussion into your activity; perhaps with questions and getting people to record their responses. For good resources on building dialogue visit the Dialogue Academy Website[4]. Try and build up further interaction post-event, e.g. Nicola Stanley-Wall describes how posting the results of an activity on a website can direct people to other sources of information (Case Study 6.2).

Equipment – it is our experience that people like seeing equipment, for example microscopes, gel electrophoresis tanks and automatic pipettes. One of the best items we ever used was a vortex machine; an ordinary piece of laboratory kit but it attracted children like a magnet.

Keep it short and simple – don't have too much happening in your activity at the same time. Your activity should last between 10–15 minutes, to allow people to keep circling around different activities within the event. Use clear and simple language when talking to your audience, and in your written material (see Section 4.7). The sign of a well-designed activity is one which is used repeatedly. Training others to run your activity will also ensure sustainability; a good example of this is Case study 9.1 by Harrison and Shallcross, 'A Pollutants Tale' demonstration lecture.

To get a good flavour of the different types of activities which can form an event, there is a 5-minute film of an event which we ran in 2009 at The Forum in Norwich called 'Cells Alive' (see the website associated with this book).

Case study 6.4 by Annabel Cook describes the design of a hands-on activity of DNA extraction from fruit.

Case Study 6.4

Strawberry DNA
Annabel Cook

Background

My aim was to produce a reusable activity that could be used at a variety of different events which looked at one of the most basic areas relevant to genetics – DNA extraction. There were several reasons for this, including the 10th anniversary of the Human Genome Project, meaning that there was a lot more news coverage of DNA, although not all of it was accurate! I was looking for a way to get people interested in genetics, familiarising them with some of the terminology involved and making DNA relevant.

---

[4] http://www.dialogueacademy.org.uk/

Activity design

There are many DNA extraction protocols available online, but none seemed to explain what was happening and why. I wanted the activity to appeal to a broad range of ages and backgrounds. I had no specific funding and wanted it to cost as little as possible so that anyone could do it.

After some research and discussion with geneticists it became obvious that DNA extraction from fruit and vegetables should be very straightforward. Comparing existing protocols for extraction, I tried variations to determine which steps were necessary and which weren't. I was able to come up with a very simple method, using only household equipment, which could be done in approximately 10 minutes. Many people use kiwi fruit for DNA extractions, but the number of children experiencing allergic reactions is increasing. Onions were used for a while, but for indoor events this proved uncomfortable! Strawberries were chosen for various reasons but mostly because they have several sets of chromosomes (unlike humans who have just two sets) meaning plenty of DNA.

I decided to present the activity as a 'recipe' with an ingredients list and step-by-step instructions. This reinforced the idea of using household equipment and avoided any mental barriers of it 'looking like science'. On the reverse of the recipe sheet I added some information on the science behind the activity (for the recipe sheet see the book website).

In 2010 I took part in a Dialogue Academy workshop[5]. I used this opportunity to develop my basic activity and the dialogue I would use. For example, I made up small flashcards and large posters with one-line facts about DNA and genetics (e.g. 'Identical twins do not have identical DNA.') for people to read before, during and after the activity. I also had several props including a DNA double helix built from K'NEX.

This encouraged dialogue, which made a huge difference to the activity, stimulating discussion amongst families and engaging the young children.

Safety issues

1 Food allergies were the most obvious safety issue.
2 Part of the protocol involves using hand-held stick blenders to break up the flesh of the fruit. By using strawberries, which are very soft, we had the option of using a metal fork instead. This was much more suitable for young children.
3 Alcohol is used to precipitate the DNA, allowing it to be seen with the naked eye. Although any strong alcohol would work, I wanted to avoid children seeing and using alcoholic drinks for this purpose. I opted for lab-grade pure ethanol, pre-measured into plastic tubes so there was only a small volume in reach at any one time.

Activity highlights

1 The use of household equipment meant that there were no barriers to engagement as everyone taking part was familiar with the items being used.
2 By carefully choosing the facts presented on the flashcards and posters, I was also able to relate genetics to a familiar environment and surprising facts.

Problems

After running the activity several times I became aware, from a feedback sheet, that one child had thought we were extracting 'D and A'. We are now very careful to explain that it's DNA and that it's short for a long chemical name.

Resources

Volunteers to help me run the activity were my biggest resource. In planning the activity I was able to get relevant scientific input and test lots of different protocols quickly. When running it, more volunteers meant more members of the public could take part.

---

[5]The dialogue academies were free training workshops for professionals in science engagement, and were supported by the Wellcome Trust. The resources from these workshops can still be found at the following URL http://www.dialogueacademy.org.uk/

Personal gain

I found it really satisfying to give so many people more of an understanding of DNA – what it is, that everything living has it and where you find it in the organisms. I have heard months after an event that people are still talking about it and have kept their strawberry DNA!

Top tips

1 Keep the requirement for equipment to a minimum. Although I prefer to use stick blenders, I haven't always had access to power sources. There are no 'musts' for this activity, although it's much easier to be near a toilet and running water, for disposing of the mashed up strawberry mixture.

2 Be over-prepared. I prefer to have strawberries left at the end than to run out half way through. Although, if I did run out of anything, their household nature meant they were easily replaced at short notice.

3 Be imaginative and think like a non-scientist. Remember the things that fascinated you and caught your eye.

4 Let any venues know your requirements well in advance and think about the detail. Are your leads long enough to reach the power sources? Are the running water and toilets close by, or several minutes walk through busy areas and up/down stairs?

6.7 Consider your resources – consumables, equipment, expertise and people (CEEP)

Resources are also covered under Sections 6.8 on funding and 6.10 on health and safety. Many of the case studies throughout this book give valuable advice on running events and activities, and provide specific advice on health and safety related issues. However, also consider the following general advice.

Consumables – these are all the materials (including chemicals), which you need for your activity. You must ensure that you bring enough for the duration, e.g. for a DNA extraction from strawberries, which lasts 10 minutes, four participants can do it at once, which is 24 per hour and 144 during a 6-hour event. If two people run the stand, then this number can double.

Equipment – must be portable, durable and easy to use. Keep the purpose of the equipment in mind during the design of the activity. Do not bring anything to an event which you cannot afford to replace. Check to see what the excess is on the insurance for equipment.

Expertise – if you need help in specific aspects of your event or activity design, try and find appropriate people to give you advice. This could include having a professional designer doing the posters for your event. You can include the cost of this in your funding proposal.

People – you can't run an event without willing helpers. Each activity needs one/two people running it, with others who can step in while they have a break. You should have one person who is overseeing the entire event, who is not manning an activity stand, but 'trouble-shooting' for others; this will usually be the event organiser (you). Help is also really invaluable when packing and unpacking your equipment and materials for your activity as you go to and from the event.

6.8   How to get your project funded

Once you have decided on the design of your activities within the event, an important part of your project is to find the money to run it. Clearly, activities and events don't all cost the same; they can range from less than £100 for a small Cafe scientifique to over £1 million for large-scale science festivals. Note that most engagement grants are not fully economically costed (FEC), thus you are reliant on your organisation to allow you to use their infrastructure to support your grant.

Different funding sources will open up to you when you decide if you are going to do a school-based or public-based event. For example, the Royal Society Partnership Grants are designed specifically for use in schools, and the grant is paid to the school. Wellcome Trust funding, e.g. Peoples Award can be used for both school and public events and will be paid to your organisation.

6.8.1   Making an application

Once you have decided on the main aims and objectives of your project, you need to decide on the best place to obtain your funding. You can get money, and also 'in kind' support from a variety of different sources. As an initial step, visit the website of the organisation and download the application form and the guidance notes. Many websites will also have the reports of past projects which have been funded; these are really useful sources of information, not only for ideas, but also to ensure that your idea is covered by the grant conditions.

The funding sources can broadly be split into the following:
- research councils;
- learned societies, charities and trusts;
- companies;
- local government authorities;
- British Council;
- your own organisation;
- self-funded.

6.8.2   Funding from research councils

In the UK all RCUK funding bodies have Pathways to Impact, a mandatory submission alongside the research grant that covers the economic and societal impact of the research and can include public engagement. They are sent out to peer review and form part of the secondary assessment criteria, science excellence being the first, used to judge whether funding for research is to be awarded (RCUK, 2011). Research councils also have other smaller grants available for school and public communication.

6.8.3   Funding from learned societies, charities and trusts

Learned societies such as the Society for General Microbiology (SGM) and charities such as the Wellcome Trust offer public engagement grants. A report

Table 6.7 Examples of UK learned science societies.

Name of society	Web address
British Society for Immunology	www.immunology.org/
Society for General Microbiology	www.sgm.ac.uk/
Society for Applied Microbiology	www.sfam.org.uk/
Biochemical Society	www.biochemistry.org/
British Pharmacological Society	www.bps.ac.uk/view/index.html
Society for Endocrinology	www.endocrinology.org/
The Physiological Society	www.physoc.org/
British Ecological Society	www.britishecologicalsociety.org/
The Marine Biological Association	www.mba.ac.uk/
Royal Entomological Society	www.royensoc.co.uk/
The Nutrition Society	www.nutritionsociety.org/
British Association for Lung Research	www.balr.co.uk/
British Mycological Society	www.britmycolsoc.org.uk/
Institute of Physics	www.iop.org/
The Royal Academy of Engineering	www.raeng.org.uk/
Royal Society of Chemistry	www.rsc.org/
Institute of Mathematics and its Applications	www.ima.org.uk/
The Royal Statistical Society	www.rss.org.uk/site/cms/ contentChapterView.asp?chapter=1

by the Society for Biology (2011) identified the different reasons for this financial support for public engagement activities as being:

- to win support for science;
- to develop skills and inspire learning;
- to fulfil their charitable remit;
- to promote more efficient, dynamic and sustainable economies;
- to benefit members by providing opportunities to enhance their careers and the impact of their research.

Table 6.7 provides a list of different UK science societies and Table 6.8 lists UK organisations that provide funding for science communication for public and school science communication events.

6.8.4 Applying to companies

Different companies give money for a variety of reasons and these include, but are not limited to:

- creating goodwill in the community;
- to be associated with causes that relate to their business;
- to build good relations with employees;
- because they have always given.

Some companies have funding schemes, so it's worth spending time looking at their websites and finding out about them, your eligibility and the process of application. Some companies have very clear policies on giving and you

Table 6.8 Examples of funding for science communication projects.

Funding source	Name of award	Amount	School/public	Eligibility	Award process	Deadlines	Web address
Wellcome Trust	Peoples Award	Up to and including £30 000	School and public	Open to anyone with a good idea for engaging people with developments in biomedical science.	One application form to fill in, which is peer reviewed, a decision on funding is made by a funding committee	Four deadlines, in previous years they have been February, May, July and October	http://www.wellcome. ac.uk/Funding/Public-engagement/ Grants/People-Awards/index.htm
Wellcome Trust	Society Awards	Above £30 000	Public	Open to anyone with a good idea for engaging people with developments in biomedical science.	A preliminary application followed by a full application. This is peer reviewed and applicants are invited for interview	Usually two deadlines per year	http://www.wellcome. ac.uk/Funding/Public-engagement/ index.htm
Royal Society	Partnership grants	Up to £3000	School	The application comes from the School, but the scientist is heavily involved in the application process	There is one online form to fill out, with your school partner. The award is decided by a funding committee.	There are several deadlines per year	http://royalsociety.org/ education/partnership/
The Society for General Microbiology	Promoting Microbiology to the Public	Up to and including £1000	Public	Members of the Society for General Microbiology	An application form, followed by an evaluation report on completion of the project.	No specified deadline	http://www.socgen microbiol.org.uk/grants/ dtf.cfm

(continued)

Table 6.8 (*Continued*)

Funding source	Name of award	Amount	School/public	Eligibility	Award process	Deadlines	Web address
BBSRC	Public Engagement Awards	Up to and including £5000	Public	BBSRC-sponsored researchers, BBSRC-sponsored institutes and BBSRC local coordinators	An application form, a decision is made by the funding committee	Check website for details	http://www.bbsrc.ac.uk/society/pe-strategy-and-funding.aspx
The British Ecological Society	Public Engagement Awards	Up to and including £2000	Public	Individuals and organisations who organise science communication events in ecology	An application form which can be sent in electronically	Check website for details, but it's awarded once per year	http://www.britishecologicalsociety.org/grants/policy/peg.php
Paul Hamlyn Foundation	Education and Learning Open Grants Scheme	A range of awards	School and Public	Formally constituted organisations, individuals are not eligible	A two-stage application process	No closing dates	http://www.phf.org.uk/page.asp?id=85
AstraZeneca Science Teaching Trust	Improving learning and teaching of science in the UK	A range of awards	School	Check website for details	Check website for details	Check website for details, but it's awarded once per year	http://www.azteachscience.co.uk/the-trust/what-we-offer.aspx
The Royal Society of Chemistry	Promoting the public awareness of chemistry and the chemical sciences	A range from £250 to £5000	Public	RSC members and member groups to support public activities but other applications will be considered	An application form is available from the website	Applications are accepted all year round. Funds will be allocated on a first come, first served basis from January to December each year	http://www.rsc.org/ScienceAndTechnology/Funding/SmallGrants.asp

Organisation	Purpose	Amount	Public/School	Eligibility	Application	Deadline	Website
Big Lottery Fund	Supports community groups and projects that improve health, education and the environment	A range of awards, including Awards for All and Heritage Lottery Fund	Public	Not-for-profit group, parish or town council, school or health body	An application form is available from the web site	No deadlines, you will be informed with eight weeks of the outcome	http://www.biglottery fund.org.uk/
Institute of Physics	Improving engagement with physics	Up to £1000	Public	Individuals and organisations	An application form is available on the website	Check website for details	http://www.iop.org/activity/outreach/
Royal Academy of Engineering	Education	-	School	Organisations (schools)	An application form is available on the website	Check website for details	http://www.raeng.org.uk/education/default.htm
Waitrose Community Matters	Supports good causes	£1000	Public/School	Organisations	See website for details. The good cause is voted for by shoppers	No deadlines	http://www.waitrose.com/content/waitrose/en/home/inspiration/community_matters.html
Small Awards from Science and Technology Facilities Council	Provides funds for small, local or pilot projects promoting STFC science and technology	£500–£10 000		Open to researchers funded by any of the research councils	An application form is available on the website	Check website for details	http://www.stfc.ac.uk/Public+and+Schools/1396.aspx

must read them to make sure your application is appropriate. Top tips for applying include the following.

1 Always write your proposal down on paper, with a clear aim and objective, description of your activity, the geographical location, the audience towards whom it is targeted and the timeframe for the project.
2 Don't send the same covering letter and proposal to hundreds of companies; it's much better to research the companies and target those that are most suitable and fit your needs.
3 State clearly what you would like the company to contribute. This might be money, but it can also be services, materials, advertising space or even training.
4 Make it clear how the company will benefit from supporting you, e.g. their logo on leaflets and other promotional material.
5 Always try and find someone within the company to address your covering letter to. Linked-In might be a good way to make connections with people.
6 The *UK Guide to Company Giving* is a book published annually by the Directory of Social Change. This is a really good place to begin as it has many contact details.
7 Follow up your initial contact with an email or phone call.
8 Do not give up, even if your initial contacts are rejected.
9 Don't ignore local companies either large or small; they may be able to provide materials or support; for an example, see Box 6.3.

Box 6.3   Asking companies for support

In 2006, the British Science Festival came to Norwich, and in collaboration with the Norwich Castle Museum, I ran an event called 'Under the Microscope'. One of the activities was creating art collages inspired by what the children saw under the microscope. We approached our local department store, a family run business called Jarrold, with details about our event, asking for a donation of art materials. They kindly supplied a box of art materials which supported this event and also others which we ran in subsequent years.

6.8.5 Funding from your own organisation

You may find that your own organisation will support engagement activities both with schools and the public. School engagement may well be funded through an 'outreach office'. These outreach funds might have target schools in mind (e.g. those schools falling behind the national level of attainment in core subjects), so you must check that your planned audience fits with their agenda.

6.8.6 Self-funding

If all other funding streams fail, you can always consider funding the project yourself. This would allow you to evaluate the project prior to making a subsequent funding application stronger. Remember some communication activities don't require money, just your time, energy and a good idea. For a good example read the case study on teaching the scientific method by Niamh Ní Bhriain (Case study 10.1).

6.8.7   Grant-finding websites

There are some websites which keep lists of grant awarding bodies, some of these websites you have to subscribe to, while others are free. It takes quite a lot of time to go through these funding sources, but you can get funding ideas you may have not previously considered. See Table 6.9 for some examples.

6.8.8   The funding process

Each funding body will have a different application form, different deadlines and different procedures for applying and subsequent handling of the application. You will need to check the individual websites in order to make sure you are aware of this important information. While all application forms are different, they often have the following points in common and you should consider these in detail before making your application;

- clear aims and objectives;
- clear target audience;
- good design;
- realistic timeframe;
- clear costings;
- clear evaluation strategy;
- clear dissemination potential.

6.8.9   Setting your aims and objectives

Section 5.2 gives you good information on setting your aims and objectives.

6.8.10   Realistic timeframe

Projects last for varying amounts of time, from a few weeks to years. When considering the timeframe for your own project, you must first decide how much time you will have to devote to the project. If you need help in delivering it, then ensure you apply for a grant which allows you to cost in staff time. Your final evaluation report should be included within the timeframe. It is often very useful to construct a table of Key Milestones with dates, to allow you to effectively plan and manage your project.

6.8.11   Costing projects

You must check the guidance documents for the grant you are applying for, to find out what they will and won't fund. For example, many grants will not fund the purchase of equipment or staff time. Check if you will need to pay VAT on items that you purchase. Remember that most engagement grants are not FEC, so your organisation will be contributing towards:

- office space;
- use of computer and telephone;
- use of finance office;
- your time and that of other staff members.

All of these can be put into grants as your organisation's financial contribution to the project.

If your grant allows the purchase of pieces of equipment or to make use of other services, make sure that you have included quotes from the suppliers. If

Table 6.9 Details of grant finding websites.

Name	Description	Free or fee	Web address
GRANTnet	Helps community and voluntary groups, sports and other clubs, schools and social enterprises and small businesses to find suitable funding.	Free	http://www.grantnet.com/index.aspx?pid=BEE1857D-B60D-4550-8D48-1C4F9D24A1A7&
Funding Central	This site is funded by the Cabinet Office for Civil Society. It guides you through thousands of different grants by asking a set of simple questions about what you are trying to fund.	Free, but you will need to register before use	http://www.fundingcentral.org.uk/Default.aspx
Government Funding	This site is supported by the Directory of Social Change. You can browse different funding sources or search for specific themes.	There is a fee to use this site	http://www.governmentfunding.org.uk/Default.aspx
Easy Fundraising	This is a site where you can register your club or society. When people shop online with different retailers, the retailers will donate a percentage of your spend to your chosen good cause. Schools can register, as can sports teams and youth groups.	Free, but you need to register	http://www.easyfundraising.org.uk/how-it-works/and http://www.easyfundraising.org.uk/register-a-cause/

you can apply for staff costs, ensure that this has been costed properly; your organisations research office, which deals with research grant applications, should be able to help with this.

Make sure that you include funds for the evaluation and dissemination of your project. This could include the cost of setting up a website, the printing of final project reports or presenting your project at a conference. Finally, ensure that your costs are linked clearly to each part of the project.

6.8.12   The grant process
Grants from different organisations will be handled in different ways, but they have some steps in common:
1 deadline for applications is announced;
2 applications are received and sent out to review;
3 referees send back their written comments;
4 awarding panel meets;
5 decisions are announced.

Be aware that not all engagement grants are sent out for review, some decisions about funding will be made by the awarding panel without a review process. Unlike research grants, you will not often be allowed to respond to referee's comments. Some of the larger awards (e.g. Wellcome Trust Society Awards) require a preliminary expression of interest, which is reviewed. Applicants are then invited to submit a full proposal, which is then assessed and applicants invited to interview before a funding decision is made.

6.8.13   Deadlines
1 Some organisations accept applications at any time, others will have deadlines several times a year, for example the Royal Society Partnership Grants has four deadlines per year, the actual project has to be started within the 3-month period leading up to the next funding round. Some organisations (e.g. the Wellcome Trust) may call for a special funding round, for example for projects specifically dealing with genes and health. It is wise to check funding sites regularly, or to set up an email alert if the website allows this.
2 Some applications are still paper-based, others are done electronically – make sure you know which format is required.
3 Always make sure that you have had the correct authorisation from your own organisation before you apply for a grant.

6.8.14   Referees
Burnet (2010) makes the important point that as science communication is interdisciplinary you may not get a referee who is an expert in science communication. You may or may not be allowed to nominate referees for your proposal. Unlike research grant applications, where you will always have an opportunity to respond to referee's comments, you may or may not be allowed to respond to referee's comments for a science communication project proposal.

6.8.15 Awarding panels

Frank Burnet (2010) offers the following advice on the basis of his experience of membership of awarding panels:

- policy varies about whether you can discover the membership of a particular awarding panel; do if you can, it's useful in ensuring your application is likely to hit the right buttons;
- each panel has its own dynamic, but most see themselves as being strongly guided by the views of the expert referees and will only go against their collective view in exceptional circumstances;
- officers from the funding body attend panels, usually as observers, and they often have data about the outcomes of previous work of yours that they have funded; so be sure you have submitted strong final reports and other documents;
- panels usually have at least a rough idea of how much money they have to award in that funding round and generally wish to ensure that a significant number of projects are funded. They, therefore, tend to look more critically at applications that cross particular cash thresholds. So there can be an advantage in staying just below, for example the five or six figure barrier if this is possible given your plan (and you don't make it too glaringly obvious; avoid £9999 and its equivalents);
- panels do sometimes have it within their power to partially fund work, particularly when it has interdependent stages or when they feel that not all costs are justified. However, this is an unusual occurrence; most funders don't negotiate.

—Reproduced with permission from Burnet (2010).

6.8.16 Some final hints

1 Try building up your experience with smaller grants from learned societies.
2 Do a joint application with a more experienced colleague.
3 Don't give up, keep trying for funding if your first application is rejected (Case study 6.2).

6.9 Top tips for successful marketing

An often forgotten aspect of designing a public science communication event is how you intend to market it, but this is clearly crucial if you want the right audience to attend. There is no question that this is one of the most difficult things to get right, too few people at your event and you will feel deflated, too many and you will feel overwhelmed and your audience will feel frustrated. Good marketing requires forethought and hard work. Think about holding your event at a Museum or Science Centre where they may be able to help you with a marketing strategy. They will keep email lists of regular users and be in a position to place the event on their website and distribute flyers.

1 Make sure that any grant application you make has a budget for the marketing of your event(s).
2 Write a press release for the event and give it to your communications officer, who will be able to release it at the right time. They will often give you help in writing it as well. Remember the five W's of a press

release – what, why, who, when and where. The press release should be picked up by the local press and local radio. You may be asked for a phone or radio interview. DO GO! This need not be a scary experience and you will be able to reach a wider audience more effectively. If you manage to get a slot on local TV, then that's brilliant.

3 Make sure that your event has a good 'hook' which will draw people in. One of our most successful event titles was 'Poo, Pee and Puke' which was very attractive to a younger audience (see Section 6.5 on hooks).

4 If you are at a university, make sure the event appears on the university homepage; again, your communication officer should help you with this.

5 Design an attractive flyer with clear information; use your press release as a basis. Make sure that the flyer includes where and when the event is being held (a map might be useful), and also the cost of the event. Distribute flyers in the area where your target audience lives. For examples of flyers, visit the book website.

6 Ask to place an ad in a local free magazine, if such exists. We have used this strategy effectively in the past. We have had the cover of a free magazine and taken advantage of their wide distribution. This does cost more money, though, so make sure you include it as a cost in the funding proposal (see book website for an example of a magazine cover).

7 Give flyers to schools in your target audience area, these can be distributed to children in their book bags, or pinned up on notice boards.

8 Leaflets can be distributed in other places such as local cubs and scout groups as well as sports centres.

9 Ask your library if you can put leaflets in their information area, or if possible to display a small poster of the event.

10 Even if your event is not being held in a science centre, but you have one locally, ask if your event can be placed on their website and if they will distribute leaflets for you to their visitors (perhaps upon entry).

11 Other good places to place leaflets include museums, art centres, zoos, tourist information shops and adult learning centres.

12 We've also given leaflets to parents in school playgrounds and parks.

13 Try using social media, set up a Facebook page and get people to Twitter about your event – or perhaps even set up a discussion through a Blog.

14 Try and get your event onto 'What's on . . . ?' pages of local newspapers and websites.

15 Hold your event where you will get 'passing people', for example a city centre. We regularly use an area in Norwich called the Forum, which is a large, open area in front of the city library.

6.10   Health and safety

In the UK all places of work must adhere to the Health and Safety at Work Act of 1974, employers are responsible for the health and safety of staff and when in a school, the pupils. The Management of Health and Safety at Work Regulations, 1999, requires the assessment of risks during practical procedures and trips and the steps taken to reduce risk must be identified. The

Control of Substances Hazardous to Health (CoSHH) regulations must also be adhered to; CoSHH looks specifically at the hazards of chemicals used during procedures, the steps needed to reduce risk and the steps taken if an accident does occur.

The health and safety of the participants of your science communication project and of yourself and your helpers is of prime importance. As the organiser of the event, it is your responsibility to make sure you have completed the appropriate risk assessment and that you have insurance.

The information provided here is intended to be a guide and we accept no liability for your events. You must always check to make sure information is correct.

6.10.1 Risk assessments

The risk assessment is a crucial part of any science communication activity. The law does not expect you to eliminate all risks, but to protect people as 'far as is reasonably practicable' (Health and Safety at Work Protection Act, 1974).

There are three main parts to a risk assessment:

1 identify the hazards;
2 identify steps to reduce the hazards;
3 steps to be taken if an accident occurs.

If you are doing the event or activity in a school, then the school may do the risk assessment for you. It is really important that you check that this is the case. If the school does not provide you with this support then you need to do your own risk assessment and make sure that it is given to the school in advance of the event or activity. You must ensure that the risk assessment is acceptable to the school, so check it with the school ahead of time. For practical risk assessments, many schools in England, Wales and Northern Ireland use the Consortium of Local Education Authorities for the Provision of Science Services (CLEAPSS), which is an advisory service for practical science and technology. In Scotland there is a similar organisation called Scottish Schools Education Resource Centre (SSERC).

If you are doing an event in a public place (e.g. museum, science centre or community centre), then you must ensure that you have given the risk assessment to the appropriate person. If another organisation is doing an activity at an event which you are organising, you must ensure that they also complete a risk assessment and that it is given to you in advance.

Always ask for advice and support from your outreach officer, contact teacher or colleagues if you are unsure what needs to be included. It is always a good idea to regularly review your risk assessments for activities, to capture any changes which may have been made.

The risk assessment form can be any format, as long as it contains the relevant information. The Health and Safety Executive website (see Section 6.10.9 for websites) has a blank template which you can use. We have also provided a form and an example on the book website which you can tailor for your own use.

6.10.2   Working with microorganisms

In the UK the use of microorganisms in schools also comes under CoSHH regulations. There are a surprising number of microorganisms which can be used in schools, providing that the risk is minimised with good laboratory practice. The Society for General Microbiology has a comprehensive site for working with microorganisms in schools, which includes a list of those which can be taken into schools. This list was compiled with the Association for Science Education (ASE), CLEAPSS, SSERC, Health and Safety Executive (HSE), Microbiology in Schools Advisory Committee (MISAC), Society for Applied Microbiology, Society for General Microbiology (SGM), National Centre for Biotechnology Education (NCBE), Science & Plants in Schools (SAPS), the Wellcome Trust and the educational suppliers Philip Harris and Blades Biological.

6.10.3   Working with DNA

The regulations for working with DNA are specific to each country, so you must check the regulations where you live. The NCBE have an excellent document which explains what can and can't be done in school laboratories in the UK; the link is provided in the useful websites section (see Section 6.10.9). For example, you cannot use cheek cells for DNA extraction in Northern Ireland.

6.10.4   Electrical testing of equipment

All electrical equipment must be tested for safety through portable appliance testing (PAT). This is a legal requirement through the Electricity at Work Regulations. Your technical staff should be able to help you with this. For further information on PAT testing see the useful websites list at the end of this chapter. Equipment needs to be tested on an annual basis.

6.10.5   Working with children and vulnerable adults

The law on working with children and vulnerable adults varies between different countries and also changes quite regularly. This means that you must check the law in your own country. In most countries you need to be vetted.

In England, Wales and Northern Ireland you will need an enhanced Criminal Records Bureau (CRB) check. You apply for a CRB through the organisation for whom you are doing the event (e.g. STEMNET), or a school, you may find that you will need several separate CRB checks (e.g. for different schools), as they are not transferable. If you work in a university, your Human Resources Department may help you arrange a CRB. In Scotland you need to register with the Protecting Vulnerable Groups (PVG) and in the Republic of Ireland, you need to be vetted through the Garda. You could include the cost of obtaining CRB checks in any funding applications.

6.10.6   Public liability insurance

To run a public event you need to have public liability insurance. Your university should be covered for outreach work with schools and the public, but you will need to find out what you are covered for and what the policy number is. Sometimes you may also be asked to produce the certificate. A good place to start to find this information is your outreach office.

Be aware that you might need to get extra cover for pieces of equipment. In addition, if you are travelling to a school or public event in your own car, you need to make sure that you have business insurance (your workplace will also require you to write a risk assessment associated with your travel).

It is worth noting that if you are a STEM Ambassador, then you can get your activity or event covered through STEMNET providing that you supply them with the event details in advance. If you are running events on behalf of another organisation (e.g. a learned society), then they may also cover you, but you must check this.

6.10.7   Permission letters

If you are planning an activity at a school, then it is a good idea to talk to your contact teacher about a permission letter which a parent or guardian needs to sign before the child can take part. The school will almost certainly be doing this anyway, but it is a good idea to get involved. I always ensure that details of the activities are written in a letter, to ensure the parent or guardian is aware of the activities to be undertaken. This is especially important for activities that include:

- tasting, e.g. salt, lemon and sugar as part of an activity on senses;
- extraction of DNA from cheek cells[6];
- fingerprints.

If these types of activities are taking place in a public event, obtain verbal permission from the parent/carer or guardian for the child to do the activity.

It is also illegal to take pictures or videos of children without prior parental consent. If you want to take pictures or a video of your school event, please contact the school for advice and they will be able to send out a permission letter. The school may also request to screen pictures and videos.

If you want to take pictures or videos of public events involving children, then you must still ask for written consent. There is a form which you can use on the book website.

6.10.8   Work experience students

If you take on a work experience student, either at the end of Year 10 or in the sixth form, this must go through your official channels. You must make sure that health and safety issues as well as insurance are covered. Your Personnel or Human Resources Department should be able to help you with this.

---

[6] CLEAPSS information states that experiments in schools which involve saliva are not allowed to be done in Northern Ireland. Elsewhere in the UK it is acceptable providing that Good Laboratory Practice is maintained. Please check what the status is in your own country.

Table 6.10 Details of other guides for getting started with science communication.

Guides	Source	Description	Website
A Guide to Good Practice in Public Engagement with Physics	Institute of Physics IOP April 2011	Whether you're completely new to public engagement or have had some experience and want to become more involved, this guide is a summary of the processes involved in developing, delivering and evaluating an activity, or volunteering for events organised by others	http://www.iop.org/publications/iop/2011/file_50861.pdf
Design a Project for Public Understanding of Science	Institut de Recherche pour le Développement (IRD) 2006	Offers essential information for carrying out projects for public understanding of science and technology for those who have little or no experience of science outreach activities	http://www.latitudesciences.ird.fr/outils/guide/ird_handbook_project_public_understanding_sc.pdf
Taking Science to the People	Dr Frank Burnet UWE, Bristol, UK January 2010	Guide designed for science communicators, scientists or researchers in any field, seeking to engage with a wider audience with their work and it's social implications	http://frankburnet.com/taking-science-to-people/
Guide: Why and how to Communicate your Research	Dr Frank Burnet UWE, Bristol, UK January	A concise 26-page guide designed to help scientists, engineers and technologists develop effective, attention-grabbing ways of communicating what they do and why they do it to public audiences	http://frankburnet.com/why-and-how-to-communicate-your-research/
The Engaging Researcher	Vitae and the Beacons for Public Engagement have joined forces to develop this booklet	This booklet highlights some of the many ways you can engage the public, offers practical tips for getting started and explores how public engagement can benefit you, your research and the public with whom you engage	http://www.vitae.ac.uk/CMS/files/upload/The_engaging_researcher_2010.pdf
Partnership for Public Awareness Good practice Guide	Engineering and Physical Sciences Research Council May 2003	This guide is designed to be of particular use to people who are about to start planning an activity, and perhaps applying for an award or grant to help fund the project	http://www.epsrc.ac.uk/SiteCollectionDocuments/form-notes/ppe-goodpracticeguide.pdf

6.10.9 Useful health and safety websites

Five Steps to Risk Assessment-Health and Safety Executive:
http://www.hse.gov.uk/risk/fivesteps.htm
For PAT testing:
www.pat-testing.info/
For CLEAPSS
http://www.cleapss.org.uk/about-cleapss
Criminal Records Bureau:
www.crb.homeoffice.gov.uk/
The Scottish Government Protecting Vulnerable Groups Scheme:
http://www.scotland.gov.uk/Topics/People/Young-People/children-
    families/pvglegislation
The Garda Vetting Unit:
http://www.garda.ie/Controller.aspx?Page=66
STEMNET:
www.stemnet.org.uk/
Scottish Schools Education Resource Centre (SSERC):
www.sserc.org.uk/
Society for General Microbiology:
http://www.microbiologyonline.org.uk/teachers/safety-information/safety-
    guidelines
http://www.microbiologyonline.org.uk/teachers/safety-information/risk-
    assessment

6.11 Concluding remarks

We hope that this chapter has provided you with the information and instilled
the confidence for you to begin your science communication projects. If you
still need further guidance, more information is provided on the book website.
Table 6.10 also provides a list of other guides that you might find useful in
your journey towards science communication and public engagement.

References

Bauer, M.W. (2008) Survey research on public understanding of science. In: *Handbook
of Public Communication of Science and Technology* (eds Bucchi, M. and Trench, B.).
Routledge International

Bultitude, K. (2010). Presenting science. In: *Introducing Science Communication* E (eds
Brake, M.L. and Weitkamp, E.). Palgrave Macmillan.

Burnet, F. (2010) Taking Science to the People. http://frankburnet.com/why-and-how-to-
communicate-your-research/ (accessed 5 May 2012)

Burns, T.W., O'Connor, D.J. and Stocklmayer, S.M. (2003) Science communication: a contemporary definition. *Public Understanding of Science* **12**, 183–202

Health and Safety at Work Protection Act (1974) http://www.legislation.gov.uk/ukpga/
1974/37 (accessed 5 May 2012)

Ipsos MORI (2009) Social Grade, A Classification Tool, Ipsos MediaCT. http://www.ipsos-mori.com/DownloadPublication/1285_MediaCT_thoughtpiece_Social_Grade_July09_V3_WEB.pdf (accessed 5 May 2012)

Public Attitudes to Science (2011) Ipsos MORI poll for the Department for Business Innovations and Skills. http://www.bis.gov.uk/policies/science/science-and-society/public-attitudes-to-science-2011 (accessed 5 May 2012)

Society for Biology (2011) Society of Biology and Member Organisation. Science Communication Project Report and Action Plan (2011). http://www.societyofbiology.org/home (accessed 5th May 2012)

CHAPTER SEVEN
Direct Public Communication

It is surprising how often people in all walks of life own that their interest in science was first aroused by attending these courses when they were young, and in recalling their impressions they almost invariably say not 'we were told' but 'we were shown' this or that.

—Sir Lawrence Bragg referring to the RI Christmas Lectures
in Taylor (1988)

7.1 Introduction

This and the following chapter specifically focus on different types of science communication events and activities which are outlined in Table 7.1. These are illustrated with case study examples that aim to:

1 showcase different ways in which scientists can engage with the public;
2 provide specific advice on running different types of activities and events;
3 share the pitfalls associated with different types of science communication;

4 provide top tips for reproducing activities and events.

7.2 Direct communication delivering information

Although the current rhetoric of science communication is for 'two-way engagement' with the public, the direct delivery of information still holds an important place in science communication. Hearing a scientist speak who is both knowledgeable and enthusiastic about their research can be an inspiring experience. Under this umbrella comes:

- lectures both traditional and demonstration style;
- public activities and events (such as those in museums and science centres);
- festivals, which can be both information and dialogue focused as many different events will make up the festival programme;
- stand-up comedy and theatre.

The advantage of direct information-based delivery is that you can reach large audience numbers. The audience does not feel pressured into taking part; they can relax and be entertained. We have to accept that sometimes

---

Science Communication: A Practical Guide for Scientists, First Edition. Laura Bowater and Kay Yeoman.
© 2013 John Wiley & Sons, Ltd. Published 2013 by John Wiley & Sons, Ltd.

Table 7.1 Types of science communication activity.[a]

	Direct communication	Indirect communication
Information	• Lectures • Festivals • Plays • Science 'stand-up' comedy • Public events and hands-on activities (e.g. at a museum or science centre)	• Television programmes • Radio programmes • Science writing, e.g. popular books and magazine articles • Open access • Websites • Information leaflets
Conversation	• Cafe scientifique • Festivals • Book clubs • Science 'stand-up' comedy • Citizen's panels • Citizen science projects • Policy formation	• Social networking, e.g. blogging and tweeting • Online opinion fora • Citizen science projects

[a]Examples provided are not limited. Some examples feature in more than one section, as they can have both information and conversation.

the audience doesn't want to spend energy actively engaging in the process of communication. They would rather sit back and enjoy being a passive absorber of knowledge, information and entertainment.

7.2.1 A focus on public lectures, both traditional and demonstration

It may seem obvious, but a public lecture is very different to a scientific talk presented at a conference, but many of us seem to forget the distinction. I have been guilty in the past in trying to adapt an existing scientific talk to a public or school audience. For me, this approach rarely works. I find it much better to start by really thinking about my target audience and constructing a new talk from scratch.

Planning is the key to giving a good public talk. Consider the following questions before you begin writing the content (Shortland and Gregory, 1991).

• Who are your audience?
• What is your main motivation for speaking? To entertain, convince, motivate or inform?
• Why should people be interested?
• Why is the subject important?
• Why am I speaking about this now?
• Why should people trust me?

If you are not introduced to the audience, make sure that you say who you are and what you do. You need a strong opening to the actual talk, perhaps pose a question, relate an anecdote or set a challenge. In your introduction outline the purpose of your talk. Regardless of your audience, the main body of the talk needs structure and a logical order. Keep the acronym KISS (Keep

It Short and Simple) in mind. Make good use of colourful images. Avoid complex terms, or if you use them, explain them. If you use numbers, think of good ways to present them. Perhaps instead of 25% say 1 in every 4. Instead of 1×10^6, use one million. For more tips on clear writing see Section 4.7.

Always ensure that you summarise your points at the end. You can't go wrong by following these three steps for the overall structure:

1 tell them what you are going to say;

2 say it;

3 tell them what you said.

You will be asked for a title, – probably in advance of you writing the talk – as it will be needed for publicity. Try and make the title inviting but also make it obvious what the talk is going to be about. If you are not clear, people may come and then be disappointed if the talk is not what they expected.

Traditional public lectures on science are a very good way of getting large audience numbers, providing that the title of the talk is attention grabbing and the talks are located in a suitable place. They are perhaps seen as a traditional, linear form of communication, but you could build in more interaction, by posing questions and actively seeking audience participation. However, do be aware that people can feel intimidated by a large audience and not feel inclined to ask questions. The environment is also important; holding a public lecture at a university may put many people off, not only from coming but also from asking questions if they do come.

Doing a public talk is a good way of getting involved with science communication for the first time, but the next step is to consider setting up a lecture series. Case study 7.1 by Dr Sarah Field shows how you can get a public lecture series going at your organisation.

Case Study 7.1

'Talking Science' – A Public Lecture Series
Sarah Field

Background
In 2006, I attended a communications skills course at the Royal Society[1] aimed at effective explanation of science both to a non-specialist audience and to your peers. After the course,

---

[1] Royal Society Communication Skills Course http://royalsociety.org/training/communication-media/communication/

Plate 1 *The Blooming Snapdragons*: a play depicting the struggle of women scientists in the early 1900s. Case study 7.6.

Plate 2 Bee inventive: 'The Red Mason Bee', a school citizen science project. Case study 10.5. *Source:* http://www.seefurtherfestival.org/exhibition/view/bee-inventive-building-perfect-bee-house (accessed 3 August 2012).

Plate 3 (a) 'Fun with Fungi' for primary schoolchildren. Case study 10.2.

Plate 3 (b) 'Fun with Fungi' for primary schoolchildren. Case study 10.2.

Plate 3 (c) 'Fun with Fungi' for primary schoolchildren. Case study 10.2.

Plate 3 (d) 'Fun with Fungi' for primary schoolchildren. Case study 10.2.

Plate 4 (a) The Mobile Family Science Laboratory Afterschool Club. Case Study 10.3. Reproduced by permission of the Wellcome Library, London.

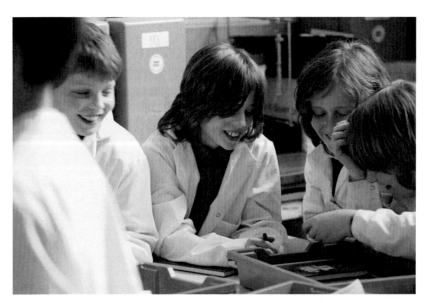

Plate 4 (b) The Mobile Family Science Laboratory Afterschool Club. Case Study 10.3. Reproduced by permission of the Wellcome Library, London.

Plate 4 (c) The Mobile Family Science Laboratory Afterschool Club. Case Study 10.3. Reproduced by permission of the Wellcome Library, London.

Plate 4 (d) The Mobile Family Science Laboratory Afterschool Club. Case Study 10.3. Reproduced by permission of the Wellcome Library, London.

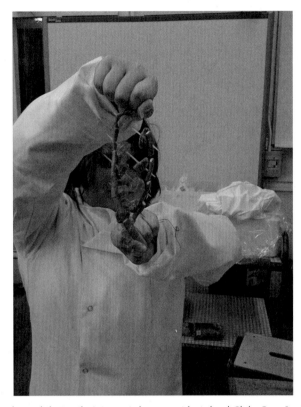

Plate 4 (e) The Mobile Family Science Laboratory Afterschool Club. Case Study 10.3.

Plate 5 The Bad Bugs Bookclub. Case study 7.8. *Source:* www.hsri.mmu.ac.uk/ badbugsbookclub (accessed August 2012).

Plate 6 Cafe scientifique, Aberdeen. Case study 7.7.

Plate 7 (a) Elements from the periodic table. Case study 7.3. A, B: Copyright 2011 Madeleine Shepherd.

Plate 7 (b) Completed Periodic Table. Case study 7.3. A, B: Copyright 2011 Madeleine Shepherd.

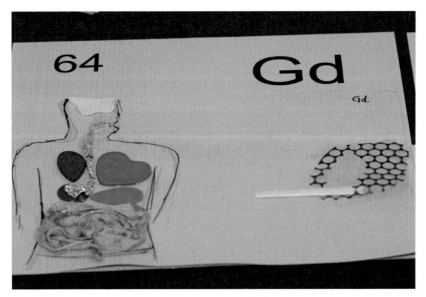

Plate 7 (c) Element from the periodic table. Case study 7.3. C, D: Copyright 2011 Matthew R. Farrow.

Blooming Snapdragons

A play depicting the struggle of women scientists in the early 1900s

Wednesday 14th July 2010

The Royal Institution of Great Britain

19:00

Followed by a post show debate "What makes a good scientist?"

Tickets cost £8, £6 concessions, £4 RI members

Further info @ www.rigb.org

www.jic.ac.uk

Plate 8 A flyer from The Blooming Snapdragons play described in Case study 7.6 that was also performed in London at the Royal Institution of Great Britain For more information about flyers see Chapter 6 section 6.9.

Plate 9 The 'Norfolk Science Past, Present and Future' event held at the Norwich Castle Museum. Case studies 4.1 and 5.2.

Plate 10 Building international links in Korea. Case study 7.4.

Plate 10 (*Continued*)

Plate 11 (a) 'Magnificent Microbes' at Dundee Sensations Science Centre. Case study 6.2.
Source: http://www.sgm.ac.uk/pubs/micro_today/pdf/081009.pdf (August 2012).

Plate 11 (b) Survival of the Fittest: an interactive exhibit that took place at the Forum in Norwich. This event was designed to celebrate the anniversaries associated with Charles Darwin by showing how microbes evolve rapidly in response to their changing environment. Case study 6.2.

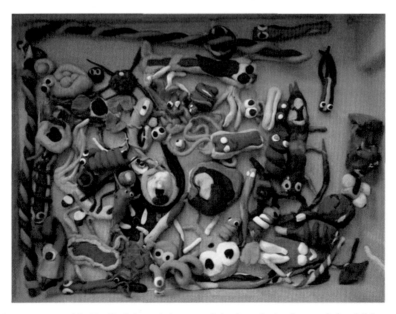

Plate 11 (c) Mobile Family Science Laboratory Selection of microbes made by children.

Plate 12 Dr Robert D. Wells (FASEB President 2003-04) addresses Vice President Richard Cheney. Flanking Dr Wells are Nobel Laureates (left to right) Drs Al Gilman (University of Texas Southwestern Medical School), Sid Altman (Yale University), Sherwood Rowland (University of California Irvine), Thomas Cech (Howard Hughes Medical Institute, President), followed by Mr Pat White (FASEB Legislative Officer), two staffers from Vice President Cheney's office, Vice President Cheney and the Honorable Bob Michel (former House Minority Leader). The meeting was held in the West Wing of the White House in the Roosevelt conference room; note the portrait of President Teddy Roosevelt above the fireplace and his Nobel Prize above the lamp on the far wall. Case study 7.9. *Source:* http://www.faseb.org/LinkClick.aspx?fileticket=6CJbV%2BQu%2Fl0%3D&tabid=390 (August 2012).

Plate 13 The images used by the children taking part in the SAW Trust activity. Case study 10.6. (a) Cyanidin molecule, reproduced by permission of Melissa Dokarry. (b) Aphid feeding on leaf by Volker Steger. *Source:* http://www.sciencephoto.com/. Reproduced with permission. (c) Lavender leaf SEM by Eye of Science. *Source:* http://www.sciencephoto .com/. Reproduced with permission. (d) Viola flower. Photo of viola flower by Andrew Davis, JIC Photography.

Plate 14 'Science in a Suitcase', Worstead Festival. Case study 7.5.

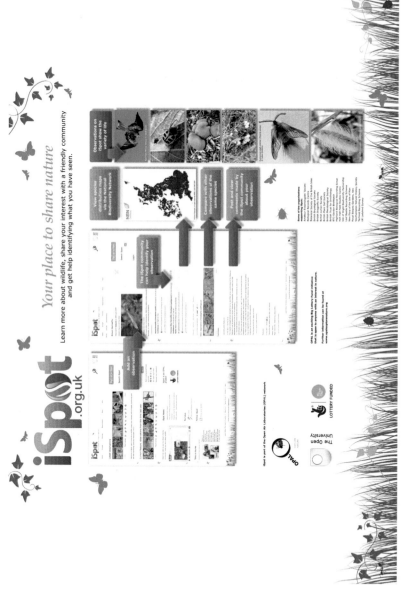

Plate 15 (a) The iSpot Citizen Science Project. Case study 8.5.

Plate 15 (b) iSpot and the *Euonymous* leaf notcher. Case study 8.5.

I became interested in the idea that science should be much more accessible. I felt that people were likely to go to talks about the history of their city or by a famous local author, but may feel excluded from finding out what scientists in their local community were researching. I decided to organise a series of public lectures that would communicate the world-class science being undertaken where I work at the University of East Anglia (UEA), not only across the different schools within the university but also to the wider community.

Developing the 'Talking Science' lecture series

In order to develop my initial idea, I discussed it with a number of work colleagues as well as staff from the university's communications office. The UEA communications office pointed out a lot of things I needed to consider and gave me contacts for internal and external publicity. These discussions enabled me to develop a proposal for a series of science talks that would be open to the public and held once a semester. I needed to get some funding for the lecture series, so I pitched this proposal to the Science Executive Committee at UEA, who agreed to provide a nominal budget to cover the publicity and a wine reception to follow each talk.

I set up a specific email address for the event and an IT technician helped me to create a mailing list and to manage enquiries.

The actual talks in the lecture series were to be given by different members of the Science Faculty to a general public audience. The talks aimed to be 45 minutes, allowing time for questions afterwards. The initial events were held at the university and the wine and nibbles reception that followed the event gave the audience an opportunity to talk to the speaker.

Running the 'Talking Science' lecture series

There was a large audience for the first pilot talk 'How Stable is Planet Earth?' about climate change, which was very encouraging. I then organised a series of three lectures to be held on the UEA campus, entitled:
* 'The Mathematics of Infinity';
* 'Will we ever cure cancer?';
* 'Cloning The Smell of the Seaside – and other amazing things that bacteria do to the atmosphere'.

Although 'The Mathematics of Infinity' was very well attended, I was disappointed with the number of attendees at the other two talks, especially 'The Smell of the Seaside', which was a very entertaining and informative lecture. I was hoping that audience numbers would increase if people attended a lecture on a topic that interested them and found it accessible and engaging, then they might attend other talks on subjects they knew less well. I had advertised the maths lecture through the Norwich Maths Society and realised that its success in attracting an audience had been due to this targeting of local interest groups.

Undeterred, I organised a series of three lectures for the next academic year, even though I would be on maternity leave for most of it, and hoped that attendance would improve. Although these talks did attract a reasonable audience, I didn't feel that we were really reaching non-university people, so I decided to move the venue to the Assembly House, a nineteenth century meeting venue which provided a beautiful and central city location for the talks and proved very successful.

Moving the venue from the university to a city centre location worked really well and lead to an increase in audience numbers.

Highlights of the 'Talking Science' lecture series

A key outcome for initiating this series of talks was to give the public an opportunity to talk to scientists on an equal footing. Throughout the series of talks it was clear that the audience always felt confident about asking the speaker questions.

Challenges with the project

It was more challenging to recruit speakers than I had anticipated. Initially I approached a couple of people I knew and asked them to deliver a lecture. I also asked the Heads of the Science Schools to recommend people or ask for volunteers. This tactic proved unsuccessful and I found the direct approach more fruitful. I now try to keep abreast of the research that is generating publicity across the Faculty of Science and I try to link UEA research with science which is making headlines.

Top tips

1 Location is very important; UEA is a campus university on the outskirts of Norwich. Although my initial thoughts were that it would be great to get people on to campus, it isn't so inviting on a cold dark February evening. I also wondered if the fact that people hadn't prepaid for a ticket to attend the lecture meant that they were less inclined to venture on to campus. The facilities on campus in the evening are very student orientated whereas a city centre location provides plenty of bars and restaurants.

2 The time of the event is also worth considering carefully; many university talks tend to run straight after work. If you are aiming at a public audience, you need to give people time to 'get home, turn round and get out'.

3 The title of the talk can entice or discourage people to attend, it needs to be informative and user friendly, for example 'Realistic modelling and the synthesis of talking faces' might put people off when the actual talk covered elements of film animation. This could have attracted a young and diverse audience if the title had been less daunting. It can be difficult to convince a senior scientist that a public friendly title does not mean that you are 'dumbing down' their subject.

4 It is really important to publicise your talk as widely as possible:
   - most universities or research institutes will have a press office that will publish a press release;
   - consider whether there is a campus newsletter or local papers that will run a piece on your event;
   - generate a mailing list from people who attended the first event, and use this to circulate information about future talks. Choosing a topical and crowd-pleasing first speaker will help kick this off; we chose climate change as our first topic as this was relevant to all people and also headline research at UEA.

5 A few emails can also go a long way. Local science centres may have mailing lists for their events and will be happy to advertise yours.

6 Local interest groups can be very useful. When organising a lecture on the plight of bumble bees, a few emails to local bee-keeping societies, naturalist groups and the council biodiversity officer generated a lot of interest. People tend to pass emails on to others they think might be interested.

Other important considerations

Safety is not an issue with this sort of event unless the speaker wants to include a demonstration that involves safety hazards or public volunteers. It is worth checking that the speaker is happy with the audio-visuals they are using. An external venue will probably have a technician present, which will be covered in your hire costs. When using university rooms, many lecturers will know the systems available, but if your talk is out of university hours, the presence of an audio-visual technician may be an extra cost, but this is worth the expense for peace of mind.

The demonstration lecture

The demonstration lecture, which uses live experiments to illustrate scientific concepts, has been in existence for over three centuries. Robert Hooke was appointed as 'Curator of Experiments' to the Royal Society in 1662, and offered to produce for each meeting three to four considerable experiments (Jardine, 2003). However, it is John Theophilus Desaguliers, assistant to Isaac Newton and also one-time Curator of Experiments at the Royal Society, who is credited with its popularisation (Taylor, 1988).

> Without Observations and Experiments our natural Philosophy could only be a Science of Terms and an unintelligible jargon.
>
> —*J.T. Desaguliers (1763)*

In 1799, The Royal Institution (RI) appointed Thomas Garnett, a trained physician, to deliver a popular course on experimental philosophy and also a more serious course of study on the same subject (Berman, 1978). The popular lectures were held twice a week in the afternoon and the more serious course of study three times a week in the evenings. Humphrey Davy was appointed as Garnett's assistant, and after Garnett's resignation in 1801, after an undisclosed dispute, Davy became Lecturer in Chemistry. The RI lecture theatre, the first of its kind to be designed for lecturing purposes, became the backdrop for the popular demonstration lectures designed first by Davy and then by Michael Faraday. Examples include the Friday Evening Discourses started in 1826 by Michael Faraday; they were superb examples of demonstration lectures and they were incredibly popular with the public. Faraday also instigated the Royal Institution Christmas lectures, still in existence today and he himself did 19 Christmas lectures between 1827 and 1860 (Royal Institution, 2011).

The modern demonstration lecture still has the power to be exciting and memorable and these types of lectures work equally well for a school audience. Case study 9.1 by Tim Harrison and Dudley Shallcross shows how demonstration lectures can be tailored to suit different school audiences combining both education and entertainment.

Case study 7.2 by Dr Stephen Ashworth shows how demonstration lectures can be designed for a very particular audience, that of the Women's Institute (WI).

Case Study 7.2

Portable Demonstration Lectures Illustrating Aspects of Research
Stephen Ashworth

Background
I applied to the Engineering and Physical Sciences Research Council (EPSRC) for funds for equipment to put on a series of demonstration lectures which would illustrate aspects of

my research.[2] The EPSRC stipulated that the activity should communicate EPSRC-funded research to the public. I proposed to develop one lecture per year over the course of the 3-year grant. Initially my audience was the Women's Institute (WI). This was driven by a desire to avoid complications of CRB checks, fitting in with the school curricula and having to leave work during the day to give my presentations. In addition, the audience is ready made: these groups have monthly speaker meetings. I got in touch with the Women's Institute and asked whether they would be able to advertise my demonstration lecture. They replied positively, but first I had to pass an audition. Every so often the WI runs an auditions day for potential speakers. Each performer is given about half an hour to give a slightly shortened taster of their presentation and the assembled audience is asked to give feedback. The feedback is then sifted by a committee and the successful speakers are listed (for a fee) in the WI yearbook. As a direct result of the auditions day I was asked to visit a local WI at short notice as the speaker that had been booked pulled out (short notice in this case being about six weeks). At the end of the presentation one of the audience came up to me and said 'Much better than hats!'

Design of the demonstration lectures
Each demonstration had to be designed so that:
- it was portable;
- there were no harmful residues after the demonstration;
- I could set everything up within one hour, give the presentation (about an hour) and pack up again as quickly as possible.

Portability meant that some experiments had to be done on a small scale, but a movie camera and a data projector (purchased through the original grant) enabled the small-scale experiments to be visible by all. Whenever possible I try to involve the audience by using volunteers to help with experiments or providing an activity that involved the whole group.

Safety considerations
Given that these are generally small-scale experiments that are under my control, I am careful to ensure my safety and that of audience members when working up demonstrations. The audio-visual requirements mean that there are a lot of trailing leads, so I ensure that these are taped down so there is no trip hazard. Fortunately there are no pungent smells produced in any of the reactions. I also use a laser in one of the presentations. The equipment I use is inherently safe as long as one does not stare into the beam and I am careful to minimise that risk.

Highlights and problems
- I have now done many of these events. Sometimes things fail spectacularly; sometimes something breaks or gets left behind. With experience, it is possible to gloss over the missing parts. Occasionally there is a fantastic audience and there is a good reaction, a real buzz. Other times it is really hard work from start to finish. I can usually tell when I crack the first (weak) jokes how things are going to go. The presentations are generally a combination of PowerPoint slides and demonstrations.
- There are some tricky demonstrations that don't always work, so I have taken videos of these and hidden them in the PowerPoint so that I can call them up if necessary.
- Every once in a while a prop breaks – sometimes I am able to fix it on site or at least jury rig it for the occasion. Every so often a key prop is left at home. This has happened to me fairly recently and I was able to hide the relevant slide in the presentation so the audience was none the wiser.
- If you are asking the audience to take part in a demonstration, any instructions have to be clear and the task relatively easy. One of mine involves sorting balls into different tubes to create a histogram. The balls are different colours and I have had some interesting (wrong) results from audience members who have not distinguished between some of the colours so well.

---

[2] http://www.uea.ac.uk/~c021/ppu/welcome.html

Moving forward

I received a second EPSRC grant to develop another three lectures in a 3-year period, but now I have moved away from lectures which hinge on an aspect of research, as this is no longer supported by the research council.

I have also started to branch out from interested adult groups and have begun to be more involved with schools. I have also given performances at SciFest Africa 2008, 2009 and 2011.

Personal gain

Confidence, experience, a pragmatic attitude to things that go wrong on stage and the possibility to indulge myself doing simple, enjoyable experiments over and over again!

Top tips

1 Practise, practise, practise . . . at least for the first one you do. I found at the start that I had a specific order in which I wanted to present the material, it was logical and made (at least to me) most sense. Having run through the presentation in order, it became obvious that there was too much 'dead time' in rearranging props. With a different running order that 'dead time' can be minimised.

2 The other aspect of the presentation that I practised was setting up and packing away. Don't neglect the time it will take for both of these activities. Try to minimise the amount of disassembly that has to be done. Disassembly will be a compromise, of course, given the amount of space available for transportation.

3 If at all possible use your own equipment. Lots of venues are very helpful 'Oh, yes – we have a screen/microphone/projector/camera.' It is only on site that you find that your plug does not fit their equipment or vice versa. I don't usually have the leisure to investigate the venue beforehand, so having my own equipment and knowing how it works together is a big advantage.

4 Do your first 'show' in front of a 'friendly' audience if possible. I was fortunate enough to have some colleagues sitting in who gave me lots of pointers and helped make my first presentations (and all subsequent ones) slicker.

7.2.2 A focus on museum and science centre communication

Museums enable us to explore culture through objects (Pedretti, 2008). Science museums grew out of private collections and objects in 'curiosity cabinets' during the Renaissance (Alexander and Alexander, 2008). These collections, gathered by wealthy individuals whilst travelling, included fossils, rocks, animal and plant specimens. They became important reservoirs of material for scientific research and have maintained this role into the twenty-first century. Botanical gardens first appeared in universities: Pisa in 1543 and Oxford in 1620. The first natural history museum, the Ashmolean, was founded in 1683 at Oxford University. This was followed 70 years later by The British Museum, opened in 1753, after Parliament bought the natural science collection of Sir Hans Sloane (Alexander and Alexander, 2008). The number of museums rapidly expanded during the nineteenth century. Sir Richard Owen, head of the natural history collection at the British Museum, which included Sloane's collection, campaigned for a new building to be located on a different site, and in 1881 the Natural History Museum, designed to be a 'Cathedral to

Nature' was opened. Richard Owen has been credited for making museums accessible to the Victorian public, but in the twentieth century they had a reputation for being dull and out of touch with a modern technological society. Museums have changed over time and have moved away from their image as dusty repositories of old collections. Objects are being placed in social contexts and there is more emphasis on interactive education and the discussion of societal issues (Pedretti, 2008).

Science centres have a different historical background and a different founding philosophy. During the seventeenth century, the scientific method was developed, based around Francis Bacon's idea of inductive reasoning. Bacon proposed a 'Museum of Discoveries', but it wasn't until 1794 that the first science centre, Le Conservatoire National des Arts et Metiers, was opened in Paris (Fors, 2007). In contrast to museums, science centres are more focused towards scientific laws and the process of discovery; they have interactive rather than passive exhibits. A forerunner to many of the modern science centres was the Children's Gallery, which was opened in the Science Museum in1931. The Gallery had working models with buttons to push and handles to wind and it inspired many to have careers in science and technology (Caulton, 1998). In 1969, the Exploratorium was opened in San Francisco, the brainchild of Frank Oppenheimer. The Exploratorium became the template for many other science centres across the world. Oppenheimer wanted people to 'discover' science through experimentation. Thus science centre exhibits are designed to be interactive; something will happen when the visitor manipulates the exhibit, which in turn invites further activity on the part of the visitor. The first UK science centre was the Exploratory, opened in Bristol in 1987, and then the Science Museum opened 'Launch Pad' which is still very popular. Subsequently there was an explosion of science centres opening across the UK funded by the Millennium Commission. However, science centres in the UK face an uncertain future, smaller ones began to close and in 2007, the House of Commons Science and Technology Committee began an inquiry into whether science centres should be publically funded. The main criticism was that the centres could not demonstrate longitudinal impact. Government funding was withdrawn from UK science centres, and many more have since closed, including our own Inspire Discovery Centre in Norwich. Another criticism levelled at science centres is that children are just 'playing'. The power of play in learning is not in dispute. As scientists we are curious, and we sometimes wonder 'What would happen if...?' We understand that useful insights can arise from those 'Friday afternoon' experiments where we set up an investigation because we were wondering 'What would happen...?'

Museums and science centres provide opportunities for 'informal learning' (see Chapter 4). The UK Association for Science and Discovery Centres[3] (ASDC) brings together over 60 different science organisations. These organisations include museums, science centres, learned societies, university

---

Table 7.2 Science centres in the UK.

Location	Name	Web address
Birmingham	ThinkTank	www.thinktank.ac/
Bristol	At-Bristol	www.at-bristol.org.uk/
Cardiff	Techniquest	www.techniquest.org/start/
Dundee	Sensations	www.sensation.org.uk/
Edinburgh	Our Dynamic Earth	www.dynamicearth.co.uk/
London	Launch Pad	www.sciencemuseum.org.uk/visitmuseum/galleries/launchpad.aspx
Newcastle	The Centre for Life	www.life.org.uk/

departments and environmental organisations. The benefits of the ASDC to these organisations include:

- lobbying and advocacy;
- bringing staff together;
- funding, PR and other opportunities;
- collaborative projects;
- raising profiles;
- advice and support.

A few of the larger science centres with their web addresses are detailed in Table 7.2. Their websites often contain resources for schools as well as online shops for science kits. If you are considering running your own activity or event, it is worth visiting the science centre websites.

The 'Taking Part National Survey of Participation in Sport and Culture'[4] (2011) showed that 47.5% of adults and 68.8% of children had visited a museum, gallery or archive. Those citizens who live in rural areas are just as likely to visit these places as those who live in urban areas. However, those in a higher socioeconomic group visit more frequently. In 2010, the British Museum had nearly six million visitors, and the Natural History Museum just under five million. Museums and science centres are excellent venues for holding science events. Staff are experienced and well trained; they are valuable sources of advice regarding both design and accessibility of activities within events. In fact, you can extend your own event by directing people to the existing exhibits, or perhaps whole events can be themed around a few different objects or objects on loan. For example, in 2007, I ran an event at the Norwich Castle Museum called 'The Skin You're In'; the idea for the event came from an object which the museum had been loaned. This object was a book bound in human leather. Facilities are generally excellent and museums and science centres often have extensive mailing lists that allow you to market

---

[4] The Taking Part Survey is commissioned by the Department for Culture media and Sport in partnership with the Arts Council, English Heritage, Sport England, and the Museums, Archives and Libraries Council. http://www.dcms.gov.uk/images/research/Taking_Part_Y6_Release.pdf http://www.guardian.co.uk/news/datablog/2011/feb/23/british-tourist-attractions-visitor-figures

your event. There are a number of examples of events and activities in this book which were held either at museums or sciences centres:

- 'In Your Element' (Case study 7.3) was held at The National Museum of Scotland in Edinburgh;
- 'The Oomycetes' (Case study 4.1) was held at the Norwich Castle Museum;
- 'Living or Lifeless' (Case study 10.7) was held in conjunction with the Norwich Castle Museum;
- 'Magnificent Microbes' (Case study 6.2) was held at the Sensation Science Centre in Dundee.

Case Study 7.3

In Your Element!
Elizabeth Stevenson

Background
To celebrate International Year of Chemistry 2011, one of the activities which I designed was called 'In Your Element!' This activity involved creating a large periodic table of the elements and was delivered, over a weekend, in the National Museum of Scotland in Edinburgh. The Museum is a very popular venue for visitors, both local and international, and is particularly popular with family groups, therefore we had no difficulty in completing the periodic table within the 2-day duration of the event.

'In Your Element' activity design
Participants were invited to select an element, find out some basic information about its physical and chemical properties and uses by:
- chatting to facilitators;
- using the fantastic book *The Periodic Table, Elements with Style* (created by Basher and written by Adrian Dingle).

Participants were then invited to interpret the properties of the element using arts and craft materials supplied by us. The activity was facilitated by a combination of staff (me), postgraduate students from the University of Edinburgh School of Chemistry, and members of an Edinburgh-based craft group called Craft Reactor. The completed table measured approximately 4 m × 2 m.

Funding and resources
I raise funding for public engagement activities from various sources. Craft materials were purchased from craft stores and in addition we used items such as lentils, beans and aluminium foil.

Event highlights
The periodic table of the elements is fundamental to chemistry and its layout is immediately recognisable and attracted much attention, even from visitors who did not craft an interpretation of an element. The activity provided a useful platform in which to highlight:
- the role of chemistry in our everyday life;

- the fact that every solid, liquid and gas is composed of elements from the periodic table (more often in combination with other elements);
- the systematic arrangement of the elements in the periodic table.

By approaching the activity from an arts and crafts perspective, we broadened the appeal of the activity and we were delighted about the variety of interpretations of the elements and their properties – from very literal to extremely abstract.

Safety considerations

- We were located in a spacious open area in the venue.
- Museum attendants were in close proximity and all children were accompanied by parents, so CRB checks were not required in this instance.
- The Museum has strict guidelines about the behaviour of facilitators, i.e. no eating, drinking or using a mobile phone while facilitating an activity; this helped to ensure a high level of professionalism during the event.
- We provided permission slips for parents to sign before we took photographs.
- It was important to carry out a risk assessment on the activity regardless of the fact that we were not doing laboratory experiments. I viewed the risk assessment procedure as a useful planning tool.
- By carefully considering the activity, we were able to plan the resource requirements and the layout effectively. For example, we used rounded child scissors and did not use sewing needles for our activities. We used PVA glue in squeezy bottles (to minimize the risk of glue being spilled), we did not use loose glitter as it tends to be very difficult to sweep up, and we definitely did not use finger paints as I am sure the Museum would not appreciate little coloured fingerprints decorating their artefacts!
- We carefully considered how participants would interact with the exhibit and the craft tables. All of the facilitators wore 'In Your Element!' sashes so they were easily distinguishable from the participants.

Activity evaluation

An important component of any event is to evaluate its effectiveness so that we can learn from the experience.

We did not use formal questionnaires, but were able to make observations on the way in which participants engaged with the activity:

- How many people participated in the activity?
- What was the demographic of the audience – family groups, teenage groups, older children, adults?
- Did family groups and other groups work together or individually?
- Were the craft pieces representative of the elements?
- Did the participants engage with the facilitators?
- Did the facilitators enjoy and engage well with the activity?

I was really pleased with the way in which participants engaged with the activity and put considerable effort into their interpretation of the elements. Participants readily engaged with the facilitators, asking questions about the element which they had selected, about the periodic table and about chemistry and the research areas of the postgraduate facilitators. Most groups spent 30 minutes or more at the activity. Participants valued the experience of contributing to a larger work and many families and individuals took photographs of their exhibit, particularly with cameras on mobile phones. The postgraduate students had received a briefing about the activity and engaged well with the participants. This particular group of postgraduates had not previously been involved in science communication activities, they thoroughly enjoyed the experience and were motivated to be involved in future events. As a bonus, the completed periodic table can be dismantled into two carrier bags! I was able to reconstruct it on my return to the School of Chemistry and I invited staff and students to a viewing. I was very pleased with the response, particularly from PhD students and administrative staff.

Event improvements
I would definitely run the activity again. I have experience in running science communication events and had given careful consideration to the activity and the audience beforehand. I would extend the activity by producing supplementary information on some of the more obscure elements to complement the information supplied by the books and facilitators. In addition, it would be interesting to deliver the activity with a different emphasis, perhaps to a craft group or other special interest group or with an adult audience.

Personal gain
I personally found it very satisfying to see the periodic table being populated with imaginative interpretations of the elements and that participants, including Museum staff, really engaged with the activity. Because of the informality, visitors were confident to chat about their knowledge of chemistry, ask questions and spend time talking about chemistry, which was the main purpose.

Top tips
1 Choose the venue carefully. The museum marketed our activity and in addition, because the museum is a public venue with no entrance charge, we were assured of an audience for the event.
2 Think about the materials used for the activity including any safety implications associated with them (e.g. child-proof scissors).
3 Think about housekeeping during the event and the role of the facilitators. Perhaps use T-shirts or sashes to identify those helping.
4 Be clear about the purpose of the activity and ensure the facilitators have a shared understanding of the purpose.
5 Relax and enjoy the event, and address with a smile any issues that arise.

Websites
http://www.chemistry2011.org/
http://craftreactor.com/

7.2.3 A focus on designing large-scale public events
If you already have some experience in taking part in public science communication events, you might want to take that next step and plan your own. This is good, as you can control the theme and provide opportunities for colleagues to showcase their research to a public audience. There are some points to consider:
- be clear about the aims and objectives of your event;
- ensure that the activities within the event conform to the aims and objectives, and are not just a random set of activities thrown together;
- be clear about how you will evaluate separate activities and the event as a whole;
- get as much help as you can from others to design and run activities;
- ensure you have a suitable level of funding;
- ensure that you have considered safety implications.

Throughout this book there are sections that cover aim and objective setting, design and delivery (see Figure 6.4).

Dr Sheila Dargan helped to organise a science communication event, but the fact that it was attached to a neuroscience conference in Korea was an additional challenge (Case study 7.4).

Building International Links Through Scientific Engagement

Sheila Dargan

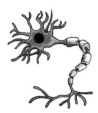

Background

Public engagement events that involve 'hands-on neuroscience' are now widespread in the UK, thanks to organisations such as the British Neuroscience Association (BNA), city neuroscience groups and university initiatives that support widening participation and science outreach. However, international neuroscience communication initiatives are much less common.

In 2008, I was privileged to be invited to contribute to the organisation of a 1-day science communication event to be held alongside the 3rd Korea–UK joint symposium on Neuroscience in Seoul, South Korea. The organisation team comprised seven UK-based researchers (myself, Kei Cho, Graham Collingridge, Jihoon Jo, Daniel Whitcomb, Heon Seok and Bryony Winters), a team of researchers based at the universities in Seoul and the Korean Foundation for the Advancement of Science and Creativity (KOFAC).

The event design

The purpose of our event was to boost public engagement with neuroscience in Korea and build positive relationships with Korean researchers. When running any type of science communication event it is important to consider the short- and long-term impact, so think about why you are putting on the event and what it is you want to achieve. The neuroscience communication day comprised interactive zones, public talks on neuroscience topics (e.g. memory, schizophrenia), a poster display and information stands. The event attracted a profoundly mixed audience of over 500 people, including families, pupils from local schools, university students and professors. It particularly attracted students of university age who would be at university if they were not doing their compulsory two years national service. These students really valued the chance to attend the free public lectures and find out more about the opportunities available to them both in Korea and the UK.

Interactive zones were situated in a large foyer area and a series of public neuroscience talks was held in a side room with audio-visual facilities. Each of the interactive science zones was manned by a pair of researchers (one from the UK and one from Korea) to promote interaction between international researchers as well as providing visitors with the opportunity to interact in the language of their choice. Hands-on activities included:
- a giant brain poster on which visitors could stick neurons they had made from play dough and scoobies;
- a 'tiny-tots area' with colouring activities;
- computer-based cognition tests (aimed at teenagers and adults);
- brain models.

Many of our hands-on activities (e.g. the Stroop test) were available in English and Korean so that visitors could perform them in either language.

More than 20 posters were displayed, which were created specifically for the event by PhD students and postdoctoral scientists, to explain the scientific mechanisms underlying common neurological disorders/diseases to a lay audience. Some of the posters, which focused on research techniques, were purposely situated by a confocal microscope used by visitors to look at fluorescent neurons. Researchers, again both English and Korean, were present in this area to promote discussion on current neuroscience research and answer any questions the visitors had. Freebies were also available at activity stands (e.g. model brains and pens) for visitors to take away.

Preparation for the event

A lot of preparation was required ahead of the event. We, the UK-based research team, held several meetings to discuss the types of hands-on activities we wanted to include, the type of audience we were likely to attract and which responsibilities each member of our team was going to take on (e.g. preparing posters, buying materials, obtaining funding, liaising with teams in Korea, etc). We also discussed more logistical issues such as how we were going to get our materials over to Seoul (e.g. what could we ship, what could we carry with us and what could we buy once there?). The Korean-speaking members of our UK-based research team spent a huge amount of time organising logistics with the events company in Seoul (e.g. booking the venue, arranging for delivery of poster boards and tables, event marketing) and our professors (Kei Cho and Graham Collingridge) organised all of the funding, including money to cover travel and accommodation.

The British Neuroscience Association and UK city neuroscience groups were represented by Anne Cooke (Bristol Neuroscience), who provided information about neuroscience research and engagement in the UK. There were also representatives from Korean neuroscience organisations, which enabled visitors to discuss local opportunities and initiatives.

We worked very well together as a team because each of us brought very different skills to the group and contributed equally. My main contributions to the event were to:

1 advise the UK-based team on the type of neuroscience activities that work well with lay audiences;
2 organise the hands-on activities and purchase material for the interactive zone;
3 run a training session for the bilingual Korean researchers (once in Korea).

We decided to fly over to Korea a few days before the actual event to allow time to recover from jet lag, familiarise ourselves with the venue and iron out any last-minute problems. Running the training session prior to the event was particularly helpful because, in addition to showing the Korean researchers what we wanted them to do, it provided the opportunity to get to know one another, ask questions and work out how to put the brain models together! This definitely increased the confidence of everyone involved and helped us function as an effective, friendly and enthusiastic international team on the day. Visiting the venue in advance was especially good for us non-bilingual English volunteers because it meant we could deal with silly things like finding our way around the building (and locating the toilets), which saved time on the day of the event.

Evaluation

- Being the first event of its kind in Korea, we wanted to generate a friendly, informal atmosphere and were reluctant to bombard visitors with feedback forms. We therefore decided to collect as much feedback as we could in an informal manner (making notes during the day on any comments we overheard). Based on the comments we collated at the end of the day the event was very well received.
- It is a good idea to retain any feedback you collect and write a brief reflection on things that went well and things that could be improved so you can come back to these notes if you want to run a similar event in the future (because you will have forgotten the details by then).

- At our event, university-age students and adults (without children) spent a lot of time at the information stands and the public talks, whereas the adults with children and the school groups spent the majority of their time at the hands-on activity stations. This kind of information can help you target activities to your predicted audience.
- Our event attracted an unprecedented interest from the national media with 15 articles in different newspapers and coverage on national TV, which is additional evidence of its popularity and impact.

Based on the overall feedback we received, we could not have done much to improve the broad portfolio of activities we put on, but if we wanted to attract more local schools we would reconsider the timing of the event and publicise it to schools much earlier in the year, so they can plan it into their schedules. The venue we chose was appropriate for the event. It was a good size, had good audio-visual facilities and IT support, and was readily accessible to our target audience. These are all important factors to consider before selecting a venue.

Top tips

1 If you have enough people willing to participate, I would strongly advise recruiting more volunteers than you actually need. This allows you to rotate your volunteers around different activities, which stops them getting bored. Having extra people around will also enable you to schedule breaks so your volunteers can go and rest for half an hour or so to re-energise.

2 If you have enough funding, it is also worth buying bottles of water for your volunteers, because talking constantly is thirsty work! If you are not paying your volunteers, it is nice to think about giving them something small as a thank you for helping out – maybe a souvenir of the event, an official certificate or a free lunch? This will of course depend on your budget.

3 Stationary or fluffy bugs you can stick on computers, are very good attractors for getting visitors to approach your stand and if you can stamp a web address on your free items the visitors can look up information about your organisation at a later date (personally I always keep good freebies, whereas leaflets tend to get binned quite readily!).

Personal gain

I have always enjoyed participating and organising science communication events in the UK and this opportunity presented additional challenges, for example training Korean researchers in strategies for engaging lay audiences with neuroscience so they can run their own successful events in the future. I did have some concerns that once the training session was over, those of us who did not speak Korean would have little to contribute on the day, but to our delight the visitors were overly keen for us to speak to them in English. Some of the visitors were bilingual and wanted to talk with us directly about neuroscience research and engagement in the UK. Others wanted us to speak to them in English (occasionally with an instantaneous translation into Korean which was possible thanks to our wonderful bilingual helpers), just to practice listening to native English speakers.

The most beneficial outcome for me was that I was invited to give a short talk on my own research at the main conference. This was obviously an honour and a very valuable experience for me because it was my first oral communication at an international conference. So, why not get involved in science communication – you never know what opportunities it will provide!

7.2.4   A focus on festivals

Festivals are generally big affairs, with several organisations working in partnership to deliver them, for example the Cheltenham Science Festival have a full-time team which organise this annual event in the calendar. There is a remarkable array of festivals throughout the UK, not only in big cities but also in smaller towns (Table 7.3). Many of them are held in March, to coincide with National Science and Engineering Week, but as you can see from Table 7.3, you can go to one at almost any time of the year as long as you are willing to travel!

The British Science Association festival is the longest running festival; it has been in existence since 1831 and changes its location in the UK every year. It has hosted many famous debates, including one held in Oxford in 1860 where Thomas Huxley and Sam Wilberforce argued over Darwin's theory of evolution by natural selection.

Festivals require many separate events and activities, so there are a variety of opportunities for involvement with different levels of commitment. Many will also run a school programme alongside the public programme, so there should be something to suit all. Festivals are an ideal way to get started with science communication. My own public science communication portfolio began after I organised an event for the 2006 British Science Association Festival. Similarly, Dr Ken Farquhar (Case study 7.5) used the schools programme from the same event to develop his innovative street theatre shows.

Science festivals incorporate many different genres of art and expression, including (but not limited to), stand–up comedy, theatre, art, dance, poetry, readings, craft activities and films. All will have a science focus (Buckley and Hordijenko, 2011). They are not just a UK phenomenon, but occur throughout the world and they draw scientists and citizens together in face-to-face communication.

7.3   Information through conversation

Dialogue with the public over science issues is seen as essential if new technology is to be adopted and if trust is to be maintained. Scientists can take part in deliberative panels and citizen juries; however, these types of activity tend to be organised by the research councils and other grant awarding bodies. Scientists are invited to attend. It is unlikely (but not impossible) that you would chose to set up your own jury or panel, if you were beginning your science communication journey.

There are a variety of other ways in which direct face-to-face dialogue can be initiated; it does not have to be within a formal deliberative panel, or within a citizen jury. Other imaginative ways aimed at providing an atmosphere conducive to discussion are:

- scientific plays;
- Café scientifique;
- book clubs.

Table 7.3 Science festivals in the UK.

Festival	Organisation	Location	Timing	Link
British Science Festival	British Science Association	Various cities around the UK	September	www.britishscienceassociation.org/web/
Edinburgh Science Festival	Edinburgh International Science Festival	Edinburgh	March-April	www.sciencefestival.co.uk/
Glasgow Science Festival	University of Glasgow	Glasgow	Summer	www.glasgowsciencefestival.org.uk/
Cambridge Science Festival	University of Cambridge	Cambridge	March	http://comms.group.cam.ac.uk/sciencefestival/
Cheltenham Science Festival	Cheltenham Festivals	Cheltenham	June	www.cheltenhamfestivals.com/science
Oxfordshire Science Festival		Oxford	February-March	www.oxfordshiresciencefestival.co.uk/
Moray Science Festival	University of the Highlands and Islands Moray College	Moray College	March	www.moray.ac.uk/moray-college/about/science-festivals.php
Newcastle Science Festival	Partnership between different groups	Newcastle	March	www.newcastlesciencefest.com/
Festival of Science and Technology	City of York Festivals	York	March	www.yorkfestivals.com/metadot/index.pl?iid=2749&isa=Category
Inverness Monster Science Festival		Inverness	June	www.monsterfest.co.uk/
Fife Science Festival	Partnership between different groups	Fife	March	www.fifesciencefestival.org.uk/about.html
Dundee Science Festival		Dundee	October-November	www.dundeesciencefestival.org/
Otley Science Festival		Otley	November	otleysciencefestival.co.uk/about/
Kent Festival of Science		Canterbury	July	www.sciencefestival.org.uk/
Manchester Science Festival		Manchester	October	www.manchestersciencefestival.com/
London Science Festival		London	October	www.londonsciencefestival.com/
Wrexham Science Festival		Wrexham	July	www.wrexhamsf.com/en/

Of course many of these activities can and do form part of much larger science festivals. Scientific plays are certainly a big undertaking and will involve considerable planning and a large team. However, Café scientifiques and book groups are more manageable for the beginner.

Another example of the public getting directly involved with scientific research is citizen science (Section 8.7). However, as many of these projects tend to be done remotely, perhaps using data collection websites (e.g. iSpot), we have discussed these in more detail in Chapter 8.

7.3.1   A focus on science in the theatre

Scientists are aware of the innate drama in science. There are races to finding results – such as solving the structure of DNA; rivalries – such as that between Isaac Newton and Gottfried Leibniz; as well as close partnerships – such as that between Marie and Pierre Curie. Book authors have captured this drama and used it as central themes in their literature. However, playwrights have also recognised the inherent, powerful appeal of watching scientists and their scientific endeavour, breakthroughs and discoveries, as well as the impact these have on society. Drama is a key component of the theatre and it is synonymous with entertainment. There are several famous plays which feature science, one of the most significant is Karel Capek's *RUR: Rossum's Universal Robots*, which was performed in 1921. In this play the term 'robot' was used for the first time. The plot of the play is simple and has been reworked in different forms ever since; robots are created, which while initially serving mankind, eventually rise up and destroy the human race.

The theatre can also be a powerful medium for educating and communicating with the public. As Djerassi (2007) asks, 'Why not use drama to smuggle (with a substantial dose of theatricality) important information generally not on the stage into the minds of a general public?'

Science can be portrayed in drama in different ways and in different subgenres of theatre. Hook and Brake (2010) recognised that there is either theatre that is demonstrating and communicating science or theatre that is drawing on science and scientific principles to create drama. Barbacci (2004) also recognised that there were distinct subgenres and I have reinterpreted his classifications and added my own analysis. I agree that Science in the Theatre can be divided into subgenres as it is apparent that science crops up in theatre and in drama in a number of different guises. What is also clear is that particular pieces of drama such as plays, musicals or physical theatre can often sit comfortably within more than one subgenre. So my description of these subgenres is as follows:

The Science Performance – is an uber-lecture with science. These shows are used to educate by demonstrating scientific principles and scientific theories in a visual and entertaining way. This type of science performance can translate easily into street theatre.

The Science as Drama – the scientific breakthrough, discovery or principle is the basis for the drama.

The Ethical Science Drama – the ethical issues surrounding science discoveries or breakthroughs are the basis for the drama.

The Science in Drama – have some scientific themes or content but the science is a supporting role.

The Scientist Drama – the decisions, the relationships and the life of the scientist is the basis for the drama.

Science as performance

There is little consensus about how frequently science appears on stage. Stephen-Barr suggests that Michael Frayn's play *Copenhagen* was the start of a surge of new plays about science (Shephard-Barr, 2006). But Djessari (2007) refutes this idea stating that 'the number of commercially produced plays is small, and there is no evidence that it is growing.' In 2011, I attended the Edinburgh festival fringe. This festival takes place every year throughout August and it is the ideal venue to find and experience these science drama subgenres. It is one of the largest arts festivals in the world and it comprises a wealth of theatre genres including cabaret, comedy, children shows, musicals, operas, dance, events and theatre. Based on my own experience as a visitor to the festival in 2011, and drawing on recent reviews of science drama, it is a clear that examples of each science-drama subgenre can be found, although some are easier to locate than others. Sometimes it can be challenging to place dramas neatly into one subgenre. In the following descriptions, the decisions on where to place the dramas are ours, but you are free to disagree. In addition, we have also tried to indicate where scientists have had an input.

The 'science performance' was clearly evident in the children's section with Marty Jopson as the one-man presenter of *Inventions Going Bang*, an interactive science show with bangs, mess and explosions. There was also the free *Mad Science Dangerous Family Show* delivered by Out of Control production where the drama is the live experiment undertaken with comedy elements and slapstick. Interestingly both shows were found in the children's section and were aimed at a family audience. The shows clearly set out to amuse and entertain, but they also aimed to educate and facilitate understanding by demonstrating scientific principles. A quick search on Google using 'science' and 'entertainer' as search terms brings up a plethora of different entertainers who provide visual science shows.[5] Some of these performers are actors who have found science as a genre that they can work into a show. Others have trained as scientists and channelled their passion and energy into entertaining others using physical theatre. One such scientist turned street performer is Dr Ken Farquhar (Case study 7.5).

---

[5] A website that houses a directory of these science performers can be found at the following website http://www.scients.co.uk/performers

Science in a Suitcase
Ken Farquhar

Background

In 2006 I was involved in the schools programme for the British Science Association festival in Norwich. This presented an ideal opportunity to take science into locations and to audiences that may have never experienced science, performance theatre and certainly not science theatre. Although the focus of the festival was in Norwich, my ambition was to take the festival to other parts of Norfolk. I drew upon my own experiences in science outreach and as a street performer, converting the suitcase full of props and my six foot unicycle with fire torches into a series of very 'experi-mental', exciting and comedic routines.

Development and funding of the project

I collaborated with Ian Walker, another street performer who had an amazing likeness to Albert Einstein and the ability to produce some of the most amazing 'Heath Robinson' props that are were so bad they were good. Most of the props were either specially designed for the shows, off-the-shelf toys or modifications. There are plenty of places on the web where you can draw inspiration. A lot of time was spent researching science content for the show and improvising with props before a script was constructed. Having established a balance of material and put it in a running order, we tried to link each part so the routines ran into each other smoothly. We set about learning the script, running through material before three days of final rehearsals.

Before an application of funding could be made, letters of support and agreements in principle were needed from all the venues. I received some great advice from the Norfolk County Council arts advisor[6] and was pointed in the direction of the Arts Council's grants for the arts which funds short tours. The performances were funded with agreements in principle from venues or local organisations[7]. Other small grants were also applied for[8] and support in kind came from the host organisations,[9] as the performances were part of their marketing campaign. The funding and revenue paid for props, costumes, some marketing, show development, artist's fees, rehearsal, overheads and evaluation. Prepublicity came from the Norwich City Council, the British Science Association Festival, South Norfolk Council and Norfolk County Council. Post-publicity included coverage in the Eastern Evening News, Eastern Daily Press, Radio Norfolk and BBC News 24.

Delivery of Science in a Suitcase

Science in a Suitcase toured for three weeks in September 2006 performing two to three 1-hour theatre or two to three 45-minute street shows per day. The tour included

---

[6]Mary Muir, Arts Officer Norfolk County Council, email mary.muir@norfolk.gov.uk.
[7]South Norfolk Council, Sea Change Arts, Profit Share From Norwich Arts Centre &, Sheringham Little Theatre
[8]Norfolk County Council small arts fund, Great Yarmouth Borough Arts Scheme
[9]BA Festival, UEA, Norwich City Council

performances in two theatres, five street theatre venues, a sport centre, a country park and even a stately home. With respect to the audience numbers, the Norwich Arts Centre was sold out and Sheringham Little Theatre was nearly half full. The country parks attracted mainly families. At weekends our street show supported other activities which also happened to be in the street and had been organised for various schools (Lab in a Lorry) so we managed to pick up passing audience members.

Evaluation

There were two distinct audiences; those who were watching the performance and those that were funders or potential funders for future shows. I tried to get a cross section of feedback inviting various contacts to review the show as well as the organiser of each venue. It was difficult to get direct feedback from members of the public unless they were quoted in the media. Since the introduction of free survey databases, I try and get funders to fill out my online survey and in the future will look at social networks to get feedback from members of the public.

Benefits from the project

This project produced a high-quality physical theatre show with a science focus that was accessible to and engaged diverse audiences in a variety of venues. Performances drew crowds in areas unaccustomed to science street entertainment. The show transferred to the stage selling out for shows in small regional theatres. The show was the highlight of a residential week for Sea Change Trust who were working with a small group of 'hard to reach' young people. There was keen interest from many media outlets during the 2006 British Science Association Festival and I was invited to take part in the best of festival showcase and to appear on www.ScienceLive.org. Positive feedback from all the organisations involved with this project has been overwhelming. The publicity, resources, internet links, reviews and DVDs, etc., has generated interest from other arts organisations and educational establishments. For example, since this project I have been working with maths buskers helping their participants develop their street theatre skills.

This has created a lasting legacy for the activity, as *Science in a Suitcase* has developed, improved and diversified its street science activities.

Problems to overcome

- Performing so many shows was a strain on the vocal chords. The acoustics in theatres vary, but on the street background noise can force you to raise your voice, which can appear aggressive to the audience. I have since purchased a headset radio microphone and amplification system and converted it for street use.
- I think working with a director may have developed the show a lot faster by speeding up decisions about what is/isn't funny science, and technical positioning on stage. It would have been interesting to have had a third party view the material and this is something to consider in the future.

Safety issues

As a member of Equity[10] I receive free public liability insurance up to £10 million. As a science outreach communicator working in schools I already had a CRB. This requirement has increased since 2006, but a CRB is not necessarily a prerequisite for the many different organisations that book the street show. Safety issues have been written into a risk assessment and method statement. As a performance with lots of props and experiments, it is essential to be aware of potential hazards around you. I try to think about the consequence of each part of the show on my work colleague, the audience and innocent passers-by.

---

[10]www.equity.org.uk

Websites
Making props
http://www.instructables.com/
http://makerfaire.com/

Evaluation
http://www.surveymonkey.com/s/KH9WZSY

Maths buskers
http://www.mathsbusking.com

Top tips for performing street science activities

1 Practice your routines on friends and family before hitting the streets.
2 Practice and refine what you are going to say before you say it. Be happy, confident and develop opening interactions to build your crowd. Initial lines that flatter or are intriguing (e.g. 'You look like an intelligent family' or 'When was the last time you used an appliance of science?') can lead to further discussions and build interested audiences.
3 Break your presentation(s) into a series of easy to remember narrative journeys.
4 Try to hold your audiences interest for a short time at first and slowly build this as you gain experience. Some routines naturally develop the curiosity of a larger crowd; others ebb and flow (away). Play with the presentation order, building to a grand finale – you need to sell this as something well worth hanging around for.
5 Location can be difficult and often surprising. Look for large areas where people meet/shop. You can do impromptu shows pretty much anywhere, but may need to seek permission from the local authority. Again, experiment.
6 Stay safe, consider potential hazards, for example pedestrians, vehicles, animals, babies in pushchairs and even individuals behaving strangely or aggressively.
7 There are many reasons a show doesn't work; don't take it personally, take a break and try again later.
8 Be aware of the weather – no one likes to stop when it's wet or very hot. When it's hot try working in the shade or inviting your audience in to the shade.
9 Performing is an iterative process. Reflect on where you were successful and build on it. Remember a street show doesn't just happen – it takes a lot of time to perfect.
10 Have someone watch you in action and give you feedback.
11 Attracting a crowd for a street show is a difficult skill which takes a lot of practice. Street performers take years to establish their act and are not confined by their material. The challenge is to balance the science content with an entertaining performance. This may change depending on the venue, for example compare the audience of a science festival with one at an arts festival like Glastonbury. I recommend experimenting with style, content, type and length of show for each different audience.

The Science as Drama and the Ethical Science Drama

The Selfish Gene: The Musical also played at the 2011 Edinburgh festival. This new musical was written by Dino Kazamia and Jonathan Salway with music from Richard Macklin, and is a clear-cut example of 'science as drama'. As a premise, successfully setting Richard Dawkins popular science book *The Selfish Gene* as a musical theatre production may appear unlikely, but it worked. This musical illustrates the science of genetics explained by a Professor of Genetics (most probably Dawkins) and illustrated by the slightly dysfunctional Adamson family through physical theatre, drama and song. The family interactions introduce the audience to the evolutionary ideas of the battle of the sexes, altruism, and the selfishness of genes and their urge to survive. This musical played to critical acclaim from reviewers and the public.

Carl Djerassi is a professor of Chemistry at Stanford University. He has written several dramas that he defines as 'science-in-theatre' but they can also be viewed as 'science as drama'. These plays include *Oxygen* (2001), written in collaboration with Nobel laureate Roald Hoffman, that centres on the discovery of oxygen and the centenary of the Nobel Prize in 2001. In addition he wrote *Calculus* (2004) based on a dispute over the invention of calculus, between Newton and Leibniz. One of his most recent plays *Phallacy* focuses on the different approaches used by scientists and art historians to date and provide provenance for an art object, in this case a putative Roman bronze in a major European museum (Djaressi, 2007). Djerassi acknowledges the difficulties of being a scientist and an unknown name as far as the world of theatre is concerned. He also recounts how *Oxygen* has been translated into nine different languages and has had more than 30 independent productions, but he acknowledges that the majority were in university theatres. *Oxygen* has entered other literary realms; it has been released by the publisher, Wiley, and badged as a science history book written in 'all dialogic form' (Djerassi 2007)[11].

Djerassi also wrote *An Immaculate Misconception* (1997) which introduced the audience to the biomedical science known as intracytoplasmic sperm injection or ICSI. This is a fertilisation technique whereby a single sperm is directly injected into a woman's egg under the microscope, followed by reinsertion of the embryo into the uterus. ICSI is demystified, portrayed on a video screen and also given a human element. This play neatly crosses subgenres as it also highlights the ethical issues inherent in IVF techniques. Another play that crosses subgenres is *Copenhagen* written by Michael Frayn and based on a meeting between two physicists, Niels Bohr and Werner Heisenberg in 1941. It portrays the science of quantum mechanics, the uncertainty principle and the military consequence of this research – nuclear weapons. The science is firmly centred within the play, but the drama also crosses genres as it raises ethical questions, for example the morality of scientists and whether scientists were acting unethically when they developed nuclear weapons.

Science in Drama

A play that highlights the impact that scientific discovery can have on society is Brecht's *Life of Galileo,* which is a play that touches on the discovery of the scientific evidence that showed that the Earth revolved around the Sun; the heliocentric model of the solar system. This play also focused on the impact of this evidence because it was viewed as a heresy by the Catholic Church.

---

[11] http://www.djerassi.com/ provides details and information about Djaressi's work.

This discovery also impacted on the life of others, including the failure of Galileo's daughter's marriage because of his teaching. However, this play also exposes Galileo as a pragmatist who is driven by self interest as he renounces his belief in his scientific discovery after he is threatened with torture by the Catholic Church. Galileo, is also the subject of another new play written by Nic Young and shown at the Edinburgh Fringe, *The Trials of Galileo*. This play also focuses on the struggle of science against the church with the spotlight on the trial of Galileo. The Catholic Church condemned the book he had written, *Dialogue on the Two Chief World Systems*, which outlined his heliocentric model of the solar system. As this play focuses on a specific point in the life of a scientist, it could also be viewed as a play that fits into the 'scientist as drama' subgenre.

When it comes to 'science in drama,' the Edinburgh Fringe festival offered the gothic horror *Frankenstein* as well as a couple of dramas that had infectious diseases as a supporting role. These included *The Infection Monologues*, which explores the reality of living with HIV today. In contrast, the *Ten Plagues musical* was inspired by the London Plague of 1665, which was used as a backdrop to the journey of one man facing the reality of inhabiting a world where you may be alive one moment and cut down by plague in the next. Interestingly, the festival also offered *Your Last Breath* as a 'science in drama' experience. This play is from the Curious Directive theatre company. It is set in Norway and combines four separate stories from distinct time periods. We meet Charles, a young cartographer, who sets out for the inhospitable climes of northern Norway in 1876. We also encounter extreme-skier Anna who sees her heart stop for three hours when she is trapped under Norwegian ice in 1999, and in 2011 a businesswoman, Freija, who travels to Norway to scatter her father's ashes. Finally in 2034, we meet a young man who explains that Anna's accident has managed to revolutionise modern healthcare.

The majority of these dramas are fairly new offerings that have received critical acclaim, but have not received financial success. An exception to this is *Arcadia,* one of the most successful plays that brings science to the theatre. It was written by Tom Stoppard and performed for the first time in 1993. It is a play that introduces the complex concepts of chaos theory, entropy and the underlying order in seemingly disordered events. *Arcadia* has been called 'a masterpiece' – but it is even more than that. The play stirs the most basic and profound questions humans can ask. How should we live with the knowledge that extinction is certain – not just of ourselves, but of our species' (Hari, 2009).[12] The play was also used as a text for A-level in the Associated Examining Board.[13] In 1994, the play received the Laurence Olivier Award for Best New Play, has been a critically acclaimed hit on both sides of the Atlantic and has consistently attracted healthy audiences.

---

[12] For a useful resume of the play and a discussion of its plot see http://www.independent.co.uk/arts-entertainment/theatre-dance/features/is-tom-stoppards-arcadia-the-greatest-play-of-our-age-1688852.html

[13] now subsumed into AQA (Assessment and Qualifications Alliance)

The Scientist Drama

A play that fits into the 'scientist drama' subgenre is the new Alan Alda play *Radiance: The Passion of Marie Curie*. It tells the story of twice Nobel Prize winning scientist, Marie Curie, who discovered the elements radium and polonium. The play is set in the period between her receiving the two Nobel Prizes and it highlights the issues associated with being a female scientist at the turn of the century. This theme is also echoed in Case study 7.6 by Dr Dee Rawsthorne which outlines the development of a play *Blooming Snapdragons* which focuses on the stories of a group of women scientists who worked for the plant geneticist, William Bateson.

Case Study 7.6

Blooming Snapdragons
Dee Rawsthorne

Background

The John Innes Centre (JIC) in Norwich celebrated its centenary in 2010. As part of this celebration, a play was commissioned for both a sixth form and an adult audience. The play told the stories of a group of women scientists who worked for the plant geneticist William Bateson. *Blooming Snapdragons* is a play that explores both the themes of genetics and the role of female scientists in a historical and a contemporary workplace. This is achieved through the interwoven lives of the two contemporary female biologists, Jo and Adi, and the historical characters who were Bateson's female colleagues. Bateson rediscovered the work of Mendel and invented the word 'genetics'. He became the first director of the John Innes Horticultural Institution that was originally founded in London in 1910, and later became the John Innes Centre, currently part of the Norwich Research Park (NRP). Before Bateson moved to the John Innes Horticultural Institution, he ran a laboratory in the Balfour Centre in Cambridge University. What was unique about this laboratory was that it was the only one within Cambridge that welcomed women scientists. It had been set up for the women who were teaching and studying within the women's colleges. In contrast to the success of Bateson, the story of Bateson's women colleagues was uncovered in the John Innes Centre archives where intriguing snippets of information alluded to a group of women who were vital for the research and insights that Bateson and his laboratory produced. However, their contribution, their discoveries and their presence had almost passed unrecorded and unrecognised. This vital and eccentric group of female scientists became the historical basis for the play.

Required resources

The original budget for the event was £5000 financed through the account set aside for JIC centenary celebrations. We chose to use professionals and commissioned a playwright, Liz Rothschild, whose previous work included a one-woman play *Breaking the Silence* that focused on the Nobel Prize winning environmentalist Rachel Carson. As well as being a playwright, Liz Rothschild is also an actor and starred in our play, *Blooming Snapdragons,* with Syreeta Kumar; they played two modern biologists, and Liz also played the historical female characters. We hired a director, Sue Mayo, who worked on the script with Liz, organised the auditions, managed the sound recordings and directed the play. To gain an insight into the historical characters, Rothschild drew on the archives of the JIC and Cambridge University. In addition she also drew on the work of Marsha Richmond at Wayne State University, Detroit, Michigan, who researches Bateson's laboratory. Rothschild also spent time talking to contemporary women who were willing to share their experience and expertise of their roles as female scientists in today's laboratories. Rehearsals took place within the science community at the JIC, which the actors said helped them immerse themselves in the scientific culture. Although we could have held/run the play at the JIC on the NRP, we chose to hire a city centre venue to attract a wider audience. We were able to hire props from a local theatre and used scientific equipment and props that we had on site. Although we would have liked to have employed a stage manager and a sound designer, our budget couldn't cover this expense, but the final version still worked well and effectively. We were able to use our Communications Department to produce publicity material and contacted local schools about the play. The evaluation was undertaken in-house.

The positive outcomes

The final product, the play itself, far exceeded our expectations. It effectively combined a historical story with the contemporary issues that still affect women scientists today. This clearly evoked a positive response in the audience who engaged with the themes, performance, performers and the stories of the characters.

It was a very positive way to combine science and the arts in a project that allowed a positive exchange of knowledge and culture. The scientists that took part in the project were given an insight into the research, creativity and craftsmanship of producing a drama. At the same time, the artists – the actors, directors and playwright – found that they developed an understanding of how the scientific world works, the difficulties and pressures of long hours and repetition, but also the passion and the inspiration that fuels each and every scientific breakthrough and discovery.

What didn't go as well as expected

A decision was made to perform the play during science week, and to attract schools we offered free attendance. However, audience numbers were disappointing. In retrospect this may reflect the fact that although we produced the publicity material, we did not have enough money to spend on effective publicity. In addition, we realised that schools found it quite hard to pitch the play: was it a drama for drama pupils or was it a science project that would appeal to science pupils? Although we offered the play for free to schools and the public to open it up to as many people as possible, this may have had the opposite effect as people didn't value the experience and were much more likely to cancel their bookings at the last minute because it was free.

Summary of the experience

One of the key outcomes that I gained personally from undertaking this project was an understanding of just how powerful drama can be as a medium to convey a story or a message. This drama provoked a strong reaction within several members of the audience who contributed to the group discussions and took the time to write to us after the event with their comments and feedback. There was one comment in particular from a female scientist that left a strong impression with me. She contacted us afterwards by email and wrote 'For me, it was a good reminder of my passion for my subject, and has helped me to decide to stay on this career path.'

Presenting genetics through the stories of the female scientists depicted within this production appealed to a wide audience including sixth form pupils. Feedback from one teacher stated 'The performance and the subject matter was brilliantly conceived and performed. For the pupils to experience drama so intimately and a Q&A session afterwards was fantastic. They were very chatty on the way back!'

I have also gained a real insight into the arts world and a sincere respect for the craftsmanship of the practitioners. I am aware of how hard they work and the commitment they show to research the backgrounds and develop the characters to bring them to life for their audience, I have great respect for their dedication and attention to detail. Indeed, many audience members were surprised to find out that Adi was an actor and not a scientist. The effect the experience had on the actors' perception of science and scientists was unexpected, but very welcome.

This project was very different to the science communication events that I had previously designed, organised and presented. In many ways it has taken me outside of my comfort zone and introduced me to a new medium for communicating science to a wider audience. However, it is one of the projects that figures highly on my list of achievements and I am very proud of it.

Contemplating commissioning a drama?

Consider the following.

1 To begin with you need to have an initial idea or a suggestion for a story, but as soon as you have that initial idea then you should have the confidence to pursue it.

2 Seek out and use professionals to help you to bring your idea to fruition. A good place to start is a script writer or a playwright whose work you have either seen before and admired or someone who has written something similar in the past.

3 Recognise that there are costs involved in undertaking an arts production such as a drama. These may include commissioning the work; hiring a venue; employing the actors, directors and stage manager, props and costumes. You should be prepared to allocate funds to invest in good publicity so that your project has the best chance for success.

4 Consider the issues associated with commissioning a performance: make sure you have public liability insurance and discuss and agree copyright issues associated with the drama. In our case we agreed that the John Innes Centre and the writer Liz Rothschild would both hold the copyright.

5 Be prepared to have open discussions between scientists and artists in order to share expertise and knowledge, but also to ensure that the science is represented appropriately within the drama and the drama works effectively to represent the science.

6 To attract a wider audience to your production, consider rebranding your event as an arts project or a drama about science, as opposed to a science communication event using drama.

7 Consider organising a tour instead of several one-off performances. A tour may increase your audience numbers as it builds momentum and publicity, but it is also an easier format for employing actors and the associated cast.

8 Finally, recognise that audiences are hard to reach and you should not judge the success of your event by the audience numbers, which may be disappointing. The success can be judged by looking at the impact that the performance has on the audience members that do attend, and who are moved, educated, inspired or influenced in some way.

Reference

http://www.jic.ac.uk/corporate/friends/events/bloomingsnapdragons.htm

It is fair to say that if you are a scientist who is seeking ways to engage with an audience, science dramas present real challenges and hurdles that have to be overcome and it is not an easy choice. However, it is not impossible, but a successful scientist playwright is an uncommon occurrence. On the other hand, the scientist turned street performer crops up on a more regular basis. Dee Rawsthorne's case study (7.6) shows that it is possible to develop a piece of drama that communicates science, but it involved collaborating with professional playwrights, directors and actors. Lastly, scientists as a reference source for playwrights and authors are not outside the realms of possibility.

Stand-up scientists!

Science and comedy may not seem to be natural cohabitants within the same sentence, but comedy has existed with science for some time, for example in TV sit-coms such as *Red Dwarf* and *Big Bang Theory*. There are comedy science cartoons, for example *Dexter's Laboratory,* and books such as Douglas Adams' *Hitchhikers Guide to the Galaxy*. But there is now a new phenomenon that is beginning to attract the attention of scientists, the public and commentators. Tom Chilvers reporting in the *Telegraph* has noticed that 'a strange thing has happened to stand-up comedy recently: it has started to find science funny'. Chilvers describes how this is an interesting movement away from laughing *at* the eccentric scientist working all hours in his laboratory, to laughing *with* scientists talking about life in their laboratory[14]. There has been a realisation that scientists and science can be humorous and that this humour can be used to communicate science to the public. In the UK, physics graduate Dara O'Briain is a household name, not because of science but because he is a well-known comedian who has developed a successful TV career and has used science in his stand-up comedy routines. Dara O'Briain is not the only scientist to have crossed over into comedy; a quick search on YouTube for scientists and comedy brings up a diverse range of videos of scientists doing stand-up comedy routines in front of engaged audiences. Simon Singh identifies these as events that have moved beyond listening to scientists in lectures to 'discussing with scientists and celebrating science'. This realisation that science offers potential for comedy is apparent in BBC Radio 4's *The Infinite Monkey Cage* comedy, now in its fifth series. The show is hosted by physicist Brian Cox and comedian Robin Ince. The show takes a science subject, two guest scientists with an interest in the science subject and a guest comedian. The outcome is a humorous and accessible conversation with science at its heart. The success of this radio show has led its presenters to take it out on the road with fellow science advocates and science writers, Ben Goldacre and Simon Singh. *Uncaged Monkeys – A Night of Science and Wonder* is the first UK national comedy tour that celebrates science (Kumar and Ince, 2011). Reviews for the show included 'very funny and very intelligent people talking about science. It's just like stand-up comedy, only with more slide shows and black holes' (Walker, 2011). This is the latest 'variety' comedy show to attract a following and it follows the successful annual event of *The Return of Nine Lessons and Carols for Godless People*. This variety performance combines short mini-lectures by scientists such as

---

[14] http://www.telegraph.co.uk/culture/theatre/comedy/4985420/Science-doesnt-make-good-comedy-You-must-be-joking-.-.-..html

Richard Dawkins, with music and stand-up comedy routines combining science and well-known scientists. Robin Ince played a key role in developing this show, partly in response to the concern that scientists as well as comedians are beginning to feel the pressure of censorship, and the concerns that freedom of expression and scientific views can be viewed as offensive and leave scientists vulnerable to libel suits.

Scientists can be found standing up for science through stand-up comedy as part of *The Bright Club,* a monthly variety night founded in 2009. This is a collaborative project with University College London's (UCL) head of public engagement, comedy promoters One Green Firework and music promoters Duel in the Deep. UCL is a Beacon for Public Engagement and *The Bright Club* has emerged as a successful output. It provides the opportunity for academics (who can be scientists) to take their research and turn it in to a comedy routine. The success of *The Bright Club* can be seen in its emergence in different cities and venues throughout the UK (Box 7.1). In addition they have also expanded into *The Bright Club Music.*

Box 7.1   Think about stand-up comedy

If you are a scientist and you feel that you might like to try stand-up, then it's worth exploring whether you have a local Bright Club. The Bright Club website contains contact information. If you do not have a local Bright Club in your area then contact the UCL public engagement team who will offer support to help you set up your own Bright Club.

publicengagement@ucl.ac.uk
http://brightclub.org/

7.3.2   A focus on Café scientifique

In this section Dr Richard Bowater explores the history of Cafe scientifique and how you can get involved in taking part or running one yourself.

Informal meetings are an attractive option to discuss important contemporary scientific issues with lay audiences. Over the past 20 years, many scientific researchers from across the globe have engaged in such discussions, often referred to as Café scientifiques. With suggestions as to how Café scientifique events are organised, the following section makes it clear that the only important rule to follow when organising such events is to make sure that they are informative and fun for the presenter(s) and audience! During the latter part of the twentieth century, it was realised that there was an appetite for people from all walks of life to meet together to discuss the latest scientific ideas that impact upon society. This idea led to the development of the Café scientifique, which was launched in Leeds in the UK in 1998 (Dallas, 2006). This title developed from the Café philosophique movement that had started in France earlier in the same decade (Clery, 2003). These informal gatherings began to appear in various guises and locations, and scientists soon appreciated the enthusiasm for such meetings among the general population.

In 2001, a Wellcome Trust grant provided financial support to spread the word about Café scientifique (Clery, 2003). An internet address was developed[15] which continues to be updated on a regular basis. The internet site is an extremely useful resource that provides listings of past and future events as well as information about how to locate your local café or develop your own. With the help of motivated organisers, Café scientifiques became established rapidly across the UK. Similar ideas have developed worldwide, and they are now held on a regular basis in most geographical areas where there is a significant presence of scientific researchers.

There is no such thing as a standard Café scientifique, with every event having its individual style. These types of informal discussions have various names that range from Café scientifique (or Science cafés) to Science Exchange and any similar combination of related words (Dallas, 2006). For an example, see Case study 7.7 by Dr Kenneth Skeldon.

Aims and objectives of Café scientifique events

Café scientifique events provide a forum for the discussion of current scientific issues for anyone who is interested, but they do not have a remit to promote science. Rather, they aim to discuss and question, in an open-minded manner, the principles and consequences of scientific research. It is important that the audience at Café scientifique events believe the discussion has no hidden agenda, either for or against science.

To ensure that Café scientifique events are accessible to the widest possible audience, they are kept deliberately informal. Speakers are encouraged to target their discussions appropriately, allowing plenty of time for clarification of complex ideas. Within any audience there is likely to be a cross-section of society with different scientific backgrounds so speakers (or other participants) must be knowledgeable enough to discuss the topic across a range of levels.

Each café undertakes its own organisation and is not necessarily connected to other cafés, though the majority will take their lead from the successful events that have been organised in other locations. Cafés are usually run by local, volunteer hosts who have a passion to discuss contemporary science. Importantly, these events are not always located in cafés; other popular venues include museums, galleries, bookshops and bars.

Café scientifique events can discuss any aspect of science, with the main point being that they need to generate interest within the local community. Topics discussed at these events often focus on points that are controversial, political or problematic in some specific way (see Section 6.5). Favoured topics include those that are relevant to everyday life, especially those in the news at the time of discussion. However, there is often a focus on science that is studied by scientists that are local to the area of the café and this can certainly be a good way to encourage specific speakers to deliver a talk.

Since the topics for discussion are decided by local organisers, the schedule of talks will often be driven by whoever is available at any particular time. It can be relatively straightforward to recruit local scientists to talk about the subject of their research or a topic that they have written about. Sometimes, though, the most interesting discussions will be about scientific issues that are

---

[15] www.cafescientifique.org

currently in the news, and it can be more difficult to recruit speakers for these. In this situation, it is important to remember that speakers at the events do not, necessarily, need to be world-leading experts in the topic being discussed. All that is needed is for them to be confident to lead a discussion of the topic, and then the audience members can dictate the direction of the discussion.

With such a flexible structure for Café scientifique events, concerns have been raised that they could be taken over by someone with a specific pro-science or anti-science agenda, particularly with political motivations in mind. However, experience has shown that this does not happen as long as there are a group of people involved in organising the local events. A number of reasons have been proposed to explain this lack of bias in established Cafés, but the simplest explanation seems to be that such events would not retain interest across scientifically interested communities. Ultimately, the audience members make up their own minds about the discussion, and experience has shown that if one person tries to impose themselves upon a Café discussion, then the audience respond by ignoring them.

People attend for all sorts of reasons and some attend every month because they are fascinated by science in general, whereas others will only turn up if they are interested in that event's topic. To ensure that enough people continue to return to make the events worthwhile for everyone, it is useful to allow potential audience members to have some input in developing the schedule of talks.

Organising Café scientifique events

If you are a scientist who is keen to engage in Café scientifique talks, then the easiest way to do this is to participate in events that are organised in your local area. To check if such events are already organised, look on the Café scientifique website[16]. If there are no such events in your area, then you may wish to organise a local group. The most effective way to do this is to identify several like-minded individuals and share the workload described below. Experienced café organisers will always be willing to offer advice and guidance, but local organisers should always be willing to alter the format to something that is appropriate to their own circumstances. This section provides some suggestions as to how Café scientifique events are organised, taking its lead from other articles that have outlined the basic requirements (Clery, 2003; Dallas, 2006; Grand, 2008, 2011). One of the main points to appreciate is that there is no single set format for these types of events. If you wish to develop a different type of Café scientifique, then the best advice is to go with whatever you think will work best for your location and circumstances.

The most important detail to arrange for any café is a venue. It must be an appropriate size and should be conveniently located for its target audience. On average, the number of people that attend these events is up to 50, so the venue should be able to cope with up to this number, but should not be so large that people will be sparsely spread around. If the events become really popular then a larger venue can always be arranged at a later date.

---

[16] www.cafescientifique.org

The venue is usually provided free, because the audience buy drinks and/or food. Typically, Café scientifique meetings are arranged during a quiet period for their venue, usually early evening. Ideally, they should occur at regular intervals, though the exact schedule should retain some flexibility; it is usual for the majority of events to run approximately once a month.

In deciding upon the venue, it is important to remember that the aim is to encourage attendance from a wide range of people and, thus, the events should be free to attend. By their nature the events are low cost to run because the organisers are volunteers who are involved for fun or their own professional or personal development. Speakers are not paid and the largest cost may involve their travel expenses. Small costs can usually be covered by asking for voluntary contributions from the audience members during the evening. If costs do build up for any particular event (e.g. because of significant travel expenses or a need for other running costs), then local companies or individuals are often willing to provide some of the necessary funds. However, if funds are accepted from anyone it is important that the independence and impartiality of the café is maintained.

Once the venue and approximate timings are decided, the next step is to arrange the schedule of talks. The organiser(s) identify a scientific topic and appropriate speaker, usually an academic scientist from a local university, college, or an author of science books. Sometimes the topic will be decided upon first, followed by a search for an appropriate speaker. However, it is more common that willing speakers are identified and allowed to decide the topic through discussion with the organisers. As a schedule of talks develops then it is necessary to arrange for appropriate administration of the talk (Table 7.4). It is worth spending time on advertising the event especially for a new Café scientifique. Attract your audience to the event using email and displaying posters in appropriate locations (Section 6.9).

Administrative details for organising the events

As with all public events, it is important that general health and safety issues are evaluated ahead of the meeting (Table 7.4). Since most meetings take place in cafés or bars, any requirement for a license or insurance coverage is likely to be provided by the owner or tenant of the establishment. However, once planning for your event takes shape, you must confirm with the owner that everything complies with local regulations.

You also need to focus on safety issues that are more directly relevant to the specific event being organised. As plans for the event develop, review whether any CoSHH or equivalent documentation needs to be prepared and assessed (Section 6.10). It is a good idea to review whether the benefits obtained from undertaking experiments during the event are worth the extra administrative effort that will be required to run them. On the plus side, once an experiment has been conducted within a Café scientifique environment, then it will be much more straightforward to undertake it again, even if the location alters.

Generally, the presenters at the Café scientifique don't need to undergo CRB checks (or their equivalent), because if any young children are present they are likely to be accompanied by parents or other responsible adults. However,

Table 7.4 Checklist of requirements to run a successful Café scientifique.

Initial Planning for the event (>3 months before the date)

√ Outline the topic and the aims for the event
√ Identify the location
√ Liaise with the host contact at the location to:
  Agree the date and format
  Identify available facilities
  Discuss likely attendees and their background knowledge (if known)
  Review insurance coverage and health and safety issues
  Source potential mailing lists
  Identify any funding requirements (and source funding if needed!)

√ Liaise with the host contact at the location to confirm plans are still on track to host the event
√ Advertise to the relevant audience
√ Begin to prepare activities if required

√ Finalise any activities and presentations required for the event

√ Arrive at the location in plenty of time
√ Deliver the activity/presentation
√ Interact with the audience, providing lots of opportunities for questions and informal discussion
√ Pass on contact details to answer further queries
√ Obtain feedback

√ Evaluate and reflect on the experience
√ Report to sponsors/funders if required
√ Share good practice and potential pitfalls with colleagues
√ Prepare for your next event!!

remember that this is a potential issue to consider if special circumstances mean that the discussion will be targeted at young children.

As for all science communication events, the event should be evaluated, particularly if funding was obtained for it, because the funder is likely to require a report. Guidelines for successful evaluation are provided in Chapter 5. Remember that Café scientifique events are meant to be informal and therefore any evaluation should have a low impact upon the audience. Whatever type of evaluation is chosen, it should take no more than a couple of minutes of each person's time and it should not be onerous.

Format of the event

Since Café scientifique events usually take place in establishments serving drinks and food, the audience are likely to be seated around tables. Most cafés last about 90 minutes, though it is wise to book the venue for at least two hours so discussion is not curtailed if it becomes really engrossing for the audience.

The events are broken down into three parts:
1 an introductory talk by the speaker(s);
2 break;
3 a discussion that encourages participation from the audience.

Most Café scientifique organisers discourage speakers from using projectors and computer presentation software, such as PowerPoint, mainly because the simplicity of just talking provides more intimacy between speaker and audience, allowing the audience to concentrate on what is being said. Often, this is also practical as many venues will not have appropriate facilities or, indeed, the space to introduce them. This type of discussion may be unusual and daunting for many speakers; however, most speakers who attempt this approach find it liberating and empowering. Any material that is really important for people to see can be printed and put on tables, or exhibits provided that can viewed during the event. Ultimately, if the speaker believes that a presentation is essential, then this can accommodated if appropriate facilities can easily be made available.

The events start with an introduction to the topic being discussed. This should not be too long, with 30–40 minutes being ideal. This means it really has to be an introduction; speakers should not try to cover everything. After the introductory talk there should be a short break, during which the audience can top up their food and drinks. The interval is important because it allows audience members to discuss the talk and identify questions that they would like to ask. Finally there is a discussion between attendees and the speaker or organiser(s). To start the discussion it is often appropriate for the speaker to pose questions to the audience, encouraging them to continue the dialogue. Speakers (and organisers) should be ready to answer a wide range of questions, many of which will not be easy to predict. Some of the audience may not have studied science since they left school, so questions can come from all angles. Most of the interchange will be between audience and speaker, but the audience can start debating among themselves. This might arise from the evening's topic, from the field in general or from people's wider experiences.

At some point the discussion will peter out, or the allocated time will be used up. At an appropriate time the organiser(s) should close the event, thanking the speakers and any sponsors and ensuring that the audience are informed about future events. Once the official part of the event has finished, it is helpful if speaker(s) can stay behind for further informal discussions with anyone who is interested.

Top Tips for running successful Café scientifique events

1 An over-arching aim of these events is that they should be accessible to anyone, regardless of their background.

2 The most popular subjects for Cafés are those where the audience feel they have some experience to contribute. Thus, subjects such as medicine, psychology and genetics and, more recently, climate change and biodiversity are of interest.

3 Another aid to a successful event is if the speaker has a controversial idea to express and sticks to the argument all evening.

4 Within the Cafés, questions come from a broad context, often much wider than are usually addressed by scientific researchers. The nature of Café scientifique events makes speakers think on their feet and the format provokes the kind of discussion and debate that make the evenings unexpected and unpredictable. Although most

speakers are initially wary of this arrangement, by the end of the event they have usually enjoyed themselves.

5 Props can work well to engage and entertain. These could be circulated during the talk or the audience could view them during the interval. The latter works particularly well if the audience are able to view exhibits or simple scientific experiments, such as microscope slides.

6 For organisers of Café Scientifique events, a unique characteristic of the event is to appreciate that changing the venue changes the tone and nature of the discussion.

7 The events provide an opportunity to listen to what people have to say about the issues that arise from their professional life, a chance to connect with intelligent and engaged members of the public and an easy, informal and inclusive way to let the light in on their professional work. It is a format that can actively encourage two-way communication.

8 Café Scientifique events make science more democratic and, thus, they are an ideal way to ensure that scientific knowledge and understanding is encouraged within the wider community.

Case Study 7.7

Community Cafe Science in Aberdeen and Aberdeenshire[17]

Kenneth Skeldon

Overview

In January 2009 the University of Aberdeen launched its first science cafe as part of its Public Engagement with Science Strategy. The aim was to create an informal city centre environment where university researchers and guest speakers can come face to face with the public to discuss a range of topics. The series echoed the format started by Duncan Dallas in Leeds in the late 1990s – an approach hallmarked by informality, promotion of discussion and avoidance of lecture visuals. Because of the popularity of the Aberdeen cafe – audiences of over 100 people are not unusual – it was decided to expand and diversify the series both geographically and thematically. As of January 2012, over 90 speakers and 4000 visitors have come together across 85 separate cafe events across Aberdeen city and rural Aberdeenshire.

Concept and venue

The Aberdeen city cafe series is held at Waterstones book store within their Costa Coffee bar. This busy city centre venue is convenient for public transport and well known to locals. The city programme has ten sessions per year from January through to October held in the middle Wednesday of the month and running from 7 to 9 p.m. The evening format is departed for the August session, which is held on a Saturday afternoon in the children's section of the

[17]Acknowledgements. Many thanks to Ann Grand for sharing some of her experiences in these events and for providing permission to use the images contained within this article.

book store, just before the end of school summer holidays and targeted to a family audience. The cafe series is planned in advance over two seasons of five events. Flyers and marketing materials are produced and distributed at least one month in advance of the first session in each season. This is essential in marketing events and building word of mouth publicity.

During a typical event, visitors take their seats in the normal cafe set-up with minimal physical separation from the guest speaker. The speaker opens with an introductory talk lasting 25 minutes or so, followed by a short break when visitors can chat directly with the speaker and obtain refreshments. A small number of people leave at this point – those with time constraints but who still wish to hear a brief overview on a current scientific topic. Following the break, an audience discussion begins. A wireless microphone system is used to ensure adequate audibility of the speaker and audience members. Topics are selected primarily but not exclusively from areas of the natural sciences and often reflect subjects of topical interest or highlight areas of research priority. The initial audience number estimate was 30–50 based on cafes held elsewhere, and the UK average attendance of around 45. However, the first session in January 2009 drew an audience of 175 – over four times the number of cafe table covers. Subsequent sessions throughout 2009 attracted an average audience of 100 making the Aberdeen science cafe one of the best attended in the UK. Such large audiences can reduce intimacy and had a bearing on the decision to expand the number of cafe programmes.

In 2010, 'Friends of Cafe Sci' was introduced whereby visitors could save 10% on any purchases made at Waterstones tills during cafe events. This was in response to concerns raised by Waterstones that the increased footfall due to the science cafe programme is recordable (for sales purposes), yet with little takings to show for it. The discount scheme together with reading recommendations by speakers and occasional programming of authors with current books, have helped address this issue. The events are all free to attend, in accordance with one of the principle pillars of the University of Aberdeen's Public Engagement with Research Strategy; namely that core opportunities should exist for our researchers to engage with the public and have minimal barriers, including entry charges.

Audiences

The evening audience is largely adult but varies with the topic. For example, evaluations demonstrate that a larger number of young adult females attend diet and nutrition events while more teenagers tend to come to physics and astronomy events. Originally, around half of the audience was made up of people with no connection to the University. The other half are what might be called friendlies: people associated in some way with the speaker's activities or the University. However as the series continued, the true public proportion increased. Currently, approximately 15% of audiences are made up of regular visitors. Interestingly, the single biggest response when asked at an event how people heard about it is now 'word of mouth'. In terms of the cost per audience member, after factoring in presenter expenses and publicity costs, the average cost per session is around £150. Using average audience numbers for 2010/11 the approximate cost per audience member is £1.50. This makes the series essentially sustainable. An option to do this is through voluntary audience donation. The Aberdeen series do not employ this however, preferring to raise the running costs in other ways. From 2009 the cafe costs have been met by a Scottish Government science engagement grant.

Evaluation and development

Evaluation is crucial to the successful evolution of any science communication activity. Within the Aberdeen cafe programmes, feedback from the audience and presenters is sought in a variety of ways. From the beginning, simple comments slips are given to the audience to rate the speaker, topic, format and venue. These slips also capture email addresses for those wishing to join the regional email distribution list for cafe science updates. Assisted questionnaires are also used during the session break. These capture more detailed comments given to support staff and allow specific questions to be probed. In addition, a simple show-of-hands evaluation is conducted ahead of the discussion part of each event. With this approach, the response to four specific questions is measured with 100% audience return and an accuracy of about 10% through speed counting of hands in the air. Together, these methods of evaluation have allowed the refining of the event format and technical issues. Feedback is also used as a form of crowd sourcing, influencing the choice of topics in future programmes.

The cafe sessions are regularly advertised in local newspapers, thanks to the efforts of the university communications officers. A number of sessions have attracted radio and television coverage, including national coverage on BBC and STV news. Evaluations also demonstrate that a sizeable fraction of the audience find out about events through the media, although as indicated above, the most sizeable mode of communication is now word of mouth. The associated press releases are important because they afford an opportunity for presenters to talk with the media about their area of research, which in itself is a public engagement mechanism.

Expansion of the cafe concept in Aberdeen

A creative and rapid expansion of the cafe series in Aberdeen and Aberdeenshire has taken place over the past 24 months. Currently, eight strands take place throughout the year, amounting to some 35 events. An example is Cafe Controversial, founded in 2010 to tackle science and society issues that attract strong viewpoints. Another example is Cafe MED, created in late 2009 to bring medical researchers, clinicians and the public together and offer a balanced discussion of the pathways that take medical research to treatments and cures. This concept is still a growth area for through-the-year activity, anchoring to recurring events in the science engagement calendar while providing ongoing community engagement during the months in between. For example, March cafes take place in National Science and Engineering Week – taking advantage of additional publicity and cross marketing.

Embedding community cafes in the social calendars of our publics

Regularity is one of the most important aspects of the cafe series. The fact that they take place throughout the year means they can become part of the social and cultural offering within the region. This is best achieved when the series are forward planned, to at least a few events in advance, as with the Aberdeen city series. All the programmes therefore share a common website and each has a dedicated flyer and marketing strategy. Individual cafe series always take place on approximately the same day of the month, e.g. the middle Wednesday, and different series occupy different days of the week. This approach results in an overall programme where, regardless of the month, there will be at least one event taking place. There are occasions where up to four series are running concurrently but they will all vary thematically, take place on different weekday evenings or at different geographical locations. Current or recent programmes within our series include:

- Cafe scientifique Aberdeen (core programme of 10 events per year January–October);
- Cafe scientifique Aberdeenshire (various venues targeting rural community groups – normally four events per year March–June);
- Cafe MED (five events per year January–May);
- Cafe controversial (four events per year September–December);
- Cafe light (four events per year September–December);
- Cafe cosmos (an initiative to re-open the UK's most northerly permanent planetarium, four events per year March–June);
- Cafe connect (an ambitious road-show cafe which piloted in July 2011 and covered over 1500 miles in 10 days).

Related series include our Discovery Gallery programme, new in 2011 (Aberdeen Art Gallery, four events fortnightly from July – August) and our planned launch of 'Cafe philosophique Aberdeen' in 2012. Our cafe programme has a website (www.cafescience aberdeen.co.uk) and a presence on the UK Cafe scientifique website (www.cafe scientifique.org) where details of cafes all across the UK are listed.[18]

7.3.3 A focus on book clubs

Book clubs or as they are known by their other, common aliases – book groups, reading groups or book discussion groups – come in all sorts of shapes, sizes and genres, and they can offer an opportunity for taking part in discussions about scientific issues. What all book groups have in common is a group of people that meet on a regular basis to informally discuss and offer opinions about a book that they have agreed to read prior to the meeting. There is no such thing as a typical book group. It is probably not an exaggeration to say that the only thing that members are guaranteed to have in common is an interest in reading and literature. Members come from all walks of life and joining a book group offers a fantastic opportunity to meet and interact with different people. The forum can be a virtual online forum, an informal group of friends who meet regularly in each other's homes on a rotating basis, or it may be a slightly more formal arrangement where individuals join a book club that is held in a library, a bookshop or a local bar or cafe. As well as your regular generic fiction book club there has been a recent emergence of different genres in the book club world. It is possible to join a club that focuses on Scandinavian detective stories, translated fiction, science fiction or historical fiction.

You may currently be a member of a club which has quite a broad remit and is willing to discuss different modern literature. If so, then an easy first step would be to suggest a novel that contains scientific themes and you can offer to lead the discussion. This can provide you with the opportunity to discuss scientific themes in an informal setting. Many novels offer opportunities to discuss science and it may be that you simply need to change your mind set slightly and to actively seek out opportunities to bring science and the impact of science into your discussions. Alternatively suggest a book which has science as the main theme. With a little bit of research and by drawing on your knowledge, you have an excellent opportunity to discuss science with others in an informal and conversational environment. If you are not

---

[18] The University of Aberdeen is grateful to the Scottish Government Office of the Chief Scientific Adviser for its ongoing support of our programmes. Cafe Scientifique Aberdeen was made possible through funding from their 2008/2009 science engagement scheme and the expansion to cafe scientifique Aberdeenshire and cafe MED was supported through their 2009/2010 scheme. Further embellishment to our cafe programmes took place with support awarded by the 2010/11 and 2011/12 schemes. We are grateful to Waterstones, Woodend Barn Arts Centre, the Fraserburgh Lighthouse Museum cafe, Aberdeen College Planetarium, Suttie Building, Aberdeen Art Gallery, Satrosphere Science Centre and the Acorn Centre in Inverurie for venue and infrastructure support across our series. All series are led by the University of Aberdeen run in conjuction with partners including NHS Grampian, TechFest-SetPoint, Aberdeen College and Aberdeen Museums and Galleries.

a member of a book club then there are two possible approaches to bring science to the book club arena:

1 join an established book club;
2 set up your own book club.

Join an established book club

There are a plethora of book clubs. In the UK it was estimated that there were at least 50 000 in 2005,[19] and there is a good chance that there is at least one near you. In recent years we have seen the advent of book clubs which have science as a theme. They include the online Guardian Science Book Club that invites readers to read reviews as well as actively reading and discussing science literature. The books chosen can be popular contemporary science literature such as *God and the New Physics* by Paul Davies. However, they have also reviewed seminal texts such as the *Double Helix* by James D. Watson, first published in 1968.[20] Other science-themed book clubs include the Fiction Lab, hosted by Jennifer Rohn and housed at The Royal Institution.[21] If you would like to take a slightly broader approach to joining a book club and would prefer one that is slightly less of a busman's holiday, then look for a book club that it is near you. A first place to start may be to ask your friends or work colleagues who may be members of a book club. Alternatively, ask at your local bookshop or check out the various websites that provide directories of book clubs including those in your local area. These websites also contain suggested reading lists as well as tips and information about where to buy or obtain books.[22]

Start your own book club

If you are thinking about starting your own book club with a science genre, then there are several key points that you need to consider.

1 Decide on the type of book club that you feel would be appropriate. For example do you want one that is light touch; it is a social occasion that includes discussing a book? Alternatively, would you prefer a book club where there is serious analysis of the text? Would you prefer to have one text read by all members at the same time, or would you prefer a pool of books that people can dip into?
2 Having decided on the type and style of your book club, then next step is to recruit members. In general there is an optimal number of members and this depends on the type of book club as well as the location. In general it tends to work best with 6–15 members. Recruitment is usually quite straightforward and your friends and colleagues are a very good starting point. If you can rustle up a core group of three or four members and ask

---

[19] http://www.guardian.co.uk/uk/2005/feb/12/books.booksnews

[20] http://www.guardian.co.uk/science/2009/feb/08/life-unauthorised-biography-richard-forteyhttp://www.guardian.co.uk/science/series/science-book-club

[21] http://www.rigb.org/contentControl?section=5443&action=detail

[22] An example of a website that offers information on local book groups as well as book reviews and suggested reading material is http://bookgroup.info/041205/index.php

them to invite a couple of their friends or colleagues who may be interested in joining, you will have a club before you know it. Alternatively consider advertising in your local library or book store.

3 Think about the scheduling of your club and consider where you would like it to take place. Traditionally most book clubs take place at the same time on a monthly basis, for example the last Tuesday of every month, but it can be less frequent than this depending on how busy you all are. It is a good idea to schedule a set time that you stick to rather than trying to arrange a time that suits everyone on a month-by-month basis. Location is also a consideration. Book clubs can be held on a rotational basis in member's homes. This works well if your club has members who know each other or have social links. Holding them in member's homes are often social occasions where members can eat, drink and catch up on news before launching into a discussion of the book. If this doesn't seem like a good idea, perhaps because you are unclear about whom the membership will be, consider using alternative venues. Suggestions include your local bookshop, library, cafe, wine bar or pub.

4 Finally, if you are keen to stick to a genre then it may be easiest if you start the club with a reading list that you have established prior to the first meeting. It should contain enough suggested books for the first year. Consider using books that have been suggested and reviewed in other science book clubs and keep your ears and eyes open for other recommendations. You may like to be the main organiser or host of the club and feel that you would like your role to include deciding on the reading list. It may be that as it evolves and members become more confident in its genre, then you could consider encouraging others to suggest books.

Professor Joanna Verran of the Manchester Metropolitan University had the idea to start a book club when she realised that there was an opportunity to launch a book club that had infectious diseases as its premise. She describes setting up the Bad Bugs Book Club in Case study 7.8.

Case Study 7.8

The Bad Bugs Book Club
Jo Verran

Background

The Bad Bugs Book Club comprises scientists and non-scientists and we read novels where infectious disease plays some role in the plot. We usually try to couple each book club meeting with an additional activity that is relevant to the content of the novel and/or the time of year, for example World AIDS Day, Manchester Science Festival, National Science and Engineering Week and World Malaria Day. We post meeting reports, reading guides as well as discussion points and other information on our website (see websites). Hopefully this encourages others to set up their own book clubs, feedback to us or participate in our book club remotely. The intended audience for the club is therefore primarily a literary one, interested in science (microbiology) and in discussion. The additional activities, such as film screening, guided walks and community events attract a wider audience, including families.

Structure of the book club

The first event took place in April 2009; we tend to meet about six times per year. Typically around eight people meet for each book club, but there are more than eight members on our mailing list. Each meeting follows a similar format. We talk about the book plot and characters, and whether we like them or not. After general discussion, we focus particularly on the way in which science in general, and infectious disease in particular, is represented in terms of accuracy and in terms of potential value in communicating information and ideas about microbiology to the general public. We also try to think of learning and or extension activities that could complement the reading, particularly for use in education. It is interesting to see how different book club members favour different books: we are going to compile a 'good bad bugs book list'!

Examples of book club sessions

Examples of activities within the event include:
- a screening of the movie *Outbreak* coupled with a book club meeting on the novel *The Hot Zone* by Richard Preston;
- a guided walk around Manchester to illustrate some of the locations and issues contemporary with *Mary Barton* by Elizabeth Gaskell;
- the production of a community quilt, commemorating World AIDS Day, displayed during a meeting to discuss *Dorian* by Will Self;
- a community music event *Malaria Migrations,* preceding *The Calcutta Chromosome* by Amitav Ghos.

Funding the book club and resources

The initial (and continuing) key resource, is my time. I had applied to the Wellcome Trust for funding, but had been unsuccessful. However, once the idea had germinated, the project continued regardless. Initially, I talked the idea through with various interested parties – friends, colleagues, professional societies before the first event. The Manchester Beacon for Public Engagement and the Society for Applied Microbiology (www.sfam.org.uk) jointly sponsored the first events (*Outbreak* screening/*Hot Zone* reading), and their enthusiastic support was invaluable, as was that of the Society for General Microbiology (www.sgm.ac.uk) who funded the book club sessions on *Dorian* by Will Self and *The Body Farm* by Patricia Cornwell. I was fortunate that the two main microbiology societies in the UK were interested in the project, and were happy to collaborate in several of the events. Other support, including some organisational help, has been provided by the Manchester Beacon for Public Engagement.

Book club highlights

Surprisingly successful meetings have focused on books that initially seemed to have little scientific content (*Dracula* by Bram Stoker), or that lacked accuracy (*The Satan Bug* by Alistair MacLean). These actually lent themselves to more discussion, whereas novels that were more overtly microbiological (*Unnatural Exposure* by Patricia Cornwell, *Hot Zone* by Richard Preston, *Andromeda Strain* by Michael Crichton) tended to elicit more technical debate.

The community events have all worked well. In particular, *Malaria Migrations* attracted around 100 members of the public on a Sunday afternoon. The event had been preceded by discussions about malaria between members of a musicians' collective and scientists. The resultant musical pieces were performed at the community event, alongside scientific presentations, student project work, information sheets, competitions and some fund-raising, before members of the book club met at another location to discuss their novel. The collaboration between microbiologists and musicians was stimulating and enjoyable for both.

Things to be aware of

The success of the activity depends on the commitment of the core members of the book club, who have been exceptionally loyal. We have always had something to say about the books we have read, and the differing opinions are interesting to hear, so the meetings themselves always proceed well. It is important to use a range of genres during the year, since the requisite microbiological content could be somewhat limiting. The use of two Patricia Cornwell novels at consecutive meetings was unfortunate (the second, the *Body Farm*, was selected to complement the forensic theme of the concurrent Manchester Science Festival), particularly for those members who were not fans of the author. It is not always possible to use the current meeting to identify date and venue of the next meeting, and the use of email lists, internet booking systems or similar, rely on the organiser having contact details of all interested parties. Early events were more heavily sponsored, and I was concerned that the transition for members from being paid for, to deciding what to pay for, would be difficult (although it wasn't). The activity is also insufficiently profiled on the internet; there is significant potential for greater, and more adventurous, participation and dissemination.

Safety issues

Safety issues are minimal. Book club meetings are held in the evenings in various licensed locations around Manchester; members are adults. Members make their own way to the venues. For some of the extension activities, safety issues may require some consideration. When transport was provided for a guided walk (to the village of Eyam, in preparation for a book club meeting on *Year of Wonders* by Geraldine Brooks), appropriate insurance was taken out. When applying for funding for the community quilt project, first aid was taken into consideration.

Personal gain

It is a very rewarding and stimulating exercise, but it does take time to organise, particularly if you are planning other activities alongside book club meetings. Overall, I have a great sense of achievement for having initiated and delivered this project. It has been really interesting discussing these novels with other members of the book club, and I have particularly enjoyed digging out and discussing the microbiology with the scientists and non-scientists in the group. The event has made me interested in the difference between scientific and creative writing – which is an area that I would like to explore in the future, and in encouraging my undergraduate students to read more widely and critically.

Top tips

1 If you are setting up a book club that has a specific focus, then you need to decide at the outset whether you will always be the 'leader', or whether each book club member will lead on a book of their choice.
2 Use a wide range of genres.
3 If you are producing reports about your books, it is important to maintain a consistent output.
4 Try to obtain funding from your subject-specific society.
5 Set up a simple communication system via the internet.

Useful websites

http://www.thereadingclub.co.uk/HowToStartBook club.html
http://www.ehow.co.uk/video_4957126_organize-book-club.html

7.4   A focus on policymakers

The 2006 Royal Society Science Communication report offered a real insight into the science communication activities of UK scientists. It also revealed the 'type of public' that these scientists want to communicate with, putting policymakers as the front-runners, with 60% of scientists indicating that they would like to communicate with this group (see Section 2.4.5). To understand how this can be facilitated, we need to consider:

* What is politics?
* What is policy?
* Who are the policymakers that scientists want to communicate with?

   Pielke (2007) offers the answer to the first question; politics 'is bargaining, negotiation and compromise in pursuit of desired ends'. Policy on the other hand, is the decision made to commit to a specific course of action. Finally, if we take a simplistic view, the answer to who are the policymakers, is anyone or any group that can influence or change policy decisions. This is illustrated in Figure 7.1 (Baron, 2010) and it indicates that policymakers are not just one large, amorphous group; instead there are different groups that can affect or change policy. These include the press, other media, the public, policymakers, managers, scientists and non-government organisations (NGOs). Scientists need to understand that each one of these groups can influence policy, as illustrated in the examples provided in Figure 7.2. This figure also highlights that scientists need to understand that these decisions:

* may not reflect the research consensus;

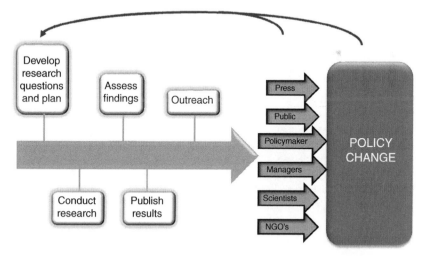

Figure 7.1 How the research pathway experienced by scientists can lead to policy change, and how policy change and pressure from different policymakers can in turn feedback and influence the research direction. This figure also describes the fact that for scientists and science to influence policy, it has to influence policymakers to catalyse a policy decision. The science cannot do this by itself. From *Escape from the Ivory Tower* by Nancy Baron. Copyright © 2010 Nancy Baron. Reproduced by permission of Island Press, Washington DC.

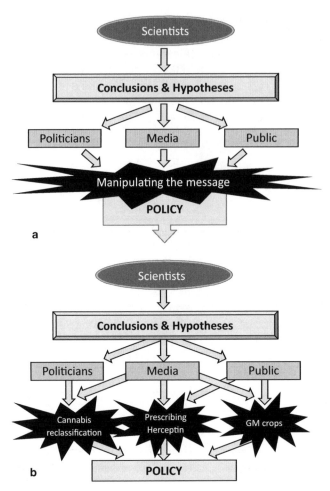

Figure 7.2 (a) How the science and the research conclusions need to pass through different groups to influence policy decisions. However it is important that scientists recognise that this message may not be fully understood by these groups and it can be manipulated because of a lack of understanding. The implication is to make the message as clear and simple as possible. Manipulation of the message can also occur if the message isn't accepted because it is contrary to common perceptions or firmly held beliefs, and is therefore difficult to accept. (b) Demonstrates how different messages that have emanated from scientists based on scientific evidence have been manipulated and the policy decisions are not supported by the scientific evidence. For example, the case of the reclassification of cannabis from Class C back to Class B was contrary to the recommendations of the Advisory Council on the Misuse of Drugs (ACMD). The policy decision was in response to a strong media campaign and a political agenda that was in opposition to the evidence. The decision to prescribe Herceptin to women with early stage breast cancer was taken before a thorough review of the efficacy, cost assessment and safety of the drug had been undertaken. The decision taken to prescribe the drug was in response to a strong media campaign and pressure from patient groups. Finally, the moratorium on GM crops within the UK was contrary to the scientific consensus but it was a reaction to the media campaign and the public outcry.

- may not be based on the scientific evidence;
- may be based on the perceptions and beliefs of the influential groups.

It is clear that scientists are no longer working in an environment where it is enough to simply communicate research findings to peers and colleagues through the peer review and publishing treadmill. Instead there have been changes in the wider society that are demanding that scientists clearly communicate their results and the evidence to a much broader audience, if we want to ensure that policy decisions are based on solid evidence. Without a doubt, it is clear that scientists cannot do this by remaining in the restricted and protected world of the laboratory, because this can enable the scientific evidence and the scientist's message to be modified, manipulated, discredited or rejected by other key players in the policy making process. Instead, scientists must venture out and learn to overcome their fear of presenting their science to the wider community. This is especially the case when it comes to interacting with policymakers in the political arena. Pielke (2007) explains how this call to share expertise is answered by scientists in different ways and he suggests that scientists fall into one of four different roles in their interaction with policymakers.

The Pure Scientist – produces and publishes relevant research information. They are content to provide it to the policy maker but in a neutral, impersonal fashion. However they have no interest or concern about what happens to the information after it has been passed on.

The Science Arbitrator – is prepared to answer questions from the policymaker and direct the policymaker to relevant sources of information on request. However they will not inform the policymaker about the correct questions to ask, and are not prepared to personally try and influence policymaker decisions.

The Honest Broker of Policy – ensures that the policymaker is provided with a wide variety of information offering different views and options to ensure that an informed decision is taken. They are also happy to provide clarification and facilitate understanding.

The Issue Advocate – generally has an issue that they are seeking to influence and holds a view that they are trying to promote to the policymaker.

The Pure Scientist and the Science Arbitrator both take a passive role. In contrast, the Issue Advocate and the Honest Broker of Policy are more active. The chances are that you will feel a natural affinity to one of these roles. However, the stance you take may be affected by your career stage, the area of science that you have chosen as your expertise, or by outside influences and events that thrust your area of science and/or yourself into the spotlight. If you are considering taking a more active role in interactions with policymakers there are a number of different training programmes for scientists (see Table 7.5).

7.4.1 Scientists and the politicians

Many scientists feel uncomfortable taking on an active role when it comes to communicating with politicians. In part, this stems from the lack of

Table 7.5 Examples of training for interaction with policymakers.

Programme	Programme details	Web address
Westminster Fellowship Scheme	3 month secondment for a student	www.rsc.org/science-activities/parliament/westminster-fellowship-scheme/index.asp
Royal Society of Chemistry	Awards 1–2 fellowships a year to work in the Parliamentary Office of Science and Technology (POST)	
Royal Society Pairing Scheme scientist and MP or civil servant	1 week in Westminster and reciprocal visits	http://royalsociety.org/training/pairing-scheme/
AAAS Science & Technology Policy Fellowships	Provide an opportunity for accomplished scientists and engineers with public policy interests to learn about, participate in, and contribute to the policy-making processes in Congress. Congressional Fellows spend one year serving on the staffs of Members of Congress or congressional committees, in legislative areas that would benefit from scientific and technical analysis and perspective	http://fellowships.aaas.org/02_Areas/02_Congressional.shtml
Leopold Leadership Fellowship	The programme offers two intensive training sessions a year apart to help Fellows gain the skills, approaches, and theoretical frameworks for translating their knowledge to action and for catalyzing change to address the world's most pressing environmental and sustainability challenges. Candidates must be midcareer academic environmental researchers who are: • based at an institution of research or higher education in North America; • conducting research related to the environment and sustainability in the biophysical sciences, the social sciences, or technical, medical, or engineering fields.	http://leopoldleadership.stanford.edu/fellowship-information
Newton's Apple	Newton's Apple, a charitable science policy foundation, was set up to help bridge the gap between scientists and policymakers. One of its major activities is a workshop programme designed to help young scientists and engineers gain a greater understanding of the processes of government and legislation.	www.newtons-apple.org.uk/

understanding of the similarities and differences that exists between the two professions – some of which are given here.

- A lack of common understanding about each other's roles. Interestingly both jobs involve research. However the scientist's role is generally to investigate a narrow area. In contrast, the politician's research involves learning a little about a wide variety of different issues.
- The scientist is free to decide on their own area of research and to adapt their research area based on their findings, hypothesis or the consensus in the scientific field. In contrast, politicians are answerable to their party, their sponsors or their constituents and therefore have less freedom to pursue their personal interests.
- Scientists' progress their work and knowledge by spending significant amounts of time focusing on their research area. It is unlikely that politicians have the opportunity to spend significant time focused on one issue or area, even if they would like to.
- Scientists are sources of detailed information, knowledge and expertise about science. Politicians are not often trained scientists, but they may be interested in scientific issues. They rely on sources of information, knowledge and expertise to make decisions that require a scientific input.
- A scientist can have a hypothesis that they will test. They will weigh the scientific and experimental evidence and if it refutes the hypothesis, they are comfortable rejecting it and testing a new one. In contrast politicians are not comfortable with this concept, which can be viewed as indecision and a lack of commitment.
- The politician may expect the scientist to be a generalist and to provide information about scientific issues that fall outside their area of expertise. A scientist may feel ill prepared to discuss issues outside their area of research.
- For scientists to succeed they need to be professional, extremely hardworking and committed to their research. The level of commitment required can only be delivered if the scientist is passionate about their role. The politician is also an extremely hardworking committed professional who is passionate about their job.

7.4.2   The passive role: the pure scientist

There will be times when a scientist becomes a passive influence on policymakers. Scientists understand the process of publishing research papers in peer-reviewed journals that have as high an impact factor as possible. The benefits of this are numerous but include ensuring research papers maximise their audience and your research is well regarded by your peers and colleagues. There are occasions when the press release for your research paper is picked up by the local or national press and it is possible that this may catch the attention of a policymaker, parliamentarian or their researchers and staffers,[23] but it is unlikely that this process will happen without media intervention. It is also a fact that there are not many trained scientists among the

---

[23] A term in the US used for researchers and information providers to policymakers

realms of the policymaker.[24] However, it is important for scientists to realise that a lack of scientific training should not be confused with a lack of interest in science or a lack of intellectual ability to understand science and scientific issues. Nor should this lack of scientific training be regarded as an excuse for scientists to avoid communication opportunities. There are many policy parliamentarians who are interested in science and scientific issues and welcome the opportunity to communicate actively with scientific experts who they respect and trust to provide accurate information. Clearly scientists have an opportunity to engage and communicate effectively with this receptive audience. There is no denying, however, that a politician will not always be a receptive audience. As with society in general, there are some politicians who, to quote Dr Huppert, MP, are 'antiscience' and 'have a set of beliefs and they will argue that regardless of the science' (Morris, 2010). As a scientist, the prospect of communicating with this particular group of policymakers is not an attractive proposition. However, at the same time scientists have to ask whether they can afford not to challenge or communicate with this group?

7.4.3 The passive role: the science arbitrator

There are opportunities to act as science arbitrators by providing information and knowledge as witnesses at Hearings called by committees. Hearings are the forum used to gather information, as well as different perspectives and points of view, from experts about specific issues as part of the decision making and legislative process. Hearing committees are also opportunities to draw attention to and raise the profile of scientific issues.[25] One of the most recent high-profile select committees that required scientists to provide evidence was the one in which Professor Phil Jones provided evidence to the UK Parliamentary Enquiry into Climategate. This committee looked at the series of events that took place at the Climate Research Unit (CRU)[26] based at the University of East Anglia. The UK Government undertook this inquiry to examine 'the two fundamental issues of events at CRU: firstly, were CRU's data and science sound, and secondly, were the university and its scientists intentionally trying to hide information?' Clearly this is not an experience that many scientists would seek out or enjoy. However, the outcomes of this enquiry were ultimately positive as the committee's conclusions were that 'the information contained in the illegally disclosed emails does not provide any evidence to discredit the scientific evidence of anthropogenic climate

---

[24] Lack of scientific background is discussed in the following articles and book by Nancy Baron http://news.sky.com/home/politics/article/15676161, and http://www.independent.co.uk/news/uk/politics/only-scientist-in-commons-alarmed-at-mps-ignorance-2041677.html Escape from the Ivory Tower by Nancy Baron

[25] For more information about the role of parliamentary committees and hearings visit http://www.youtube.com/watch?v=0f-SRugWrkw&feature=plcp&context=C3674006UDO EgsToPDskLtl81sEhqHC7fCNI0zXNsZ the following video on the Parliamentary You Tube channels outlines select committees in action http://www.youtube.com/watch?v=2feyugss YOk&feature=plcp&context=C3c1afa6UDOEgsToPDskLkNbc7_z4zXOWFNyHYOh2z

[26] http://www.youtube.com/watch?v=gPFdhxBbHX0

change'.[27] This enquiry also challenged scientists to review their processes and their practise with regard to the openness of data. The report states that '[we] recommend that the scientific community should consider changing those practices to ensure greater transparency'[28] (Secretary of State for Energy and Climate Change, 2010).

It is not just the parliamentary Select Committees that seek scientific evidence; there are also Public Bill Committees that can request written information from outside organisations such as NGOs, lobby groups and interested individuals who are members of the public. The evidence can be written evidence and/or oral evidence submitted at the beginning of the proceedings. Note that all Public Bill Meetings are public affairs, meetings are recorded by Hansard and the record is available on their website or in hard copy shortly afterwards.

In the UK, the Science and Technology Committee 'exists to ensure that Government policy and decision-making are based on good scientific and engineering advice and evidence.' It comprises a group of 11 MPs who represent different political parties in the UK parliament. A chair of the committee is also appointed. The current members come from different backgrounds and have a variety of different science qualifications with some at degree level. However, there are no members that currently hold a PhD. Nevertheless, members of the committee have been appointed to their role because they have an interest in science.[29]

The UK Houses of Parliament have published information about the role of the expert witness in Select Committees.[30] They have also produced a video released on YouTube that outlines this role. The video highlights that preparation and rehearsing the key points that you want to make are of paramount importance to ensure that you deliver them as succinctly, persuasively and as clearly as possible.

7.4.4   The honest broker of policy

Becoming an honest broker of policy takes a clear mindset as well as patience and commitment. Scientists have to be committed to making a difference, be prepared to act as a useful scientific resource, but to be apolitical and effective communicators. This role will, by its very nature, mean that you must be prepared to answer and address issues outside your area of expertise; it requires a scientist to be prepared to deliver a bottom line even if it is, at best, an informed guess. A politician is looking for someone to help them to clarify issue and to be accessible. They are also keen to know a scientist who is able to offer solutions, recommendations or suggestions about outcomes

---

[27] http://www.official-documents.gov.uk/document/cm79/7934/7934.pdf

[28] http://www.official-documents.gov.uk/document/cm79/7934/7934.pdf

[29] http://www.parliament.uk/business/committees/committees-a-z/commons-select/science-and-technology-committee/inquiries/

[30] http://www.parliament.uk/get-involved/have-your-say/take-part-in-committee-inquiries/witness/

and consequences. The scientist has to foster a wider understanding of the political and societal issues that will also impact or be impacted by scientific decisions or choices. Nancy Baron describes this as a job description for a 'go to scientist' (Baron, 2010).

7.4.5 The issue advocate

It is a fair assessment that scientists who choose to get involved with policy-makers usually do so because they have a particular issue they feel strongly enough about to overcome the activation energy associated with communication activities at this level. Generalising somewhat, scientists that become issue advocates usually do so because they:

- are seeking to raise awareness of scientific issues, because either they wish to ameliorate negative consequences or they wish to ensure positive outcomes;
- are seeking financial support for science or scientific issues.

There are several approaches that you can take to communicate with parliamentarians to influence policy. You can take the first step of contacting your local MP and arrange to meet; contact details including email addresses are easily tracked down using an initial Google search. The other alternative is to approach them in person. In the UK, every MP spends time in their local constituency and they hold clinics in accessible locations that you can attend as a scientist who is also a local constituent.

An alternative approach is to work through your science society. One of the roles of science societies is to provide:

- briefings to key stakeholders and policymakers;
- evidence and opinion, based on the most up-to-date research;
- expertise for governmental reviews and committees.

The Royal Society is one of the most active in this area and it has been providing evidence to governments and to policymakers since 1664, when it delivered a report on the state of Britain's forests to King Charles II. It has its own science policy advisory group[31] and also organises 'policylab' meetings that bring together scientists and policymakers to discuss emerging and timely scientific issues.[32] These meetings are podcast and more information about the discussions can be obtained if you sign up for regular updates about the policy work undertaken. The majority of science societies are actively seeking to play a role in policymaking decisions in the UK. These include The Royal Society of Chemistry, The Society of Biology, The Institute of Physics, and the London Mathematical Society amongst others. Scientists can join these societies for a fee (which can be tax exempt) but you may need a member of the society to nominate you.

As well as science societies, there are other campaigns and petitions that are an effective resource of advocacy for science and scientific issues. In the UK, 'Save British Science' started in 1986 as a grass roots movement of scientists who wanted to ensure better prospects for British science. The movement

---

[31] http://royalsociety.org/about-us/governance/committees/science-policy/
[32] http://royalsociety.org/policy/policylab/

SBS SAVE BRITISH SCIENCE

Basic science has given us radio and television, plastics, computers, penicillin, X-rays, transistors and microchips, lasers, nuclear power, body-scanners, the genetic code, ... All modern technology is based on discoveries made by scientists seeking an understanding of how the world works, what it is made of and what forces shape its behaviour. Basic science is uncovering the secrets of life, gaining knowledge that defeats disease, inventing new materials, understanding the Earth and its environment, looking deeper into the nature of matter and reaching towards an understanding of the Universe.

Today's basic research enlarges our conceptions of the world and our place in it and underlies tomorrow's technology, the basis of future prosperity and employment.

Yet British science is in crisis: opportunities are missed, scientists emigrate, whole areas of research are in jeopardy. The Government's support for research is declining, falling further behind that of our main industrial competitiors in Europe whose policy is to increase investment in scientific research.There is no excuse: rescue requires a rise in expenditure of only about one percent of the Government's annual revenue from North Sea oil. We can and must afford basic research, Britain's investments for the future.

**ASK YOUR MEMBER OF PARLIAMENT
TO HELP SAVE BRITISH SCIENCE
BEFORE IT IS TOO LATE**

1500 scientists
have paid for this advertisement

For information write to:
SAVE BRITISH SCIENCE
P.O. Box 241,
OXFORD, OX: jQQ
or telephone: (0863) 54993

Figure 7.3 Original advert placed in *The Times* by the 'Save British Science' Campaign. Reproduced by permission of The Campaign for Science and Engineering.

started as a fundraising event to buy an advert in *The Times* newspaper to raise concerns over the lack of funding for scientific research in the UK (Figure 7.3). Over time this movement morphed into The Campaign for Science and Engineering (CaSE). An interesting feature of the CaSE website is that it provides information about British MPs who are known to have an interest or background in science.[33] CaSE is now one of the leading advocates for science and engineering policy in the UK and is supported entirely by its members. Scientists can join CaSE for a low monthly fee and begin to take a more active role in raising the profile of science and engineering.[34] More than 20 years after the Save British Science campaign, the spectre of savage cuts to British science in 2010, was the catalyst for the new 'Science is Vital' campaign.[35] This was another grassroots movement backed by CaSE and founded by Dr Jenny Rohn. This campaign was organised both as a protest against the

[33] http://sciencecampaign.org.uk/?page_id=1543
[34] http://sciencecampaign.org.uk/
[35] http://scienceisvital.org.uk/

Government cuts to science spending and the proposal to limit the immigration of talented and skilled international scientists to work in science laboratories within the UK. The campaign included a petition that described the importance of science and engineering to the UK economy, which quickly gathered momentum and support as it circulated through the internet. The petition was handed into parliament and led to meetings between representatives of CaSE and the 'Science is Vital' team as well as key policymakers. It also led to more than 2000 people attending a rally outside the HM Treasury supporting the campaign, which attracted a lot of media attention, as well as an Early Day Motion for MPs to sign and finally a lobby, attended by supporters of the campaign and 24 cross-party MPs. The impact of this campaign as well as other evidence gathered prior to the spending review, resulted in a more beneficial settlement for science than had been expected (Brook and Leevers, 2010). Another recent campaign that has focused on a scientific issue is the 'Antibiotic Action' campaign highlighted in Box 7.2.

Box 7.2   The 'Antibiotic Action' campaign[1]

The British Society for Antimicrobial Chemotherapy (BSAC) has launched a campaign to raise awareness about the lack of new antibiotics that are reaching the marketplace and to highlight the threat that this has to the management of 'superbug strains'. The global initiative 'Antibiotic Action' includes a petition that is receiving international support from scientists, researchers, clinicians, MPs and concerned individuals. The petition calls for the government to identify opportunities to accelerate the licensing process of new antibiotics, provide incentives for industry and pharmaceutical companies to bring new antibiotics to the marketplace, and finally to provide incentives to encourage collaborations between academia and industry to accelerate the production of new and novel antibiotics. The campaign has attracted the support of MP Kevin Barron, who tabled the following question for debate in the house:

> Does this House agree that it must support and help identify initiatives that will regenerate the discovery and development of new antibiotic agents, as advocated by the UK led global initiative 'Antibiotic Action'?

In addition it has also led to an Early Day Motion number 2418:

> That this House notes the importance of the aims of the 'Antibiotic Action' Initiative of the British Society for Antimicrobial Chemotherapy that is examining ways to encourage the discovery, research and development to bring to market new antibacterial agents; further notes the need to ensure that licensing processes are streamlined to provide safe and rapid licensing of antibiotics; understands the need to stimulate industry to overcome the challenges in developing and bringing to market new antibiotics; and supports a drive for initiatives to encourage more partnership working between academia, pharmaceutical and diagnostics companies in order to maximise the conversion of discovered candidate molecules into licensed antibiotics available for use on the NHS.[2]

[1] http://antibiotic-action.com/
[2] http://antibiotic-action.com/2011/11/15/early-day-motion-prime-ministers-question/

In the USA, it is clear that the agenda for getting scientists to communicate with policymakers has moved on. There is a clear move to encourage scientists

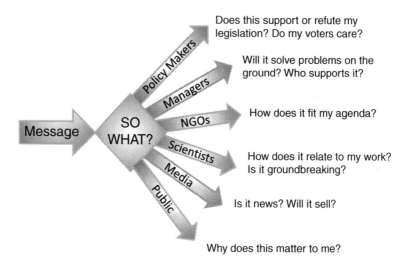

Does this support or refute my legislation? Do my voters care?

Will it solve problems on the ground? Who supports it?

How does it fit my agenda?

How does it relate to my work? Is it groundbreaking?

Is it news? Will it sell?

Why does this matter to me?

Figure 7.4 The So What Prism: Answering the question 'So What?' is the key for getting your message across effectively to your chosen group. From *Escape from the Ivory Tower* by Nancy Baron. Copyright © 2010 Nancy Baron. Reproduced by permission of Island Press, Washington DC.

to begin to take a more active role in public life: moving out of the laboratory and into public office (Russo, 2008)[36]. It has been recognised that trained scientists can offer real insight into issues such as science education, research funding and problems of national importance that require scientific insight (Dean, 2011). Recent developments have been the formation of a bipartisan political action committee called the Ben Franklin's List that aims to offer engineers and scientists the financial backing and the credibility required to win office (Dean, 2011).[37]

7.4.6   How to communicate your message effectively to policymakers

As a scientist you may contemplate communicating your research to policy-makers. There is one advantage that you have over anyone else; you are the expert. However, in many ways this is also a major disadvantage, because you know your research area in extreme detail but you cannot let the detail mask the main message that you are trying to communicate. One of the first things that you need to do is to work out what your main issue is, and to understand why it will interest politicians or policymakers. You need to recognise that your message, or the way you communicate your message, will vary depending on your audience. Therefore you need to understand who you intend your audience to be as illustrated in Figure 7.4. To pare down your research information into a clear message, you need to remove all extraneous detail and concentrate on the big picture. Nancy Baron has developed the message box (Figure 7.5) – a straightforward, simple tool that allows you to develop the

---

[36] http://www.nature.com/news/2008/080521/full/453434a.html

[37] http://www.nytimes.com/2011/08/09/science/09emily.html?_r=1&pagewanted=all

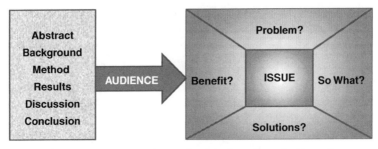

Figure 7.5 The Message Box. From *Escape from the Ivory Tower* by Nancy Baron. Copyright © 2010 Nancy Baron. Reproduced by permission of Island Press, Washington DC.

key message that is most appropriate for your audience. Among other things the message box can help you:

• explain to non-scientists what you do;
• prepare for interviews;
• refine your 30-second elevator speech for talking to policymakers;
• polish an abstract or cover letter for publication;
• write an effective press release;
• storyboard your website.

—From *Escape from the Ivory Tower* by Nancy Baron. Copyright © 2010 Nancy Baron. Reproduced by permission of Island Press, Washington DC.

The message box consists of four quadrants arrayed around a central issue. The following questions should be answered in each quadrant.

1 **Issue** – in broad terms, what is the overarching issue or topic?
2 **Problem** – what is the particular problem or piece of the issue that I am addressing?
3 **So what?** – why does this matter to my audience?
4 **Solutions** – what are the potential solutions to the problem?
5 **Benefits** – what are the potential benefits of resolving the problem?

—From *Escape from the Ivory Tower* by Nancy Baron. Copyright © 2010 Nancy Baron. Reproduced by permission of Island Press, Washington DC.

The message box does not need to be completed in a linear way. It can be started and completed in a stepwise fashion, but you can answer each of the quadrants in an order that is effective for you and the issue or problem that you are addressing. Ideally you should be aiming to get the answer to each of the four quadrants onto one side of A4. For an example of a message box applied to the 'Antibiotic Action' campaign see Box 7.3.

Box 7.3   Using the message box to analyse the BSAC, Antibiotic Action Campaign.

The Issue: lack of modern antibiotics and an increasing resistance to infections.

The Problem: there are not enough new antimicrobial compounds reaching the marketplace. At the same time we have a growing number of drug-resistant bacteria that cause devastating human diseases.

> **So What?:** this will have a major impact on global health and a return to a pre-antibiotic era where disease-causing bacteria will cause significant mortality and morbidity.
> **The Solutions:** raise awareness of the issue and appropriate use of antibiotics. Incentivise the development of new antibiotics and accelerate the licensing process for new antibiotic agents.
> **The Benefits:** avoiding a potentially significant health crisis and, consequently, improving the prospects for global health.

7.4.7 The scientist-citizen

Scientists have to remember that even when it comes to making policy decisions regarding science, a politician will not just be influenced by the scientist, or the evidence. Instead they will make their decision based on media and public opinion, the information provided by science societies as well as information provided by research funders or grant-awarding bodies. This should not be a reason for the scientist to disengage from the process. On the contrary, a scientist who has taken steps to move outside the laboratory environment and into the political realm can be regarded as a scientist-citizen because they can make an impact on policymaking decisions. Case study 7.9 is written by Professor Robert Wells and it describes his forays into the public arena of policymaking and decisions in the U.S. It outlines some of the different issues that have required advocacy and the different approaches used.

Case Study 7.9

The Scientist-Citizen
Robert D. Wells

The early years

I started my professorial career at the University of Wisconsin, Department of Biochemistry in the mid 1960s. This department was spectacular for a young, professor attempting to build a research and educational programme. It has a very long and distinguished history (over 86 years at that time) and was one of the premier departments in the US. A large amount of the basic work on many of the vitamins had been performed there, and warfarin (a widely used anticoagulant and rodenticide) had been discovered by one of my colleagues. The graduate students were talented and the educational programmes were quite solid. My research programme built quickly and I soon had a bustling lab of eager young students and postdoctoral fellows working on important DNA biochemical problems.

I quickly realised that the progress in my lab, indeed in the entire biomedical enterprise in the country, depended intimately on the availability of appropriate funding. The federal government is the only source of suitable, sustained funding to support an ongoing and vigorous programme for an extended period of time. Admittedly, some smaller and shorter term programmes may be supported with funds from state governments, private foundations, corporations, or other private sources. Thus, my dependence on the US federal government was complete. I was curious about the mechanisms whereby I received money from the National Institutes of Health (NIH), the National Science Foundation (NSF), the Department of Energy and/or other agencies in response to my grant proposals and, indeed, I was curious about the way that the agencies received money from the US Congress. Hence, the next 30–45 years were, in part, engaged in learning about this process and attempting to positively influence the support of important scientific research.

My first interfaces with congress

My serious involvement in public affairs came in the late 1980s after I had taken the Chairmanship of the Biochemistry Department in the Schools of Medicine and Dentistry at the University of Alabama at Birmingham (UAB). We rapidly grew this medical complex to over $120 million from less than a million dollars in extramural support in about ten years and were very successful in the grantsmanship arena; this included the funding of an new NIH Training Grant in Molecular and Cellular Biology, an exceedingly competitive area. When we applied for a renewal of this grant after five very successful years and received a 'fundable' priority score, we were informed that the NIH had removed all new funding from the training grant programmes; our wonderful training grant would not be funded. I was *livid*. I felt this was a wrong decision by the NIH as the education of new young scientists is the life-blood of the research enterprise.

Hence, I decided to take action. Instead of attempting to work within the NIH bureaucracy, I was advised by a UAB Vice President to meet with Congressmen to ask them to intercede on my behalf on this important educational issue. I wrote to all other NIH Training Grant Directors (about 45 in total) and solicited their support, both philosophically and financially, and was reasonably successful. I broached a lobbying company in the District of Columbia that works in the area of health affairs and they volunteered to assist with this effort. In the course of my Congressional meetings, I was introduced to Senator Mark Hatfield (Republican from Oregon) who was one of the most supportive and inspiring men I have ever met. While waiting in his outer office, I was informed by a staffer that I had ten minutes with the Senator; be out in 12 minutes since a large number of others were waiting to meet him. Upon meeting Senator Hatfield, I informed him of my mission and he let me know in a short time that he knew as much, or more, about the NIH budget as I and he fully supported my efforts. He then moved on to raise a number of mammoth issues with me such as 'Why do we have a scientifically illiterate public in the US' and 'Why do our kids score at or near the bottom of the pack compared to other enlightened countries in math and science'? We talked for the next 90 minutes and he commented toward the end of the conversation, 'Dr Wells, I don't often have the chance to talk to someone like you'. Thus, you never know how a Congressman will react.

To summarise the outcomes, Senator Hatfield made a phone call. The NIH restored the training grant funding and the UAB grant was refunded. Also, Senator Hatfield asked me to organise a consortium of biomedical scientists to meet with him and Admiral James Watkins, then Secretary of the Department of Energy, to discuss the issue of science education in the US; we met on three occasions over the next 18 months for lunch meetings in the US Capital to take initial steps toward improving these problems. This resulted in the bill (S.2114) sponsored by Senators E. Kennedy and M. Hatfield entitled 'Excellence in Mathematics, Science, and Engineering Act of 1990'. The committee was composed of Drs Robert M. Bock (University of Wisconsin), Robert T. Schimke (Stanford University), I. Bernard Weinstein (Columbia University), Daniel Nathans (Johns Hopkins University), Robert H. Burris (University Wisconsin), Dieter Soll (Yale University), and Robert L. Hill (Duke University). This act was one of the initial efforts to call attention to the growing deficiencies of science education in the US; happily, there is now a tsunami of interest in this problem.

Note: I am not anyone special. I had no special qualifications or talents for these endeavours. I simply took action after I became upset by a federal decision since I felt deeply about our financial support. I knew that the governmental support was critical for our scientific progress. Thus, *get involved, stand up and take action*. You have everything to gain and little or nothing to lose. No one is likely to 'require' you to get involved; you must be self-motivated. But remember, you are on the 'side of angels' in these efforts.

Texas and society presidencies

In 1990, I founded the Institute of Biosciences and Technology (IBT), Texas A&M University, in the Texas Medical Center, Houston. Hence, I was extremely busy hiring the faculty, building the graduate educational programme, the infrastructure, Center system, human resources, outreach, capital development, plus many other activities including my exciting research programme (http://www.ibt.tamhsc.edu/labs/cgr/wells.html). Today, the IBT is a vigorous institute comprised of ~25 research teams working on a variety of medical research projects in an 11-story ~130 000 square foot building in the heart of the Texas Medical Center. On top of all of this, in 1990, I was elected to be the President of the American Society for Biochemistry and Molecular Biology (ASBMB).

The ASBMB (founded in 1906) consists of ~12 000 members and has become an international society. My 4-year term, as President-Elect, President, and then Past President, enabled a fine overview of the broad agenda of advocacy, meetings (the large national conference and smaller, specialty symposia), and publication (*Journal of Biological Chemistry*, *Molecular and Cellular Proteomics*, *Journal of Lipid Research*, and *ASBMB Today*) interests of the Society. Each of these three areas has several staff personnel in the home office (Bethesda, MD) and diverse and multifaceted activities within their areas. Hence, I was able to support and strengthen the wonderful public affairs programmes of this society. Some of these interests included: Congressional visits, news pieces and information, fellowships for junior science policy interns, awarding of the Schachman Public Service recognition, and development of position statements and letters to congressmen. Also, we developed a DVD entitled *Meeting Your Congressman: A Guide for the Grass Roots Advocate* to train scientists on the suitable methods for advocating with government leaders. I encourage you to visit (mms://asbmbcdn001.navisite.com/ASBMB/Meeting-with-your-Congressman.wmv).

The ASBMB is one of 23 societies that make up the Federation of American Societies of Experimental Biology (FASEB). I was elected to be the President of FASEB in 2003 and served a 3-year term in office. FASEB, founded in 1912, has more than 100 000 members. This is the largest and most prestigious consortium of biomedical research societies in the US. Its mission is to enhance the ability of biomedical and life scientists, through their research, to improve the health, well-being and productivity of all people. With its home office in Bethesda, MD, it serves as the dominant spokesman for the interests of the medical scientist in the US government. Hence, this is the 'gorilla in the room' with respect to public affairs. FASEB, like ASBMB, has a broad agenda with vigorous programmes with respect to journals, publications and especially advocacy. The FASEB public affairs office in Bethesda consists of ~12 professional staff who are actively conducting the programmes working toward the goals stated above.

The White House visit

When I assumed the FASEB Presidency, I decided to attempt to have an impact on the successes of biomedical research. I knew that the US budget was initially drafted in the White House by the President and his staff before Congress began their deliberations on this document. If the President's budget for scientific research was too low, it would be exceedingly difficult for FASEB and other efforts to get the funding up to a reasonable level to sustain the research enterprise. I broached the FASEB Office of Public Affairs staff during the first weeks of my Presidency about attempting to get a meeting in the White House with President G.W. Bush to attempt to improve science funding for the future. We had two preliminary meetings to discuss this possibility over a period of two months; this type of initiative had never been attempted previously by FASEB officials. At our third meeting during the third month, Dr Howard Garrison, the FASEB Director of the Office of Public Affairs, declared 'Let's go to the White House'. 'If you [Bob Wells] can get several Nobel laureates in science involved to accompany us, I'll try to arrange the visit at the critical time when the budget is being finalized and sent to Congress'. This was a major, innovative and bold step forward.

In July–October 2003, I worked hard to solicit the involvement of four Nobel laureates who I could depend on to make a positive impression and to commit to a day in DC; simultaneously, Dr Garrison and his staff attempted to arrange a meeting with President Bush. Rep. Bob Michel, Republican from Illinois and former House Minority Leader, was invaluable in contacting the White House to make suitable arrangements. Rep. Michel has been a vigorous supporter of medical research and a good friend of FASEB. The date for the visit was arranged for 20th November 2003. Surprisingly, Mr. Bush was summoned to London on that date for a state dinner with the Queen and we rearranged to meet Vice President Richard Cheney. Numerous advisors have told us this may have been a serendipitous substitution since V.P. Cheney may have been more influential than Mr Bush.

The meeting was held in the Roosevelt Conference room of the West Wing of the White House. Our topic for the meeting was that medical and scientific research is a valuable and noble endeavour and that the US should increase its support for these programmes. The presence of Drs Gilman, Altman, Rowland and Cech insured that I had the attention of the highest level of the US administration. Our meeting went very well for a period of ~50 minutes; although Cheney's staff told me that we had ~20 minutes with the Vice President. In fact, he was quick to acknowledge that he was a grateful benefactor of medical research; his cardiac issues are well known.

We learned later that V.P. Cheney spent a full day on the US budget two days after our visit and that he moved ~$750 million in new money into the budget of the National Institutes of Health. Not bad for a day's work!

Top Tips-General principles

- Get involved – stand up and let your thoughts be known.
- In democratic forms of government, legislators *need* to know your opinions; these people, in general, are attorneys or businesspeople and they do not know what is transpiring in science. They need our input. You can provide a valuable service to your government.
- You have a right, indeed a responsibility, to meet with your representatives in a democracy. Whereas this is a time-honoured tradition in the US and is becoming more commonplace in the UK, I encourage scientists in other countries to also engage their parliamentarians. If you have any questions about the appropriateness of these actions, ask your institutional superior.
- 'The squeaky wheel gets the grease'.
- Adopt the attitude that you are advocating and/or educating for a cause, not lobbying.
- Invite the Representative to visit your workplace; they (and you) might benefit in several ways.
- Volunteer to provide information/advice in the future on science and/or science policies. You might be positively surprised at the outcome.
- Join your scientific society; many societies have an advocacy office and can provide support and training.
- Inform your home institution of your activities; many will support your efforts.
- Identify a mentor; ask for help. Some of the issues are so large that people are anxious to obtain assistance from others.

- Meeting with Congressmen is a time-honoured tradition in the US; it's an important component of the democratic enterprise.
- You can meet with Congressmen in your capital city; also, you can meet with them in their home offices. (The latter is cheaper.) Frequently, they are honoured, even charmed, to meet with you.
- Scientists are upset when a proposal is not funded; but remember, the granting agency would like to fund the grant but they have insufficient funds to fund all proposals. Attack the problem at its roots, which is the level of funding.
- Don't be misled to think that someone else is representing your interests; in all likelihood, they are *not*.
- Just because you know that scientific and/or medical research is a noble activity, don't be snookered into thinking that legislators will extend special considerations and fund these endeavours. In general, they will *not*! Remember that research funding is competing with prisons, defence, highway construction, welfare, etc. and each of these activities has their supporters who are 'knocking on the legislator's doors'.
- In the past 20 years, I have met a large number of the important US legislators. They are, in general, well meaning and dedicated people.
- No one will 'coerce you' to undertake these activities; you must be self-motivated.
- The advocacy process is enjoyable, interesting, and rewarding.
- *One person* can make a difference!

7.5   Concluding remarks

This chapter has highlighted the different ways that scientists can communicate directly with different segments of the public not only to provide information about science but also to take part in a more open and equal dialogue. Developing a successful scientific career depends on building a strong reputation amongst your fellow scientists, but remember that as you develop your expertise you are also ideally placed to provide and share your experience, knowledge and inspiration with a much wider community than just your peers. Indeed, making a difference to the scientific community doesn't just have to happen at the laboratory bench. There are many other opportunities to become a scientist-citizen and the message seems to be that each individual action and every single voice matters and can have an impact, including your own. Chapter 8 continues the theme of public science communication. In contrast to this chapter it highlights indirect interactions, which nevertheless can have dialogue built into the process.

References

Alexander, E.P. and Alexander, M. (2008) Museums in Motion. *An Introduction to the History and Functions of Museums*. AltmaMira Press

Barbacci, S. (2004) Science and Theatre: A mutifaceted relationship between pedagogical purpose and artistic expression. Paper presented at the 8th International Conference on Public Communication of Science and Technology, Barcelona, June 3–6, 2004

Baron, N. (2010) *Escape from the Ivory Tower*. Island Press, Washington DC

Berman, M. (1978) *Social Change and Scientific Organisation. The Royal Institution 1799–1844*. Heinemann Educational Books

Brook, M. and Leevers, H. (2010) Science is vital. CaSE News **65** (October), 8–9

Buckley, N. and Hordijenko, S. (2011) Science festivals, *in Successful Science Communication, Telling It Like It Is* (eds Bennett, D.J. and Jennings, R.C.). Cambridge University Press

Caulton, T. (1998) *Hands-on Exhibitions: Managing Interactive Museums and Science Centres*. Routledge

Clery, D. (2003) Bringing science to the cafés. Science **300**, 2026

Dallas, D. (2006) *Cafe scientifique—deja vu*. Cell **126** (2), 227–229

Dean, C. (2011) Groups Call for Scientists to Engage the Body Politic, The New York Times (August 8)

Djessari, C. (2007) When is 'Science on Stage' really science? *American Theatre* **24**, 96–103

Fors, V. (2007) The Missing Link in Learning in Science Centres, Doctoral Thesis, Lulea University of Technology, ISSN:1402–1544 http://epubl.luth.se/1402–1544/2006/07/

Government Response to the House of Commons Science and Technology Committee 8th Report of Session (2009–10) The disclosure of climate data from the Climatic Research Unit at the University of East Anglia September (September 2010) Secretary of State for Energy and Climate Change by Command of Her Majesty

Grand, A. (2008) Engaging through dialogue: international experiences of Café Scientifique, *in Practising Science Communication in the Information Age* (eds Holliman, R., Thomas, J., Smidt, S., *et al.*). Oxford University Press.

Grand, A. (2011) What's it like to speak at Café Scientifique? *The Bulletin of The Royal College of Pathologists* **154**, 114–116

Hari, J. (2009) Is Tom Stoppard's Arcadia the Greatest Play of our Age? The Independent (22 May)

Hook, N. and Brake, M. (2010) Science in Popular Culture, *in Introducing Science Communication* (eds, Brake, M.L. and Weitkamp, E.). Palgrave Macmillan

Jardine, L. (2003) *The Curious Life of Robert Hooke: The Man who Measured London*. Harper Perennial

Kumar, M. and Ince, R. (2011) The Science of Comedy, Daily Telegraph (1 May)

Morris, N. (2010) Only Scientist in Commons 'Alarmed' at MPs' Ignorance, The Independent (3 August)

Pedretti, E. (2008) T.Kuhn meets T.Rex: critical conversations and new directions in science centres and science museums. *Studies in Science Education* **37** (1), 1–41

Pielke, R. (2007) *The Honest Broker*. Cambridge University Press

Public Attitudes to Science (2011) Ipsos MORI poll for the Department for Business Innovations and Skills. http://www.bis.gov.uk/policies/science/science-and-society/public-attitudes-to-science-2011 (accessed 5 May 2012)

Royal Institution (2011) http://www.rigb.org/registrationControl?action=home (accessed 5 May 2012)

Russo, G. (2008) Meeting urges scientists into politics. *Nature* **453**, 434

Shepherd-Barr, K. (2006) *Science on Stage: From 'Doctor Faustus' to 'Copenhagen'*. Princeton University Press, Princeton NJ

Shortland, M. and Gregory, J. (1991) *Communicating Science, A Handbook*. Longman Scientific and Technical

Taylor, C. (1988) *The Art and Science of Lecture Demonstration*. Institute of Physics (IOP)

Walker, A. (2011) Review: Uncaged Monkeys – A Night of Science and Wonder, South Wales Echo (11 May)

CHAPTER EIGHT
Indirect Public Communication

Our species needs, and deserves, a citizenry with minds wide awake and a basic understanding of how the world works.

—Carl Sagan

8.1 Introduction

This chapter investigates examples of science communication that are generally indirect, i.e. the encounters are not usually face-to-face. Indirect communication can still include both information and conversation using a variety of different media. This media includes television, radio, print newspapers and books. This chapter also includes the growing area of citizen science and explores how you can engage in conversations with the public through Web 2.0 tools and services.

8.2 A focus on science and television

The majority of people (54%) receive most of their science knowledge through television (PAS, 2011), as shown in Figure 8.1. Thus it can be argued that science communication through TV remains the medium with the most impact. Despite this, very few scientists develop TV careers. Those who do, become highly influential.

Science crops up on television in many different guises:
- drama series such as the successful CBS drama franchise *CSI* and the BBC's *Dr Who* and *Sherlock Holmes*;
- comedy dramas such as CBS's *The Big Bang Theory*;[1]
- news reports on all local and national TV channels;
- chat shows such as the BBC's *The One Show*;
- panel shows such as the BBC's *QI*;
- science and nature TV series such as the seminal 1980s, Emmy award-winning documentary *Cosmos*, broadcast on PBS (Bucchi and Trench, 2008).

---

[1] http://www.cbs.com/shows/big_bang_theory/

Science Communication: A Practical Guide for Scientists, First Edition. Laura Bowater and Kay Yeoman.
© 2013 John Wiley & Sons, Ltd. Published 2013 by John Wiley & Sons, Ltd.

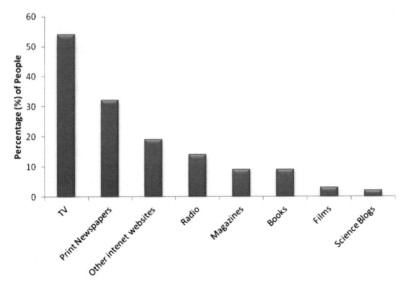

Figure 8.1 The percentage of people aged 16+ responding to the question about which were their most frequent sources of information about science. The data has been taken from PAS 2011.

Current programmes with a focus on science and nature include the factual science programmes *Horizon,* and the popular science show *Bang Goes the Theory* that are broadcast on the BBC. Other programmes with strong science content are the scenic and panoramic natural history and wildlife series such as the BBC's *Frozen Planet* and *Walking with Dinosaurs.* In fact, the longest running BBC series is *The Sky at Night* hosted by Patrick Moore, amateur astronomer and communicator extraordinaire.

Interestingly, TV coverage is not evenly spread across all scientific disciplines; content analysis studies of TV programmes have shown that nearly half of all science stories on TV have a health and medicine angle. This perhaps reflects the common interest of a wide viewership on health-related issues (Hargreaves *et al.*, 2004). More detailed, qualitative in-depth, narrative studies have also investigated particular television programmes such as *Tomorrow's World* which used to be broadcast by the BBC and the *Nova* on the PBS. These studies indicate that science is portrayed as a 'product rather than a process of inquiry'. In addition, the journey to scientific discovery and certainty is shown as orderly, predefined and ultimately sequential. It also showed that overall television tends to portray scientists as 'omniscient, authoritative and credible' (Hansen, 2009). However, viewers of *The Big Bang Theory* would struggle to recognise this perception of the scientist. Instead, viewers of this comedy drama are confronted with a group of scientists portrayed as neurotic, competitive and socially inept, who resemble the well-known stereotype of the mad scientist, albeit without the grey hair.

In general, investment in science programmes such as *Tomorrow's World, Nova* and *Horizon* tend(ed) to be commissioned by television channels paid for by the public purse: The BBC in the UK and PBS in the US. Programmes with a significant science and nature content match the remit of public broadcasting as defined by the UK's Office of Communication (Ofcom, 2004), namely to create programmes that make the 'viewers think'. Ofcom states that the role of public broadcasting is to provide programming that 'stimulate our interest in and knowledge of the arts, science, history and other topics through content that is accessible and encourages personal development' thereby promoting public participation (Ofcom, 2004). Nevertheless there are reports that suggest the amount of science programming is being squeezed across all terrestrial channels (Russell, 2010). Issues such as cuts in production budgets for specific science programmes, the concern that science programmes are too serious for the public audience (Bennett, 2009) and the emergence of other genres such as history are taking their toll on the broadcasting space for science programming (Russell, 2010). This is despite the recent PAS 2011 study indicating that half of people surveyed think that they hear and see too little science. Nevertheless, the BBC is still investing in, commissioning, producing and providing a variety of science programmes such as *Bang Goes the Theory*[2] that encourages its audience to actively undertake science experiments, as well as the *Wonders of the Solar System* in 2010[3] and the follow-up series *Wonders of the Universe*[4] in 2011, to name just a few.

Research has shown that citizens like TV science coverage as they felt programmes broke the science down into manageable chunks of information. However, despite this, 57% of people felt either not at all informed or ill informed about science (PAS, 2011). The data from the PAS 2011 study also indicated that those people who got their information from TV were least likely to feel informed about science (37%), while those who obtained information online through science blogs felt better informed (71%).

8.2.1 Scientists and television
You can catch glimpses of scientists on the TV talking about scientific breakthroughs on news programmes, occasionally as guests on talk shows and panel shows and as presenters on science and nature documentaries. Being interviewed on television can provide real challenges for scientists. Firstly, there is the necessity of effectively communicating your message while ensuring that the science is reported accurately. Secondly, and perhaps a more challenging issue, is to make sure that you convey a positive impression to a TV audience. This encompasses thinking about how you look and appear to the viewer as well as ensuring that you express your take home message in a clear and succinct manner. Scientists who seem to do this well have usually gained broadcasting experience, had media training and have rehearsed their message to create the key 'sound bite' to get their point across. Creating

---

[2] http://www.bbc.co.uk/programmes/b00lwxj1
[3] http://www.bbc.co.uk/programmes/b00qyxfb
[4] http://www.bbc.co.uk/programmes/b00zdhtg

a positive impression is also important in this visual medium. As a first step, conveying enthusiasm about your subject is really important. This can include explaining why your science matters and is interesting to the public, but it can also be helped by providing insight into your own views, concerns and opinions (Baron, 2010). Box 8.1 describes how two microbiologists have used their scientific expertise to provide TV features which have public appeal.

Box 8.1   Microbiology on TV

Dr William Rosche describes how he worked with his local TV news station Fox 23 based in Tulsa, Oklahoma. The concept was to produce a series of individual news segments with Bill and his colleagues working as a 'SWOT team'. This team visited different places around Tulsa (changing rooms in a clothing store or bus shelters are examples), swabbed them for bacteria and took the swabs back to the laboratory. The cultures were then grown on agar plates and the isolated bacteria identified. The isolated cultures were then revealed in short 3-minute news segments.

Similarly, this concept for TV material also cropped up in the UK. Microbiologist, Dr Anthony Hilton, took part in *Grime Scene Investigation*, which was an eight part television series that aired on BBC3 during 2006. In each episode a team of scientists in their mobile laboratory visited a member of the public to reveal the microbes that they were associated with. As well as *Grime Scene Investigations,* Dr Anthony Hilton has produced a series of three short segments for *The One Show* broadcast on BBC1. These segments explored whether nasty bacteria could be isolated from well-known objects in public places such as cash machines, pelican crossing buttons and computer keyboards.

http://www1.aston.ac.uk/lhs/staff/az-index/hiltonac/

Scientists have also been recruited as high profile presenters of science and nature documentaries. Interestingly there are certain scientists such as Professor Robert Winston, Sir David Attenborough and Professor Brian Cox who are now regarded as a 'marketable commodity or a brand in themselves' when it comes to attracting large audiences. In addition, the use of scientists as presenters brings credulity, credibility and authority to the programme. In addition these presenters engender trust in their audience that in turn, attracts viewers and promotes belief and acceptance in the subject matter being conveyed. It is clear that the success of scientist presenters such as Brian Cox and Alice Roberts, is in no small part due to their innate ability as communicators. But it is also because as well as being an expert in their subject, they are passionate about it and they are able to convey complex ideas in a straightforward and accessible way. However it isn't always the case that you have a TV science presenter who is an expert in that particular scientific discipline. Professor Robert Winston entered the public eye as an expert in fertility and genetics. Since this time he has morphed into a TV presenter who is comfortable fronting TV programmes such as the BBC's *Child of our Time* and *Walking with Cavemen* which are clearly outside his personal, scientific area of expertise (Bennett, 2009). In the UK there are several scientists who have come to the fore as effective communicators in their field of science thanks to their expertise and a unique collaboration between the BBC and the Open University. This has led to the production of many seminal science and nature programmes, including the natural history documentary *Frozen Planet* as well as the popular science programme *Bang goes the Theory* (see Box 8.2 for more details about the BBC and OU collaboration).

Box 8.2   Science on TV: The Open University

The Open University Works, the BBC and the Open University (OU), a distance learning higher educational institute, co-fund programmes from in-house and independent suppliers. These 'learning journey projects'[1] are broadcast across all BBC TV channels and radio stations. A priority subject area for these learning journeys includes science technology and nature, one of the five subject strands taught by the OU. The OU and the BBC commission programmes that contribute to the teaching material of the OU, enhance public understanding by highlighting key research areas and hopefully generate public interest that can be converted to students applying for OU courses. Additionally, the programmes are also moving to incorporate the latest technology, allowing content to be delivered in new and innovative ways. Academics and scientists from the OU at the start of every new project meet with the production team to ensure:

- the project can provide enough educational material to support a learning journey project;
- to assess what support can be made available at research and scripting stages;
- whether there is any need for academic involvement during filming.

Finally, for the collaboration to be successful, the OU has to be satisfied that it's appropriate for the University to be acknowledged as a co-producer in the programme and the OU Broadcast Commissioner has final sign-off on behalf of the OU. Final editorial control of the programme rests with the BBC.

---

[1] The learning journey is defined as an 'educational journey' that enhances and enriches the broadcast experience that the viewer takes after watching/listening to the programme.

www.bbc.co.uk/commissioning/briefs/tv/browse-by-genre/open-university-1

If you want to try and develop a TV aspect to your science career, then one way to see if you have the right skills is to enter 'Fame-Lab'[5] (Box 8.3).

Box 8.3   Fame Lab

If you want to develop a TV career, enter Fame Lab which began through the Cheltenham Science Festival. Fame Lab, which is now an international organisation, has produced alumni who have gone on to take part in and to present TV programmes. These people include space scientist, Maggie Aderin-Pocock, who presented *Do We Really Need the Moon?* and evolutionary biologist Simon Watt, who has worked as a presenter on programmes such as *Inside Nature's Giants*.

---

[5] http://famelab.org/

8.3   A focus on radio and science

Although it may seem a contradiction in terms, radio is both an intimate and a visual medium. In general people tend to listen to radio in isolation and as Martin Redfern states 'as a radio producer I often like to think of there being one listener and work from there' (Redfern, 2009). Radio has to be entertaining; it has to support and complement everyday life by capturing the listener's attention. It does this by 'creating a waking dream, a mental movie, using sound and the voices of people that count' (Joyce, 2010). Failure to capture attention can result in the listener pushing the off button or tuning into to an alternative radio station. The arrival of the digital age has led to a huge expansion of channels that can be received through the radio. This is expanded even further with the internet, allowing access to radio stations that broadcast throughout the world. In addition, listening to a 'live' broadcast is no longer a prerequisite, as many radio shows are recorded as podcasts which can be downloaded and enjoyed at a time that fits into your lifestyle. Radio has long recognised that science offers opportunities to inform and entertain. As early as the 1950s, radio programmes that feature science as a mainstay have been broadcast within the UK and further afield. Many contemporary science shows can be found on the programming schedules of different radio stations. In the US, the National Public Radio station broadcasts short science features as part of different shows, but in addition it produces a Friday Science Programme devoted to science. In the UK, the BBC continues to provide programmes that feature science such as the *Material World* on BBC Radio 4,[6] and *Science in Action* on the BBC World Service.[7] Both shows feature a magazine format that discusses contemporary issues and breakthroughs across all science disciplines. Such science programmes are not the just the preserve of (inter)national radio stations however. *The Naked Scientists*, is the brainchild of Dr Chris Smith a consultant virologist in the Pathology Department at the University of Cambridge. What it offers in contrast to other science shows is an interactive experience that started out on a local radio station in Cambridge. Since then it has been broadcast on different BBC Radio stations throughout the Eastern Region, as well as the ABC Radio National in Australia (Smith, 2011).

There has been an expansion in the number of radio stations that broadcast to a wide diversity of listeners. Many radio stations attract their own distinct audience. Community radio stations have sprung up and offer their listenership a programming schedule that is responsive to the interests and concerns of their audience. One such station is Future Radio, a community radio station for Norwich. It is part of the charity organisation Future Projects, a social inclusion project that aims to inspire and empower its local community. Case study 8.1 describes a science communication project that involved Future Radio as well as academics from different disciplines across the University

---

[6] http://www.bbc.co.uk/programmes/b006qyyb
[7] http://www.bbc.co.uk/programmes/p002vsnb

of East Anglia. The academics were recruited to take part in a series of radio shows that celebrated the anniversary of the birth of Charles Darwin and the publication of his seminal work, *On the Origin of Species*.

Case Study 8.1

The Charles Darwin Radio Series
Laura Bowater and Kay Yeoman

The Charles Darwin Radio Series was a collaborative project between the University of East Anglia and a community radio station based in Norwich: Future radio. Future radio is part of Future projects, a community based arts, media and education charity. It provides school inclusion projects for:
- young people, aged 13–16 years;
- post 16 support;
- its local and wider communities.

One of its mission statements is 'to educate, inform and advise on, music, media, radio training and education films/DVD's'. The radio series aired in the summer of 2009. It was written and produced to celebrate the bicentennial anniversary of Darwin's birth and the 150th anniversary of the publication of his seminal work *On the Origin of Species*. The project used the life and times of Charles Darwin and his scientific theories to provide a framework for five, weekly, half hour radio programmes that were broadcast on a Sunday afternoon at 3 pm. Each individual show had a theme:

1 Who was Charles Darwin and why did he develop his revolutionary ideas?
2 Why are scientists still studying evolution today?
3 Bacteria and evolution in action;
4 Choosing our sexual partners;
5 Effect of Darwin on today's society.

This series of programmes covered a wide variety of academic disciplines and research expertise which provided many opportunities for academics and local experts to become involved in the shows. This had an added advantage as we were able to attract academics to take part in the shows because we encouraged them to discuss their own particular area of expertise. We designed a series of questions with each of the contributors, then we invited them to come to the radio studio where they were interviewed using the agreed questions. This ensured that each interviewee was adequately prepared for the interview session, but it also allowed the interviewee to be actively involved in the design and content of the interview. Without exception all contributors were able to talk about their research interest with confidence and enthusiasm. The interviews were undertaken prior to the radio shows being aired. This allowed editing for 'ummms and ahhhs' to occur and meant that the final edited version of each interview had a professional quality. It also meant that once we had our package of edited interviews for each radio show, we were able to write a narrative that linked the interviews together to produce a programme that had a cohesive structure. Each academic and expert that we approached and asked to take part, did so with enthusiasm and good grace. Many commented that they actively enjoyed the process and welcomed the opportunity to take part in a science communication event that allowed them to chat about their own area of interest and expertise. By clearly explaining the demographic of the radio station's audience prior to the interview, all contributors had an understanding of this audience and made huge efforts to minimise scientific jargon whilst not 'dumbing down' the science.

We also had a fantastic team at the radio station who bought into the concept of the Darwin series. They provided us:

- with the studio time to undertake the interviews;
- webspace to house the podcasts and to provide opportunities for the radio listeners to get involved with questions and comments. Each radio show was individually podcast by the radio station and is still available online;
- with a radio producer who had knowledge and expertise of the radio studio and its equipment;
- with opportunities to advertise the series ahead of it's airing, which was invaluable.

The series of radio shows worked largely through the good will of all involved. However we did receive a small amount of funding from our local Beacon, CUE East which helped to cover some of the costs of studio and producer time of the community radio station.

The series of radio shows were successful because we involved a wide variety of experts who were enthusiastic, confident and well prepared. But this success came with a significant time cost. We made an initial contact with each contributor and met each one prior to the show to discuss the project, decide how they could contribute to the shows and to design a series of questions that worked for the interviewee and for the show. We also made sure that we attended each interview that took place at the radio studio. Finally we wrote and thanked each contributor after the interview recording had taken place. Contacting the interviewee after they had completed the interview also ensured that they were happy with the process as well as providing an opportunity to answer any questions that had arisen after they had taken part in the show.

The production of each radio show also took a lot of time as we were actively involved in the editing process. In addition, we wrote the transcript of the narrative that tied the individual interviews together creating a cohesive whole. Being so actively involved throughout the series meant that we had to be incredibly organised in order to fit the production of these radio shows around our 'day job'. However the amount of time and effort that we gave to the project produced a product that we are incredibly proud of.

http://www.futureradio.co.uk/charlesdarwinseries

If you are invited to take part in a radio programme it is important that you recognise that getting your science across in a factually correct way should not be your only outcome. Although this is important, 'creating a waking dream and a mental movie' also matter. As a first step you need to consider the key messages that you want your listenership to understand and learn about your science. It is worth taking the time to rehearse this message as short sound bites that roll of your tongue with ease. In addition they should also express the interest and excitement that you want to share with your audience. You need to keep your language simple and accessible but this doesn't mean that it should be dry and uninspiring. Consider building visual pictures through the use of metaphors and vivid descriptions. As scientists we are used to conveying our science in this way and we feel comfortable talking about DNA ladders and enzymes that catalyse reactions using a lock and key mechanism. What we don't feel so comfortable about is personalising our science and our research journey with anecdotes that bring the 'job of a scientist' to life. However, it is these personal anecdotes that can capture the audience's interest and provide you with the personal (and human) dimension that your audience can relate to.

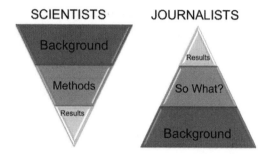

Figure 8.2 The difference between articles written by journalists and scientists. From *Escape from the Ivory Tower* by Nancy Baron. Copyright © 2010 Nancy Baron. Reproduced by permission of Island Press, Washington DC.

8.4   A focus on newspapers

According to the Public Attitudinal study of 2011, 32% of citizens receive their science information through the print newspapers (Figure 8.1). However the study also revealed that people were critical of science reporting in newspapers, and felt that they were less trustworthy sources of information compared to either TV or radio.

It is also fair to report that scientists are slightly sceptical when it comes to trusting journalists to accurately report their science story. This lack of trust stems from some of the differences that exist in the way in which articles are written (Figure 8.2), but also by other differences between the two professions:

- **timescales** – a scientist may have invested many years of careful, painstaking work prior to publishing a piece of research. In contrast the journalist wants to know the gist of the story as quickly as possible;
- **emphasis** – a scientist prioritises the evidence that underpins the conclusions. In contrast the journalist wants to understand the news angle of the conclusions;
- **absolutes** – a scientist recognises that there are no absolutes, a scientific hypothesis is only as good as the observations and results of the last experiment. A journalist recognises that their readers want absolutes and certainties;
- **personal angle** – the scientist wants to remove their personality from the research by reverting to the passive voice so that the research is the story. A journalist wants to bring the story to life by adding the voice and the personality of the scientist;
- **reputations and perspectives** – the scientist values and respects and judges the academic credibility of other scientists and his peers accordingly. Similarly a scientist judges the hypothesis and makes a balanced assessment based on the body of evidence. A journalist doesn't judge scientists in the same way. They also seek to bring all views and as many perspectives as possible to the story. A journalist tends to seek out a balanced view not based on the evidence but by presenting different opinions and interpretations of the evidence.

As a scientist it is worth remembering that if a journalist is to get your story into the newspaper ahead of other stories, they need to provide it with news appeal and this accounts for many of the differences between scientists and journalists. A good journalist also has a professional reputation to uphold and this depends on ensuring that they report your story accurately. This means ensuring that they quote you correctly and they do not over exaggerate the conclusions of your research whilst maintaining the news appeal of the story.

Throughout your career as a scientist, you might get your scientific research into the media, but it may be an irregular occurrence. There are many excellent guides that have been published and courses which you can attend to help you interact with different media and journalists; we have directed you to these sources in Box 8.4.

Box 8.4   Scientists and media

There are excellent books and chapters within books looking at how best to interact with journalists and the print media.

- *The Hands-on Guide for Science Communicators*, written by Lars Linberg Christensen and published by Springer.
- *Escape from the Ivory Tower: A Guide to Making Your Science Matter*, written by Nancy Baron and published by Island Press.
- Chapter 'Writing science' by Emma Weitkamp in *Introducing Science Communication*, edited by Mark L. Brake and Emma Weitkamp.
- *Am I Making Myself Clear? A Scientists Guide to Talking to The Public*, written by Cornelia Dean and published by Harvard University Press.
- *A Scientists Guide to Talking with the Media* by Richard Hayes and Daniel Grossman, published by Rutgers University Press.
- Science Media Centre, *How Science Works* guides,
http://www.sciencemediacentre.org/pages/publications/index.php?&showArticle=
&showAll=0&showSeries=12

Media Courses and internships

The Royal Society offers three communication and media courses:
- communication skills course helps develop written and spoken communication;
- media skills training-develops skills working with TV, radio, newspapers and other media;
- a residential course covering the two 1-day courses above.
http://royalsociety.org/training/communication-media/
The Wellcome Trust offer broadcast internships for scientists.
The British Science Association Media Fellowships bridge the gap between journalists and scientists.
http://www.britishscienceassociation.org/web/scienceinsociety/MediaFellowships/

If you have a paper about to be published that may be of interest to the media, then your communication department will help you prepare a press release. The press release is now the standard way in which scientists can inform the world of advances (Christensen, 2007). The press release is a summary of your research findings and will include these six essential pieces of information:

1 What? – what is the main point of the research?
2 When? – time of the research, publication or event;
3 Where? – location of the research;
4 Who? – who did the research?
5 Why? – why is the research news?
6 How? – how was the research done?

The most important pieces of information should go at the start of the press release. This is so the reader can read as much of the article as possible before they get bored and move on to the next story in the newspaper. As well as this, front loading the majority of the information into the beginning of the story means that when space is tight and the editor trims the bottom of your article, the key points will still remain. It is really important that the front of the story contains the key conclusions. The information contained within the press release will form the backbone of any articles written by journalists, so it's essential that the information is written clearly in the active voice and is free of jargon (see Section 4.7). Not all stories will be successful in the press. Success depends on different things, which could include some of the following:

- **timing** –is it newsworthy and have no other big political or world events taken place to push your story off the news?
- **the 'hook'**– is it interesting and to whom? Is it attention grabbing – is it the fastest, or oldest, is there a human interest angle, or is there a mystery?
- **proximity** – has the story got local appeal?
- **major discovery** – is the research ground breaking?

If the press release is picked up by local or national media, then you might be asked for a TV or radio interview, in which case the information in the preceding sections on TV and radio will be of assistance.

If you would like to do more media work, then your university or institution communications department may have a 'directory of experts', where you will be able to register yourself and your area(s) of expertise. If the press then contact your organisation to ask for someone to comment upon an issue in your area, you will then be contacted. For example, I have been asked in the past, to comment for Anglia TV about a case of fungal poisoning.

In the wake of the BSE crisis, The Royal Institution was involved in the establishment of the Science Media Centre (SMC).[8] The SMC provides advice and support for scientists working with the media, and they publish good practice guides to help scientists explain issues. The SMC also provide journalists with access to relevant experts who are able to comment on breaking science news stories.

---

[8] http://www.sciencemediacentre.org/pages/

8.5   A focus on science and writing

8.5.1   In books

Scientists should recognise that science writing isn't restricted to textbooks or academic tomes although these have a role in communicating science within a scientifically literate community. Science in writing has emerged within different genres and one such genre, popular science, has seen science rewritten and repackaged to increase its accessibility and its appeal to a public audience. In 2011, according to Nielson Bookscan[9] which tracks the sales of over 90% of all retail book purchases in the UK, popular science books accounted for sales of £4.6 million. To put this figure into context, sales of crime and thrillers topped £87.6 million whereas gardening books come in at £1.3 million[10]. These days popular science is an established genre that encompasses all scientific disciplines[11] including mathematics. In 1978, Carl Sagan's, book *Dragons of Eden* was awarded a Pulitzer Prize and since then, popular science books have entered the finalist list on a regular basis. As well as the Pulitzer Prize, popular science books attract other awards including the BBC Samuel Johnson Prize for Non Fiction and The Royal Society Winton Prize awarded specifically to popular science books. Interestingly, it has been noted that the shortlist for the Winton Prize struggles to attract female authors which arguably may reflect the fact that this genre attracts a largely male authorship.[12] The impact and reach of this genre has been enhanced by accompanying TV programmes such as Carl Sagan's *Cosmos*, Richard Leakey's *The Making of Mankind* and more recently Brian Cox and Andrew Cohen's *Wonders of the Universe*. Popular science has attracted the penmanship of eminent scientists. *A Brief History of Time* by theoretical physicist Stephen Hawkins and *The Selfish Gene* by biologist Richard Dawkins have both had several editions, can be regarded as modern classics and are still popular today. Both these books seek to make a specific science topic more attainable to a wider audience. In 2003, Mellor suggested that popular science writing could be subdivided into three separate categories:

- **narrative** – works which provide a narrative of a period of scientific discovery or the life of a scientist. Examples include Francis Crick *What Mad Pursuit* and Gina Kolata *Flu: The Story of the Great Influenza Pandemic of 1918 and the Search for the Virus That Caused It*;
- **expository** – works which seek to make a particular science subject, hypothesis or paradigm accessible to a wider audience; examples include the *Selfish Gene* and a *Brief History of Time*;
- **investigative** – works which take a particular science issue or story which may provoke public interest, concern or dissent and this is explored it in a journalistic or newsy style. A recent example is Rebecca Skloot *The*

---

[9] http://www.nielsenbookdataonline.com/bdol/

[10] http://www-958.ibm.com/software/data/cognos/manyeyes/visualizations/type-of-book

[11] A good book that highlight the different disciplines that are written about in this genre is Richard Dawkins' compilation, *The Oxford Book of Modern Science Writing*

[12] http://www.guardian.co.uk/science/blog/2011/oct/04/popular-science-books-women

Immortal Life of Henrietta Lacks and the 1960s classic book *Silent Spring* by Rachel Carson, which helped launch the environmental movement.

What all these books have in common is an author who has taken a subject that they are passionate about and have sought to share this with a wider audience. Successful popular science writing manages to capture this. It probably goes without saying that if the author doesn't find the topic interesting, then they are unlikely to make it interesting to others.

As with all intrinsic parts of our culture, science has taken its rightful place as a key element in fiction. At the beginning of the twentieth century, H.G. Wells was one of the first craftsmen to weave science into his stories. Wells came from a working class background, but from an early age he was a keen reader and he eventually found himself studying biology in the Normal School of Science (now part of Imperial College London) under the famous Victorian scientist, T.H. Huxley. Wells, however, left college without obtaining his degree after failing his third year geology exam (Aldiss, 2005). He became a science teacher and a prolific and popular author. Through his novels, such as the *Time Machine*, *War of the Worlds* and *The Island of Dr Moreau*, Wells introduces science and scientists to his readers, but he also blurred the boundaries between fact and fiction, producing influential science-fiction novels. Wells was soon followed by other scientist writers, including the physicist C.P. Snow, and the physician A.J. Cronin. Another trained scientist who was also a contemporary author was Michael Crichton, whose novels such as *Jurassic Park* and *The Lost Worlds* sparked the new genre that Brier refers to as Ficta (Brier, 2006). This genre also encompasses Robin Cook's *Coma* and *Outbreak*. They represent scientific ideas and concepts that form the critical framework to a modern thriller. In addition they draw in ethical and philosophical dimensions that appeal to a wide audience.

There has been a new subgenre of science writing that has gained prominence due to the burgeoning output of fiction with a science theme. These subgenres differ from both popular science and science fiction writing, as they are 'realistic novels that contain scientists as central characters plying their trade' (Rohn, 2010). In 1925, one of the first of these novels to be published, *Arrowsmith*, was written by Sinclair Lewis in collaboration with scientist Paul DeKruif.[13] The book describes the discovery of bacteriophage by Arrowsmith, who recognises its potential to treat bacterial disease. During the book Arrowsmith is sent to an island with plague and is determined to carry out a controlled scientific trial to test the effectiveness of his phage therapy. However, his wife succumbs to the plague and in his grief he ruins his scientific trial and gives the phage therapy to all. Paul DeKruif, who provided Lewis with the scientific detail, went on to be an author in his own right, publishing *The Microbe Hunters* in 1926 (Markel, 2001). In 2001, Dr Jennifer Rohn, a cell biologist and novelist coined the term 'lab lit' to describe this new genre (Rohn, 2006). The number of books in this genre has increased dramatically

---

[13] Paul DeKruif earned a PhD from the University of Michigan, but was fired from his position at the Rockefeller Institute for writing *Our Medicine Men*, a four part series of articles on the medical profession.

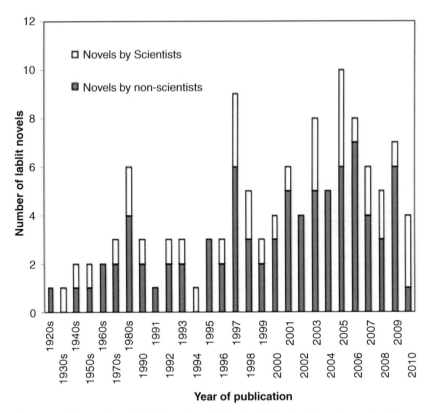

Figure 8.3 The number of lab lit novels published by both scientists and non-scientists. Reprinted by permission from Macmillan Publishers Ltd: Nature, Rohn, J. More lab in the library, 465:552, copyright 2010.

over the last couple of decades as shown in Figure 8.3, more details on lab lit can be found in Box 8.5. Rohn, who has written two lab lit novels herself, *Experimental Heart* and *The Honest Look*, wonders if 'writers may be responding to social trends that have made science and technology more palatable'. A recent example of a novel that has crossed over from the lab lit genre into the 'mainstream' bestseller list, *Solar,* was written by the bestselling author Ian McEwan and provides a day to day account of a Nobel prize-winning physicist whose scientific success is fading. It provides an account of his scientific career, outlining what maintaining a successful scientific career actually entails. Interestingly Ian McEwan studied English Literature at University; he is not a scientist. Nevertheless his account of the life of a Nobel prize-winning physicist Michael Beard is an accurate representation of a scientist's daily life that will spark recognition among the scientists who read his novel. This underlines another aspect of the lab lit genre – they can be written by both scientists and non-scientists, but they all seek to put the daily life of a scientist (from all disciplines) at the centre of the novel. If you have read any book from these particular genres, think about how you can apply elements to your own writing (see Box 8.6).

> **Box 8.5   Webzine lab lit**
>
> The webzine Lablit.com was created and is maintained by cell biologist and author Jennifer Rohn. It is dedicated to real laboratory culture and to the portrayal and perceptions of that culture – science, scientists and labs – in fiction, the media and across popular culture. The webzine is aimed at scientists and non-scientists. It contains reviews of new literature, seeks nominations for lab lit literature, and also provides links to suggested reading and reviews of lab lit. The webzine also offers opportunities to contribute reviews, essays, profiles, interviewers, and fiction to the site.
>   http://www.lablit.com/

> **Box 8.6   Writing science for a public audience**
>
> If popular science is a genre that you have dipped into, then consider the books that you have read. As with all genres it will contain books that you enjoy reading but others that you struggle to finish and wouldn't recommend to others. Take some time to think about the books that you enjoyed and try to distil the different elements of the writing that you found attractive and engaging. Do you feel that you could harness any of these elements and apply it to your own science writing for a general audience?

Writing a science textbook, popular science book or a novel is not a short-term project; it requires an initial idea for a suitable book as well as a significant investment of thought, time and effort. Next you need to covert this initial idea into a book proposal that you can use to market to publishers. Publishers are looking for:

- 'the initial sell'– What is your good idea and how can you effectively convey it to someone else? Why is it exciting and why should they publish it?
- who is likely to read your book and what evidence have you used to assess who your potential audience is likely to be?
- what makes you the ideal person to write the book?
- a table of contents that outlines the structure of the book as well as short summaries of each chapter that provide information about the likely content.

Conventionally, the next step is persuading a publisher to publish your book. It is very unlikely that you will be able to do this directly. Most publishers only work with agents, so you should consider recruiting a literary agent with an interest in your genre, who will represent you to publishers as well as negotiating a contract for a fee (which is usually a percentage of future book sales). Alternatively, thanks to the internet, there are now opportunities to self-publish books that can still muster a significant readership due to the arrival of e-books and the Kindle.

8.5.2   Alternative ways of writing for a public audience

Evidence shows that although most scientists are not successful novelists or popular authors, they can nevertheless still exercise their literary muscles and write short articles for a general audience. The key to this type of writing is that they are short pieces, approximately 700 words as opposed to a 3000 word journal article. They should be informative and written in a conversational style using humour and emotion to communicate one key point (as opposed to several points as maybe the case for a scientific paper). They should also use clear writing as described in Section 4.7. Examples of these types of articles can be found in op-Ed (opposite the editorial) pages of local and national newspapers, commentary sections in general science journals published by science societies such as *Microbiology Today* (SGM) and *Chemistry World* (RSC), departmental websites, institute newsletters, leaflets and blogs. In addition, some journals such as *Science* have a news section in the front pages. These sections are looking for science stories that are newsworthy either because they provide an unexpected answer to a longstanding question, or perhaps it suggests further questions that need to be explored. These sections can also provide opportunities to discuss the impact of recent findings on the scientific field or even speculations about the potential impacts that these findings may have on society.

8.5.3   Contributing to science writing in other ways

Authors such as Ian McEwan and his novel *Solar* require a detailed scientific knowledge and the best sources are often scientists. *Litmus* is an anthology of short stories published by Comma Press and edited by Ra Page. Each short story is a collaboration between a writer and a scientist that has a scientific discovery or a 'eureka moment' at its heart, but includes a new narrative that brings the science and the scientist to life. Each fictional story also contains a commentary afterward that adds additional impact into the science and the scientific endeavour that underpins each science breakthrough. The anthology has achieved critical acclaim and provides examples of collaborations between scientists and professional writers.

Scientific writing is also at the heart of another anthology of short stories providing opportunities for science communication: the Light Reading Project. Case study 8.2 written by Sara Fletcher describes a short story competition which invited entries from members of the public. The element that made this competition an interesting example of a science communication project is that each short story had to involve the Diamond Light Source, the UK national synchrotron facility.

Case Study 8.2

Light Reading Project

Sara Fletcher, Diamond Light Source

Background

The Light Reading project is a short fiction competition where stories must be inspired by Diamond Light Source, the UK national synchrotron facility based in Oxfordshire. Oxfordshire has a large number of local writers, but whilst it was more heavily promoted to the local community, the competition was open to everyone. Throughout the competition we were keen to emphasise that we were not looking for a particular genre; the only restrictions were a 3000 word limit, and a separate 'Flash fiction' competition for stories under 300 words. The winning stories were selected by external judges, Jenny Rohn and Anjana Anuja. The top three entries received a cash prize and will be published in a short anthology via publish-on-demand.

Structure of the competition

We tested the concept of a fiction competition initially with the 400 staff at Diamond. We received six entries, and the winner and runner-up were judged by local bookshop owner Mark Thornton and author Ben Jeapes. We were very pleased with the quality and diversity of entries, which encouraged us to go ahead with the public competition.

The main competition was between September 2011 and November 2011. It was promoted through a website (www.light-reading.org) and we produced some postcards that were distributed in local libraries, bookshops and science communication events. We also issued a press release that was picked up by local media, bloggers and *New Scientist* among others. We also directly targeted bloggers in the science communication community and held an open day for budding authors to see the facility for themselves, and ask questions on the research undertaken at the synchrotron. The communications team shortlisted 15 entries from the 55 submissions which were then passed to the external judges. The Flash fiction entries were subject to a public vote via the website, and the top five also passed to the judges.

Funding and resources

The competition was intended to be low cost, and the main resource was the time of the Diamond communications team and partner Jenny Rohn. We had a small budget for the design work for the website and postcards, and also the prize fund. There was a small cost associated with putting the anthology together and printing 100 copies for the winners and for promotional purposes.

Competition highlights

Our key criteria for success were the number and quality of competition entries. In the event we received 55 entries, which exceeded our expectations! Competition entrants included published authors, scientists and schoolchildren who had visited the facility. The entries were also very diverse, including love stories, ghost stories, comedies, murder mystery, fantasy, crime and children's fables.

Things to be aware of

We did feel that the competition reached more scientists and science communicators than established writers, which was a consequence of our own contacts being primarily in these fields. If we were to re-run the competition we would look to build relationships with individuals and organisations that are more deeply embedded in the literary community, such as publishers and book fairs.

Shortlisting entries proved a challenge as our criteria of 'original and entertaining' are very subjective. However, we were keen not to penalise stories on the basis of grammar, punctuation or scientific accuracy. It was also a considerable time investment for two people to read 55 stories up to 3000 words each, so allow plenty of time!

There is a risk that stories will not paint your institution in a positive light! We were very open about the fact that Diamond staff were involved in the shortlisting process, which reassured our CEO, but in the event these concerns were unfounded.

Personal gain

It has been fascinating reading all the stories that have come in! I'm really pleased with the number of stories, the range of people who submitted them and the imaginative plots and characters they created. We also encouraged the writers to contact us with questions about their stories, and it was interesting to read and answer these questions.

Top tips

1 On demand publishing makes this type of project possible – see e.g. lulu.com.
2 Try and find a partner in the publishing industry, or an experienced blogger with contacts to partner with.
3 Spread the word as widely as you can – the press release generated a lot of coverage.

Websites
http://light-reading.org/
http://www.lablit.com/

8.6   A focus on science advocacy

Advocacy is the act of arguing for something. Within science, this could be an argument for a cause or issue, for example for the development of GM crops or the use of embryonic stem cells in therapy. Advocacy is required where there is uncertainty about facts and differences in values (Sarewitz, 2012). As scientists, there are often occasions when we are unhappy about the way that science and scientific stories are portrayed. It is easy to feel that as a lone voice it is too difficult to counteract misconceptions or inaccuracies in science. However, you could be just the person to actively provide the scientific evidence or expertise that would enable a more enlightened debate to take place and to be brought into the public domain; in fact, you may be an expert in the field. Expertise is essential for advocacy. Incorrect facts and misconceptions coming from non-expert scientists are socially and morally irresponsible and have been deemed as acts of misconduct (Aron *et al.*, 2002).

In the UK, there are some good examples of science advocates. Dr Ben Goldacre, physician and science writer, is one. Through his *Guardian* newspaper column, blog, Twitter account and not to mention his popular science book *Bad Science*, Dr Goldacre has stood up for scientific evidence and strongly opposed scientific inaccuracy from all quarters. In the shadow of the GM debate, and 'Climategate' scientists and physicians such as Ben Goldacre have realised that there is a need to respond to misinformation, misconceptions and even deception. Some have to go to great lengths to defend their position. For example, in 2010 Dr Simon Singh won a libel case brought against him by the British Association of Chiropractors who objected to his use of language when he criticised them for defending chiropractors who used treatments on children with asthma and other conditions, for which there was insufficient evidence (Boseley, 2010).

Different groups have been developing different ways to counterbalance the effects of inaccuracies and misconceptions. In the US, a Climate Science Rapid Response team has been set up to match eminent climate scientists with law makers and the media. It promises to quickly turn around queries providing strong evidence and hard facts. The group states that they are 'advocates for science education'.[14] In the UK, Sense about Science is a charitable trust launched to 'equip people to make sense of scientific and medical claims in public discussion'. It has built up a database of more than 5000 scientists who are prepared to provide evidence and clear information as well as advice in a defence of scientific enquiry (Box 8.7). In addition Sense about Science[15] has run successful campaigns that have scientific truth at their core. These include raising awareness about the lack of scientific evidence that underpins 'Brain Gym'. This led to many schools dropping the Brain Gym programme. They have also launched an 'award winning campaign that helped patients and carers weigh up claims about treatments advertised on the internet'.

Box 8.7   Sense about Science

Sense about Science is inviting scientists to apply to join their ranks. It is seeking scientists to join their database of scientists who are willing to defend and supply scientific evidence. If you are interested in joining the first step is to go to the website and note your interest.

http://www.senseaboutscience.org/pages/offer-help-and-expertise.html

Secondly you may be interested in joining a current Sense about Science campaign.

http://www.senseaboutscience.org/pages/campaigns.html#campaigngetinvolved

If you are a young scientist who is seeking ways to get more involved in science advocacy, and want to get your voice heard, then joining the The Voice of Young Science (VoYS) programme also run through Sense about Science is an option that you should explore. This programme has been specifically set up to support early career researchers.

http://www.senseaboutscience.org/pages/voys.html

8.7   A focus on citizen science

The organisation of metropolitan and provincial field clubs and natural history societies is giving an impetus to science such as it never enjoyed before. Numbers of hard-worked men and women, wearied with brain as well as manual labour are learning to restore their lowered energies by such country outings as the pursuit of natural science involves.

—Dr J.E. Taylor FGS, FLS (1868)

8.7.1   Introduction

In writing the above in his book *Flowers, Their Origin, Shapes, Perfumes and Colours*, John Ellor Taylor sought to show how citizen involvement with

---

[14] http://climaterapidresponse.org/
[15] http://www.senseaboutscience.org/pages/about-us.html

science can benefit all. It also shows how the concept of citizen science has been with us for a remarkably long time and it could easily be argued that research was produced by the public before there was a paid 'scientific profession'. It was only when science became highly specialised and professionalised that gaps appeared between scientists and citizens.

Modern citizen science can be defined as the participation of the lay public in research; it is also known as 'Community Science'. Citizen science projects tend to take place in the fields of astronomy and ecology where amateurs often record their findings on specifically designed websites such as eBird, iSpot and citizensky. The advent of the internet has made it easier for large numbers of people to get involved with scientific research and for 'crowdsourcing', where communities perform tasks usually done by an employer or contractor. A familiar example of crowdsourcing is Wikipedia, the online encyclopaedia, built entirely from citizen contributions. The phenomenon of citizen science is not new. In 1900, the Christmas bird count was established; this has become a major source of data on US bird species in North America (Silvertown, 2011). Another long-running project that started in 1911 is the American Association for Variable Star Observers; an amateur astronomer recently recorded the 20 millionth observation. In 1952, McKeown published a classic text on Australian spiders which contained information gathered by amateur naturalists who observed the number of different species and behaviour of spiders in their gardens (Rennie and Stockylmayer, 2003).

Science is becoming increasingly expensive and the collection and analysis of large amounts of data are very costly. Developing a strategy to reduce these costs by involving a knowledgeable public is appealing. The British Trust for Ornithology (BTO) estimates that over a 50-year period, volunteers have given 1.5 million man hours towards various research projects (Bibby, 2003). Moreover, as Nov (2011) suggests, it is an excellent way to get citizens more directly involved with science and to increase scientific democracy. The director of the Division of Information and Intelligent Systems at the National Science Foundation has been quoted as saying, 'Crowdsourcing is a natural solution to many of the problems that scientists are dealing with that involve massive amounts of data' (Young, 2010).

Citizen science projects are subtlety different to other crowdsourcing projects, such as Wikipedia. There is a clear distinction between those who volunteer to do the tasks (the citizens), and those who benefit (the scientists). The final output of citizen science projects can occur a significant time after the data collection and analysis and there is no acknowledgement to each individual contributor.

8.7.2 Range of citizen science projects

An internet search of citizen science projects reveals a variety of projects, a few of which are highlighted in Table 8.1. It is interesting to note that some of these projects require little effort on the part of the volunteer (e.g. distributed computing), while others require significant time investment (e.g. Galaxy Zoo

Table 8.1 Examples of citizen science projects.

Organisation	Aim	Location	Website
Evolution MegaLab	Hypothesis driven research looking at polymorphism in banded snails in Europe	Europe	www.evolutionmegalab.org
Project Pigeon Watch	Citizens participate by counting pigeons and recording courtship behaviours observed in their neighbourhood pigeon flocks	US	www.birds.cornell.edu/pigeonwatch
British Trust for Ornithology	Independent charitable research institute which uses professional and citizen science to gather evidence about changes in wildlife populations to inform the public and environmental policymakers	UK	www.bto.org/
Bee Guardian Foundation	Education conservation organisation, where citizens can become bee guardians	UK	www.beeguardianfoundation.org/
Christmas Bird count	Annual bird survey run by the National Audubon Society	US	http://birds.audubon.org/christmas-bird-count
Open Air Laboratories (OPAL)	Involves citizens in environmental research	UK	www.opalexplorenature.org/
Swedish Species Gateway	Collects observations from the Swedish public on birds, butterflies, mammals, plants, fungi, fish and marine invertebrates	Sweden	www.artportalen.se/
The lost Ladybug Project	Website where citizens can enter observations about native and invasive ladybirds	US	www.lostladybug.org/
Monitoring nature through citizen science	Enables citizens to enter species information into a global database. The observations are used for natural resource management, scientific studies, and environmental education	Global	www.citsci.org/
National Biodiversity Network	Capturing wildlife information from citizens and recording it in datasets. The data can be used by all people	UK	www.nbn.org.uk/
Global Community Monitor	An environmental justice and human rights non-profit that empowers industrial communities to recreate a clean, healthy and sustainable environment		www.gcmonitor.org/index.php

(*continued*)

Table 8.1 (*Continued*)

Organisation	Aim	Location	Website
The Great Sunflower Project	A project which enables citizens to grow sunflowers and then observe bee pollination	US and Canada	www.greatsunflower.org/
Galaxy Zoo	A project which involves citizen volunteers to classify galaxies according to their morphology	Global	www.galaxyzoo.org/
Vital Signs	Enables citizens to monitor and record environmental conditions around Maine. Especially focused towards invasive species	US	http://drupal.org/node/694998
Citizensky	Citizen observation of the variable star epsilon Aurigae, by measuring brightness	Global	www.citizensky.org/
Zooniverse	This is a suite of citizen science projects, mainly in the area of astronomy (began with Galaxy Zoo, also includes Old Weather)	Global	www.zooniverse.org/
American Association for Variable Star Observers	Citizen observations of variable stars	Global	www.aavso.org/
Phylo	Human computing framework for comparative genomics	Global	http://phylo.cs.mcgill.ca/eng/index.html
Foldit	Protein folding	Global	http://fold.it/portal/info/science
BOINC	Distributed computing	Global	http://boinc.berkeley.edu/

and Foldit). Each project listed in Table 8.1 can be mapped onto the scientific method (Figure 8.4).

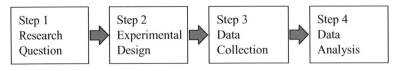

| Step 1 Research Question | Step 2 Experimental Design | Step 3 Data Collection | Step 4 Data Analysis |

Figure 8.4 Scientific method framework for citizen science.

8.7.3   Step one: the research question

Citizen scientists are unlikely to set the research question, but this can happen occasionally. A good example is the International Science Shop

Network[16] (ISSNET) that aims to provide research, usually done free of charge, on behalf of the public and the community. The actual research is usually conducted at a university by students as part of their curriculum. The Science Shops can be seen as mediators between citizens groups and research institutions. The Science Shop concept originated in The Netherlands in the 1970s and since then the practice has spread to many different countries, with investigation into noise and air pollution, as well as research into social and environmental problems.

In the Galaxy Zoo project, established in 2007 at the University of Oxford, citizens volunteer to classify galaxies according to their morphology. By 2008, more than 100 000 volunteers had classified each of the 900 000 galaxy images 38 times. There have also been 19 peer-reviewed papers using the Galaxy Zoo database (Fortson *et al.*, 2011). The creators of Galaxy Zoo, learning from the experiences of these individuals, now ensure that they supply information and analysis tools to allow people to follow-up their own research questions (Fortson et al., 2011). The Citizen Science Alliance is a new collaboration which has emerged from Galaxy Zoo. It is a collaboration between scientists, software developers, educators as well as citizens and it aims to manage internet-based citizen science projects. In an article in the *Times Higher Education*, Professor of computing science at the Open University, Darrel Ince, was quoted as saying 'citizen scientists can go beyond merely collecting data to exploring them alongside professional researchers'.

Citizens becoming research leaders is also possible. Sharon Terry became involved in molecular biology when her two children were diagnosed with a rare inherited disease, pseudoxanthoma elasticum (PXE). She wanted to find the gene responsible. Sharon and her husband collected samples from 1000 individuals suffering from disease and gave them to interested researchers. She established PXE International and now coordinates 33 laboratories and was involved in cloning the gene responsible for PXE (Terry *et al.*, 2007).

8.7.4   Step two: experimental design

Biopunk is an emerging phenomenon where citizens are conducting biotechnology research at home. This type of work would involve all steps of the scientific method, including the experimental design. In reality, most of the people doing this have considerable science experience. A good example was the development of a test for haemochromatosis by Kay Aull. She was motivated by personal reasons to develop a cheaper alternative to the test that was on offer (Wohlsen, 2011).

Citizen Science can be performed with scientists and citizens working together in the same physical space. An example of this is Manchester DIYbio (Case study 8.3); the project is now in the phase where they are building facilities to enable citizens to design experiments to answer their own research questions.

---

[16] http://www.livingknowledge.org/livingknowledge/

Manchester DIYbio

Martyn Amos, Asa Calow, Naomi Jacobs, Hwa Young Jung, Trish Linton & Jo Verran

Introduction

The biological and life sciences will have a huge influence over life in the twenty-first century, affecting areas as diverse as drug production and delivery, fuel and energy production, environmental monitoring and remediation, and security. Sustained public interest in this domain has been generated by high-profile issues such as GM food, bio-nanotechnology and synthetic biology, but until recently, the *practice* of biology has remained the preserve of well-funded commercial or academic laboratories.

However, this monopoly has recently been challenged by the emergence of the so-called 'DIYbio' movement. The increased availability of low-cost tools and techniques, combined with scientific and technological advances, has led to the development of a parallel intellectual ecology, where interested amateurs mix with established scientists to investigate the full potential of citizen science. Biology is breaking out of its traditional confines and moving into back rooms, garages and galleries.

Manchester DIYbio

MadLab is a community space for the grassroots arts, technical and scientific communities in Manchester, based (since late 2009) in a three-storey formerly disused weaver's cottage in the city centre. It hosts a wide variety of informal learning groups, workshops and one-day conferences on a variety of subjects, from computer programming and social media, through to robot building and traditional arts and crafts.

The MadLab group attended an event on synthetic biology organised by Manchester Metropolitan University, and proposed a link-up to work in this area. Although several DIY-bio groups existed in the US and Europe, we were only aware of one other in the UK (in London). We were subsequently successful in obtaining funding from the Wellcome Trust's People Award scheme, which allowed us to set up a formal Manchester DIYbio 'chapter'.

Initial engagement

We structured our activities in a staged fashion, with an initial *engagement/bootstrapping* phase, then a *development* phase. The former was used to gauge interest, attract participants and cement their involvement with a specific project.

In order to teach the basics of biological techniques, we undertook an investigation of the microbial ecology of Manchester city centre (the 'Manchester Microbe Map'). Participants collected swabs from bus stops around the city, which were then cultured and photographed at the MadLab. Several debates and discussions surrounding microbes were held during the group sessions. Academics with specialist expertise answered questions on topics such as how microbes are analysed and tested for, and how infectious diseases spread, as well as giving group members the chance to participate in the techniques of swabbing and sterile culture, and learn about the safety precautions and risk assessments necessary for this. Some members of the group also had an interest in data visualisation, and were keen to display the results in an interesting format that led to further analysis.

Risks and ethics

Interestingly, the open and in-depth discussions on the ethics of such a study – What safety precautions would we take to reduce risk? What permissions did we need to undertake such a survey? How would the resulting data be analysed? Who would we share the data and results with? – proved to be just as valuable as the data generated by Manchester Microbe Map.

Such engagement is useful, not only to the members of the public (who may not have previously considered such aspects of scientific research), but also to the academic representatives, who found it valuable to hear about issues of concern to people beyond their immediate scientific community.

Development

As the group developed, we moved into the second phase, which was concerned with setting up laboratory infrastructure, building equipment and testing it in an 'amateur' setting. Activities to date include the construction of 'homebrew' PCR machines, and investigations into microbial fuel cells. This work is still ongoing, but the eventual goal is to have a safe, dedicated, fully-functioning laboratory space to allow the group to continue and extend their work. Donations of equipment and materials from commercial companies are helping achieve this goal.

In addition to informal activities, we also held, as part of the 2011 Science Festival, the UK's first ever national DIYbio summit[17], which was attended by people from across the country, as well as Ireland, Germany and the US. The project gained much media attention and resulted in several radio interviews and press appearances.

Evaluation

A variety of techniques were used to record and evaluate the project as it progressed. MadLab has significant experience with new media technologies, and a blog, twitter feed and website[18] were used as the primary means of communication. Videos were also made of the major events, which included interviews and feedback from participants.[19]

Future plans, and reflection

The aim of founding this group is that it grows and becomes self-sustaining. At the time of writing, activities are ongoing and more events are planned, including an octopus dissection workshop and the completion of the lab.

There have been many beneficial outcomes of the project, including a better understanding of the nature of scientific research, a more nuanced appreciation of concerns held by non-scientists, the development of mutual respect between professional scientists and 'amateurs', stronger (in)formal links between community groups and academic institutions, positive press and media coverage of both the project and the participating groups, and improved public awareness of the value and benefits of citizen science.

When considering public engagement with science, it is important to consider the full range of those who may wish to engage – adults as well as children, and those who may have a strong *interest* in science, but no way in which to follow that up. We have tried to ensure that our activities appeal to as broad a cross section as possible, whilst retaining the interest of the 'enthusiastic amateur'. We hope that other groups will follow our lead and that a thriving DIYbio community will continue to grow, both in the UK and internationally.

Acknowledgements

We acknowledge support from the Wellcome Trust, Manchester Metropolitan University, Transport for Greater Manchester and Cambridge Bioscience.

---

[17] http://www.manchestersciencefestival.com/whatson/diybio-summit
[18] http://diybio.madlab.org.uk/
[19] http://vimeo.com/35156029

8.7.5 Step three: data collection

Most citizen science projects are based around data collection, and community-based monitoring (CBM) projects are a really good example. CBM is 'a process where concerned citizens, government agencies, industry, academia, community groups, and local institutions collaborate to monitor, track and respond to issues of common community [environmental] concern' (Whitelaw et al., 2003). In the UK, CBM is called Biological Monitoring, where the emphasis is on species and habitat data collection (Conrad and Hilchey, 2011). There seem to be two main types of CBM work (Science for Environment Policy, European Commission, 2011):

1 population monitoring, where non-expert citizens collect species data on birds, fish, amphibians and plants;
2 ecosystem monitoring, where processes are monitored, such as water and air pollution.

Citizen science ecological projects are flourishing, and Conrad and Hilchey (2011) suggest that this is due to several factors:

- an increase in public awareness and knowledge;
- a concern about the environment;
- worries about how governments handle ecological issues.

Conrad and Hilchey (2011) carried out a review of the literature published over a 10-year period concerning CBM citizen science projects to try and assess their effectiveness. They found that the advantages of such schemes were:

- data collection over a wide geographical area;
- increased scientific literacy;
- science democracy;
- inclusion of local people in local issues;
- tangible benefits to the environments were monitored.

The review also highlighted areas of concern about the quality of the data submitted by citizens, lack of objectivity, lack of training and problems with sample sizes. These drawbacks mean that data collected by citizens can lack credibility and that conclusions drawn from the data may not be taken seriously by governments. The review also highlights the paucity of research papers written using data collected by volunteers despite the large volume of data which exists. Conrad and Hilchey do suggest that these drawbacks are not insurmountable, but would require best practice frameworks to be written.

Other examples of citizen science involving ecological data collection include the UK Open Air Laboratories project (OPAL), a 5-year project started in 2007 with 16 different organisations headed by Imperial College. The aim is to get scientists, amateur experts, interest groups and the public together in a series of surveys, which have included water, air and climate (OPAL, 2011). Packs are freely distributed to the data collectors and provide information about how to undertake observations and measurements. For example, the Biodiversity survey investigates the invertebrate diversity found in hedges; the pack contains a field guide, workbook and a guide to common invertebrates. Results can be entered into the OPAL website and appear instantly. This allows people to compare their results to others (OPAL, 2011). OPAL also runs road shows and training sessions. Recently a data set collected by thousands of citizens across 15 countries as part of Evolution Megalab (Evolution MegaLab, 2011), has shown that observed changes in the polymorphism of banded snails is due to changing predation pressure by birds and not because of climate change (Silvertown *et al.*, 2011).

Old Weather is a Zooniverse project involving collecting historical data about the weather at sea from Royal Navy ships. The data is going towards creating better climate model projections.

8.7.6 Step four: data analysis

Distributed computing is an example of digital citizen science; it uses the idle time of computers for data analysis. Examples are seti@home, established in 1999 by the University of California, Berkley (UCB) which enables computers to run programs that search radio telescope data for signs of extraterrestrial life (Hand, 2010). In 2002, UCB released software called the Berkeley Open Infrastructure for Network Computing (BOINC). This ran many projects which allowed the public to sign up their computer's idle time. There have been significant discoveries using distributed computing, for example in August 2010 an unknown pulsar was discovered by volunteers at Einstein@home, mining data from the Arecibo Observatory.

Recently there has been a move to harness the power of gaming into science problem solving. Humans are very good at spatial arrangement, detecting patterns and solving visual problems more effectively than computers. *Phylo* is an online game for comparative genomics; sequences of human DNA are being optimised in their alignment. In the *Phylo* game, the letters of the DNA code are replaced by blocks of colour, the objective for the gamer is to produce columns of the same colour by moving blocks horizontally across the computer screen. An entire column of the same colour isn't always possible, so the gamer has to decide on the best arrangement, gaps are allowed but are penalised for in the game. The sequences come from the UCSC[20] genome browser, and they are all thought to be involved in human genetic disorders. Once the sequence alignment has been optimised by the gamers, it is fed back into the database.

Foldit is another online game, which evolved from Rosetta@home, a BOINC project. This distributed computing project harnessed the power of volunteers' computers to fold proteins into chemically stable configurations. Citizens running Rosetta@home began to report that they could see better ways in which the proteins could fold. This inspired computer scientists at Berkeley to develop a game, which they called *Foldit,* which allows online players to help the computer, but also allows them to compete with each other in folding proteins. Players accumulate points and can move through different levels in the game. Khatib *et al.* (2011), revealed that *Foldit* gamers had considerable success with providing structural models of M-PMV, a retroviral protease, which have allowed scientists to finally solve the crystal structure of this protein.

---

[20] http://genome.ucsc.edu/

8.7.7 Motivations behind citizens getting involved with research projects

Nov *et al.* (2011) researched the reasons why people volunteer to take part in three types of citizen science projects, each involving different levels of commitment. These projects were, in increasing order of commitment, Seti@home, Stardust and Weather Monitoring. The highest scoring motivation for all three citizen science projects was the aim of the actual project itself (collective motives). This was followed by personal (intrinsic) motives. These findings should impact the way that citizen science projects are designed. Nov *et al.* (2011) recommended the following:

1 projects should increase volunteer commitment by making clear the mission and goals of the project;

2 create dynamic contribution environments where volunteers can start at a low level of commitment, but work their way up to take on more tasks and higher levels of responsibility.

The founders of Galaxy Zoo realised early on in their project that their volunteers wanted to undertake meaningful tasks which would provide useful results. They found that citizens began to build communities with each other, swapping information and literature. The project represented quite a significant time investment on the part of the citizen volunteers and the scientists made sure that they were always informed about the technical, scientific and social aspects of the research (Fortson *et al.*, 2011). Scientists have found very sophisticated skills among volunteers and they are working to develop tools to enable citizens to develop their own investigations.

Organisations such as the not-for-profit Biosphere Expeditions, offer volunteers the opportunity to pay to take part in conservation projects. Two-thirds of the money paid by the volunteer goes back to the conservation project to aid sustainability (Biosphere, 2011).

One concern over harnessing the power of citizens is that they act as unpaid researchers, and therefore it could be seen as an exploitative relationship. However, this can be mitigated by clear roles undertaken by each partner in the project.

Case study 8.4 on conker tree science by Drs Michael Pocock and Darren Evans also revealed that people wanted to be involved in real science that had clear and useful goals. They also found that citizens had excellent ideas, which in turn helped them to develop new hypotheses.

Case Study 8.4

Conker Tree Science: Public Engagement and Real Research

Michael Pocock and Darren Evans

Background

'What is happening to our conker trees?' was the question we posed to thousands of citizen scientists during 2010 and 2011. This question is important because British horse-chestnut trees (also known as conker trees) *Aesculus hippocastanum* are currently under attack by an 'alien' invader. This invader is a tiny moth, the horse-chestnut leaf-mining moth *Cameraria ohridella*, which arrived in London in 2002 but is spreading rapidly across Britain (having reached Cornwall and Tyneside by 2010). The moth's caterpillars form 'mines' in the tree's leaves and the damage caused by the caterpillars can be striking; severely damaged leaves shrivel and turn brown by late summer and fall early, well before normal leaf fall in the autumn. However, the impacts on the trees are poorly understood.

Conker trees are highly valued in villages and towns, so severe leaf damage is very concerning to members of the public. In conjunction with this, scientists still have much to learn about how the moth spreads and the damage it causes, but data to answer these questions need to be collected at such large scales (across Britain and throughout the summer) that it is too costly for professional researchers to undertake this work. Considering both the public concern about their horse-chestnut trees and the need for large-scale research, we started Conker Tree Science in 2010. From the project's inception, we were clear about the dual aims of the project being public engagement with science and high-quality, hypothesis driven ecological research.

Design of the citizen science project

In Conker Tree Science we enlisted schoolchildren and other members of the public to take part in two missions and so gather data, via a website, to address two hypotheses:

1 **Mission: alien moth survey** was to test the hypothesis that the level of damage caused by the leaf-mining moth is greatest where the moth has been longest. Testing this required the observation and reporting of damage scores (a simple categorical scoring system) from horse-chestnut trees across Britain. Records could be collected and submitted at anytime throughout the summer, so providing an easy and accessible way to take part in the project. This mission also helped us (in collaboration with Forest Research) to track the continuing spread of the moth across Britain.

2 **Mission: pest controllers** tested the hypothesis that the level of parasitism by 'natural pest controllers' is greatest where the moth has been longest. The natural pest controllers that we were interested in were parasitic wasps that lay eggs inside the caterpillar, and then eat the caterpillar from the inside out. Addressing this mission required a more experimental approach, inviting members of the public to act as laboratory assistants, and it necessitated a means of assessing the accuracy of the results from members of the public, which was essential for the production of publishable quality data. The experimental protocol was straightforward but specific (pick an infested leaf during the first week of July, seal it in a plastic bag and check it for emerging insects after two weeks) and we provided a guide to the identification of the insects.

The project had a dedicated website (www.ourweboflife.org.uk/) so that people could find out more information about the problem, download detailed instructions for the missions, enter records online and view results in real time. We benefited greatly from the expertise of a website developer (at the cost of a couple of thousand pounds) to set up a reliable, secure and user-friendly website and database. In 2011, we extended the reach of the project by developing a smartphone app to allow people to upload photos, tagged with location via Global Positioning System (GPS), for the alien moth survey, although this required a more specialised team working with a larger budget.

During 2010 and 2011, nearly 10 000 people have been engaged through the project (by actively taking part and submitting records or by downloading the smartphone app) and we had almost 10 000 records submitted. We invited feedback via an open question presented to participants after they had entered records. The comments revealed overwhelming positive feedback about the project. Comments also revealed the diversity of motives for public involvement (from contributing to the search for a remedy for their own trees, to wanting to be involved in real science). Some people even contributed observations that helped us develop new hypotheses for testing.

It is easy, in retrospect, to view this relatively large project as successful, but we want to emphasise that that Conker Tree Science was developed incrementally from the success of smaller projects, each of which was valuable in its own right. With each project we gathered media and public interest, thus demonstrating the public appetite for this citizen science to funders.

We have emphasised the value that we placed on Conker Tree Science being genuine ecological research. We realised that it was essential to do three things: keep the ideas we were communicating as simple as possible (although not to be simplistic), keep the methods as straightforward as possible (and test them out before rolling them out to the general public), and ensure that the accuracy of the results were assessed. We found when validating results from a subset of the records (specifically working with schoolchildren who took part in the pest controller mission) that some aspects of the tasks were completed accurately, but some were subject to error (especially the, admittedly difficult, task of accurately identifying and counting the tiny natural pest controllers). We quantified this error and explicitly took our validation into account when analysing the results to produce reliable and publishable quality results.

Of course, the success of Conker Tree Science relied on people taking part and we achieved this in two ways: firstly by recruiting members of the public through the media and secondly enlisting volunteers to work with school classes. For the media, we found that the stories of 'our conker trees being under attack' and that 'people are needed to be scientists in their own parks and gardens' proved to be attractive. We worked hard to promote the story with the support of university press officers and we were fortunate in it being promoted in several media sources. We found our local BBC radio stations and local papers to be most receptive to taking up the story. Of course, journalists were most interested in a strong narrative, but at no time (so far!) were the media reports too far from the truth. Different journalists and interviewers picked up on different aspects of our project, so over time we had the opportunity to discuss many different stories underlying the project (the role of the public in undertaking science, the problem of invasive species, the richness of insect life, biological pest control, and so on). Enlisting volunteers to work with schoolchildren was also an effective way to recruit participants. We provided training for the Conker Tree Science volunteers and they recruited over 100 classes of children aged 8–11 to take part in the pest controllers mission. We found direct contact with teachers to be much more effective than advertising. All the teachers and children were enthusiastic about doing real research, learning about insects and being visited by real scientists.

Developing Conker Tree Science has been an enriching and enjoyable activity for us both and would have been of value even if it were only about public engagement. However, our research has also benefitted because Conker Tree Science has raised the profile of our research, opened new funding streams and provided publishable data. We believe that creatively exploiting the win–win opportunities of citizen science is hugely valuable for researchers and empowering for members of the public.

Acknowledgements
We thank Natural Environment Research Council for funding, and Nancy Jennings and Huw Jeffries for support. The smartphone app was developed by the Institute for Learning & Research Technology, University of Bristol with funding from JISC. Additional funding for pilot projects was received from the British Ecological Society, RCUK and the University of Bristol.

A criticism of citizen science projects is that they tend to lack scope in tasks and that the governance and decision-making of the project are under the control of the scientists. Stilgoe (2009) is quite dismissive about these types of project, where the science stays essentially the same. He suggests that where there is genuine citizen science, the citizen is on equal footing, the science being done is changed through the project and new knowledge created. Case study 8.4 by Pocock and Evans is an excellent example of how new research questions can come from citizens observations. Case study 8.5 by Janice Ansine shows how citizen scientists can be taken on learning journeys at the same time as contributing to natural history data collection.

Case Study 8.5

Reaching the public through iSpot: your place to share nature

Janice Ansine

Introduction

In October 2009, 6-year-old Katie Dobbins from Berkshire saw an unusual furry moth on the windowsill at home and showed it to her Dad. He helped her take a photo and posted this observation on iSpot.[21] Within 24 hours, experts confirmed it to be the euonymous leaf notcher, a species that had never been seen in the UK before.[22]

iSpot is a growing online community where anyone with an interest in nature can share their observations and get help putting names to organisms they have spotted. It was developed by the Open University (OU) and launched in June 2009 to help people build their identification skills. The premise behind the project is that unless someone can name what they see, they cannot learn about it, share their experience with others, contribute to its conservation or add to knowledge of that species. In fact, without a name, one might argue that a species is effectively invisible. iSpot uses the web and social networking and is designed for learning, building on the thrill of observing nature and the sense of achievement you get when you can identify what you have seen for the first time.

What does iSpot do and how does it work?

iSpot is a free resource aimed at everyone or anyone – from 7 to 107, or in Katie's case even younger! It appeals to those with a curiosity or casual interest, those who are more knowledgeable in wildlife identification and experts too.

iSpot is achieving its objectives by taking users on a lifelong learning journey. Our project aims are to:

- lower barriers to identification with new web and mobile software for a varied audience:
- be open to all;
- create a social network around biodiversity connecting beginners with experts;
- create a new generation of naturalists.

The website has a user-friendly interface; anyone can browse through the array of images posted by other users on the site, but registration is required to contribute. It is easy to participate: take a photograph of any plant, animal, fungus or other living organism, upload it via the observation page which will read the geographic location from EXIF data in the photograph or record the geographical location using a map provided. Then, suggest an identification with the aid of iSpot's species dictionary, or wait for other users to help.

We aim to develop identification skills and do this through our reputation system which gives social points and scores for activity on the site; this is the key factor behind how iSpot works. Users can gain social points and then scores for each of the species groups represented.

---

[21] www.iSpot.org.uk

[22] see observation posting at http://www.ispot.org.uk/node/7407

As a social network, points (recognised by stars) are awarded for making observations and comments, etc., reflecting a growth in knowledge. Importantly, scores are also awarded for species groups highlighting the expertise of a user in identification. Points are awarded when other users click 'I agree' to your ID. This reputation scoring is highlighted with icons matching the group represented and scoring can be increased depending on the rating of the user agreeing with your ID, such as an iSpot expert.

Users are so active on the site that half the observations uploaded without an ID receive a name within an hour and nearly 90% are named within a day, demonstrating the unique power of the iSpot community. This includes the keen observation and identification skills of the general public, more knowledgeable naturalists and experts from more than 70 natural history schemes and societies who all participate.

The website has a range of tools and features which further enhance the user's experience. This includes geographical mapping tools, aids to identification and information from other resources. For example, scientific and common names are checked automatically against the UK species dictionary hosted by the Natural History Museum. Once an observation has been identified, hyperlinks are shown to species' maps on the National Biodiversity Network[23] and to the appropriate species page on the Encyclopedia of Life.[24]

There are also news updates, forums offering more discussion around observations and other topics of related interest. iSpot apps for Android and iPhone are now under development and will be launched by mid 2012.

iSpot Keys are also freely available to help people identify what they find and are geared towards helping both beginners and the more experienced. The Keys are found on the iSpot front page, are accessible on computers and mobile phones and are available for a range of species. Users are able to input as much information as they can and the Key computes how closely this matches, presenting a list of possible species in order of likelihood and reaching a clearer conclusion as the user continues adding information. Users are also invited to contribute to the development of new iSpot Keys.

What have we achieved so far?

Hundreds of thousands of internet visitors simply browse iSpot to enjoy the observations and learn more about nature. Our achievements were recognized in 2010 by Wildscreen with their international Panda New Media Award, praising iSpot for taking 'maximum advantage of what the internet uniquely affords users – being both participatory and collaborative' and the fact that users 'are able to take an active role in the curation and generation of content'.

By early 2012, iSpot was a thriving community of over 17 000 registered users, who had submitted 150 000 images of over 5500 species in close to 100 000 observations. Most observations are of common species, but in addition to Katie's unique moth sighting, hundreds of rarities have been posted. In 2010, an iSpotter found a species of bee-fly that an iSpot expert quickly identified as a species not recorded in Britain before. Then, a bright yellow caterpillar was seen by a 5-year-old boy at the Titchwell Marsh reserve in Norfolk. After being posted on iSpot where it stimulated a lot of discussion, an expert confirmed it to be the sawfly *Cimbex luteus*, one of the rarest of this genus in Britain.

To conclude our starting story, Katie's moth was eventually deposited in collections at the Natural History Museum, but only after she took it to Show and Tell at her local school. However, on a larger scale, the growing numbers of observations being submitted to iSpot are contributing valuable scientific data. For example, the distribution of observations of shieldbugs reflects recent northward and inland spread of species responding to climate change such as the green shieldbug and the bug *Corizus hyoscyami*.

iSpot is part of a group of associated projects across the OU which builds on its unique reputation in distance learning, crossing from informal to formal learning while incorporating information education technology, media, communication and public outreach. For example,

---

[23]www.nbn.org.uk
[24]www.eol.org

iSpot has been the 'call to action' for the BBC/OU co-produced Radio 4 Series *Saving Species* which broadcasts to millions on a weekly basis.

The OU introductory course Neighbourhood Nature, which combines theory and practice, includes using iSpot as part of the field-based activities and helps students along a natural history learning journey. Students are badged on the site with an OU crest, which acquires a golden halo after successful course completion.

Aiming to engage as many people as possible, we have a part-time team of Biodiversity Mentors based across the UK whose work ensures that iSpot reaches all parts of the community. In three years, through the work of this team on the ground, we have directly reached nearly 60 000 members of the public through activities such as nature observation walks, bug hunts, bat walks, pond dipping, rock pooling, exhibitions and events, training days, surveys and bio blitzes.

The public's response to iSpot

Comments from users of iSpot and members of the public who have benefited from our Biodiversity Mentors, speak volumes of our impact so far and potential for the future:

> *'Just signed up and wanted to say thanks for a brilliant idea and site. Informative and addictive (in the nicest possible way)'. An iSpot user.*
>
> *'Thank you so much (yet again) for all the information. I am amazed at how easy it is to become 'obsessed' with nature... no matter how many books a novice like myself has, you cannot beat the knowledge and advice I get from everyone on iSpot.' An iSpot user.*
>
> *'Your visit was AWESOME and we loved shaking the trees and seeing what fell off into the net' Year 3 pupil, Norfolk following a session with our East of England Biodiversity Mentor.*
>
> *'I never thought that so close to becoming a teenager my son would be so excited about going out with the family to an event like this, he is so excited about getting his own detector for Christmas'. Family from the South West of England who have been on two iSpot Exploring Nature bat walks.*

Acknowledgements

iSpot and the Biodiversity Observatory are supported by a grant from the Big Lottery Fund for England as part of the OPAL project. We would like to thank the iSpot team and the iSpot user community for contributing to the success of the project to date.

The term 'citizen scientist' does have another meaning, which is the scientist not drawing the line between their work as a scientist and their responsibility towards society as a citizen (Stilgoe, 2009). As Stilgoe says 'all scientists are citizens, but not all scientists are citizen scientists'. Veronique Chable is a plant geneticist; she is trying to preserve the genetic diversity of the cauliflower. She works alongside farmers in 'participatory plant breeding', helping them go back in their records to unearth the genetic heritage of their crops. She is helping famers rediscover different cauliflower varieties with enhanced flavour (Stilgoe, 2009). This definition of a citizen scientist can be applied to scientists who involve members of the public as active partners in health research (Section 8.8).

8.8   Public involvement in health research

There is a growing movement to involve members of the public as active partners in the research process related to health. There is research in the health and clinical sciences that relies on the support of public. What makes

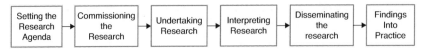

Figure 8.5 Scientific method framework for patient user involvement in research.

this research slightly different to other citizen science projects is that the public can be the 'end users' of this research. As 'end users' of clinical science and health research programmes, the public can feel stigmatised, especially if the public takes the position of the 'researched on'. In efforts to counter this, there has been a real push towards the 'researched on' becoming 'researched with' or active partners in the research process. This involvement can take place at the different stages of the research cycle (Figure 8.5).

Research between the public and the research community can also be placed on a spectrum of involvement within the research process.

CONSULTATION ↔ COLLABORATION ↔ USERCONTROL

Consultation involves meeting with the public or end users and asking for views and opinions that may or may not be acted on. Collaboration involves active partnership with the researcher and the 'end user'. Finally, user control is research driven by the end users as oppose to the research scientist. Within the UK, the Department of Health (DoH) through the National Health Service, set up the 'Consumers in NHS Research' in 1996 that in 2003 evolved into INVOLVE. The premise for INVOLVE is that engaging the 'end user' within the research process can bring clear benefits to the funder, the researcher and the consumer of the research. These benefits include:

- the 'end user' can offer a different perspective to the researcher;
- research can be identified that is viewed as a priority of the 'end user', which can help to prioritise and identify research topics that may not have been researched otherwise;
- ensuring that money is provided to fund research that is deemed to be relevant by the end user and not just the researcher;
- recruitment of peers for research projects and can facilitate access to marginalised or hard to reach groups;
- involvement within the research process can lead to a more empowered researcher who can effect change and the implementation of research results;
- an increasing political priority to involve the public in the research process especially within the NHS and the DoH.

Many funding bodies such as the Wellcome Trust and the National Institute for Health Research now require scientists to involve members of the public, who are experts by experience and may well be the end users in the research process (Hanley *et al.*, 2004). An example of public involvement in the research process is the UK Alzheimer's Society. In 2000 they established a network called Quality Research in Dementia (QRD), patients and carers work with leading scientists to set research priorities. An additional part of their role, is to review research proposals, and assess and monitor research grants (Stilgoe and Wilsdon 2009; Rose, 2003). This Research Network ensures that the Alzheimer's Society only funds research that has a significant impact on people's lives.

PXE International (Section 8.7.3) is an example of an advocacy organisation, and provides a model upon which other advocacy groups, who are involved in a range of different diseases, can get involved in translational research (Terry *et al.*, 2007).

8.9   A focus on Web 2.0 tools and services

Scientists are poised to reach more people than ever, but only if they can embrace the very technology that they have developed.

—*Nature* (2009)

The World Wide Web is a truly global communication system. From modest beginnings at CERN, Europe's particle physics laboratory, the web has flourished into a phenomenon consisting of over a billion pages, enough for several lifetimes' worth of viewing. Scientists working in academia, who teach undergraduate and postgraduate students, discourage the use of the information provided on the web, as it has not been peer-reviewed and is deemed not to be trustworthy. However, the reverse is true for the citizen. It can be perceived that web content is written by ordinary people and therefore what they write is not influenced by a hidden agenda (Russell, 2010).

The web is an attractive means of communication as you can reach a vast global audience. This is not just a one-way communication system; there is opportunity for dialogue, all contained in a virtual environment. The web is now firmly part of science communication and the tools within Web 2.0, such as forums, podcasts, blogs, twitter and wikis can be used for different purposes, including evaluation of science communication events (see Chapter 5). The suite of powerful Web 2.0 communication tools can be used by individual scientists, groups or institutions. Table 8.2 provides examples of some of the tools available for communication on Web 2.0. Despite their popularity, very few studies have been done which look at the use of social media for scholarship within science or other disciplines.

Procter *et al.* (2010) conducted a study, funded by the Research Information Network, looking at how Web 2.0 services are being used for scholarly communication by researchers across different academic disciplines. Procter *et al.* define scholarly communication as:

- conducting research, developing ideas and informal communications;
- preparing and shaping and communicating what will become formal research outputs;
- the dissemination of formal products;
- managing personal careers and research teams and research programmes;
- communicating scholarly ideas to broader communities.

—Procter *et al.* (2010)

Table 8.2 Examples of different media which can be used for science communication.

Example	Type of social networking	Description	URL
Twitter	Micro-blog	Tweeting is very different to blogging. With a tweet you are limited to 140 characters, so you have to be very clear about your communication	http://twitter.com/
Delicious	Social bookmarking	Helps you find information on the web. Common themes are linked into 'stacks'. You can search stacks or build your own which you can then share	http://delicious.com/
StumbleUpon	Social bookmarking	StumbleUpon searches through the information on the web to direct Stumblers to high quality web sites relevant to their interests	www.stumbleupon.com/
Digg	Amplifier	Digg is a place where you can share web content. It can potentially shoot your research world-wide. The Digg button can be placed on your paper and then your research can be picked up by other Digg users	http://about.digg.com/
FriendFeed	Aggregator	FriendFeed will pull conversations you and your contacts are having from all different social networking and bookmarking sites and displays them in one area	http://friendfeed.com/
YouTube	Multi-media sharing	Sharing video material	www.youtube.com/
Flickr	Multi-media sharing	Sharing image material	www.flickr.com/
Podomatic	Multi-media sharing	Audio and/or visual material shared on the web	www.podomatic.com
Ning	Social networking	Create your own social networks	http://uk.ning.com/
Facebook	Social networking	Create your own personal page or page for your organisation	http://en-gb.facebook.com/

Note that their definition of scholarly communication includes communication with a wider audience. The study consisted of an online questionnaire ($n = 1477$ respondents) followed by 56 semi-structured interviews. The results of the study split researchers into frequent users (13%), occasional users (45%) and non-users (39%). This data suggests that the majority of researchers are using Web 2.0 tools and services for scholarly communication, but few do it on a regular basis. Contrary to belief, this study indicated that the use of

Web 2.0 is not the preserve of the young; of the frequent users, 69% were aged between 35 and 64 and had more senior positions. Frequent users tended to be involved in collaborative research projects and were more likely to be male. However, they did find that among the occasional users it was more junior researchers who tended to use social networking tools, e.g. Twitter and Facebook for purposes related to research (Procter *et al.*, 2010b). This is supported by a survey ($n = 66$) by Letierce *et al.* (2010) who showed that social networking tools within Web 2.0 tended to be used by younger, early career scientists. The main motivations for using these networking services were to:

- share knowledge/study/work about their field of expertise (86%);
- to communicate about some of their research projects (80%).

Of those services that were available on Web 2.0, Twitter was their favourite service – 92% had a Twitter account. The second preference was for Facebook.

The use of Web 2.0 technology for scholarly communication is still in its infancy, but growing rapidly. It is being used, however, not to replace more traditional forms of communication, but to enhance them. Lack of formal skills is a barrier to usage, knowing what can be used for what and how. Procter *et al.* suggest that Web 2.0 could be helpful in delivering 'impact' of research to non-academic stakeholders.

8.9.1 Blogging

The Weblog (blog) seems to have evolved from online commentary (Russell, 2010). In essence a blog is a webpage created by a group or by an individual that is updated regularly (Kouper, 2010). It is a means of communication between the group/individual and an audience, as citizens reading the Blog can comment upon it. Figure 8.6 provides the lifecycle of the blog, from source to destination (Wilkins, 2008). Early in the twenty-first century, software was developed which allowed for easy and professional-looking blogs and this dramatically increased the number of people blogging. The environment in which blogging occurs is called the blogosphere. While the number of citizens who get their science information from blogs is small, at 2% (Figure 8.1), those who do, feel well informed about science (PAS, 2011).

Blogs are beneficial to science communication for a number of reasons:

1 blogging is very responsive – not only can information about a new piece of research or breakthrough be communicated immediately, but citizens can quickly comment and enter into a dialogue;
2 the readership can benefit from the expertise of the blogger;
3 blogging can help demystify science – the research in scientific papers can be placed into a broader context of other work, by the blogger;
4 blogs can cover the politics of science, not often covered by newspapers (Wilkins, 2008).

The journal *Nature* surveyed 500 science journalists and most said that they had used an idea from a scientist's blog to develop stories. Science blogs are being seen as a form of community journalism. A search of 'Science Blogs' in Google reveals a vast array, some of which are highlighted in Table 8.3.

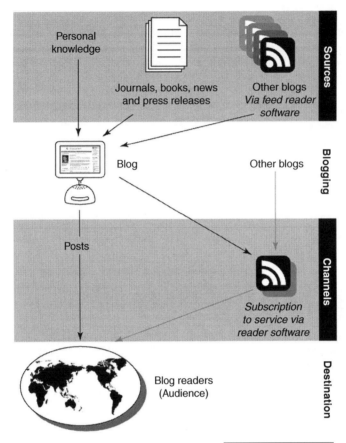

TRENDS in Ecology & Evolution

Figure 8.6 The lifecycle of a blog. Sources include other blogs, news services, journal articles, popular media and the personal knowledge of the blogger. For science blogs, posts are often based on recent science announcements and more rarely articles. Sometimes science bloggers blog about their own research. Reprinted from Trends in Ecology and Evolution, 23, Wilkins, J., The roles, reasons and restrictions of science blogs, 411–413, Copyright 2008, with permission from Elsevier.

Table 8.3 Different science blogs.

Title	URL
Microbiology Bytes	www.microbiologybytes.com/blog/
Wired Science	www.wired.com/wiredscience/
The Redfield Lab	www.zoology.ubc.ca/~redfield/index.html
A Blog around the Clock	http://scienceblogs.com/clock/
Pharyngula	http://scienceblogs.com/pharyngula/
In the Pipeline	www.corante.com/pipeline/
Panda's Thumb	www.pandasthumb.com

Technorati[25], a blog search engine returns 19 881 blogs with a 'science-tag', although many of these are classed as 'pseudo-blogs' (Bonetta, 2007). Kouper (2010) analysed 11 different science blogs to try and determine the role that blogging plays in the public engagement with science. Science blogs tend to be written by scientists, some early career, others professors; they are also written by professional science communicators and journalists. Ideas for blogs come from:

- journal articles;
- recent scientific announcements;
- personal experiences;
- commentary on other media;
- research in laboratories.

Kouper identified that the audiences of science blogs tend to have had some prior interaction with science, i.e. they were not a typical lay audience. When one blogger asked who their audience was, the answers came back as being science graduates, postdoctoral scientists and lecturers. In Kouper's study, *Wired Science* was the only blog that had a considerable readership of non-scientists. The paper concludes that scientists who write blogs should be more aware of their audience and to try and welcome non-scientists. *Pharyngula,* a science blog by Paul Z. Myers, professor of developmental biology, covers both religion and politics and it is these posts which attract the majority of his readership, at over 300 000 visitors per week (Wilkins, 2008).

Fahy and Nisbet (2011) make the point that science blogging can be very valuable and complementary to normal peer reviewed papers. Some journals even send their papers to bloggers to ensure that they are commented upon (Wilkins, 2008). Nancy Baron (2010) also makes the point that scientists should visit blogs, as they are becoming increasingly influential.

Benefits and barriers to blogging

Blogging has several personal benefits for the blogger:

- the blogger will form part of a community, with access to a network of contacts. Some bloggers have been offered jobs on the basis of their online profile (Wilkins, 2008);
- blogging scientists feel that they become better communicators and they have a broader appreciation of different scientific issues (Bonetti, 2010).
- the barrier between science and humanities can be broken down. Commentaries on blogs represent a rich resource for those studying the effect of issues on the public.

There are perceived barriers to blogging. Bloggers tend to be under 30, thus it's seen as a medium for early career scientists. Some scientists feel that they will be ridiculed by their peers if they enter the blogosphere as a blogger. Time is also an issue, a blog is meant to be frequently updated and each blog takes time to write. Some scientists dislike the nature of blogging, which tends to be opinion based and can be at odds with evidence. From a public

---

[25] http://technorati.com/

perspective, conversations can also be hard to follow. Quality can be lacking, there is no editing involved, and some blogs badged as science blogs are in fact 'pseduo-blogs'.

A caveat to science communication on the web is that it is fast moving. Facebook was little known, now it is a global phenomenon. You can't get too comfortable with one type of outlet – you need to be willing to keep up to date with new services and go where people are going.

Case study 8.6 by Joshua Howgego highlights how easy it is to set up a blog and also explores his motivations for starting one. He also talks about the benefits that blogging has bought him.

Case Study 8.6

Benchtwentyone

Joshua Howgego

Background

When I started my PhD in October 2008 I was hugely excited about doing some real science. I soon found out though, that when people say doing a PhD is hard, they are not joking. I found mastering the techniques I needed to do my research took me much longer than I had anticipated, and several months in, I found myself still loving science as a concept, but finding the day-to-day work difficult and that it often didn't work as I expected.

To keep myself from complete despair, I thought it would be good to take a much wider interest in science as a whole – the global effort to understand more about our world and universe – and put that knowledge to some sort of good use. If I could do that, I would know that even though I was a very small part of that effort, the effort itself was special and worthwhile.

I had enjoyed writing and words since I was at primary school, and I find the process of writing to be cathartic and one which is effective in crystallising my thoughts. The idea of a blog soon came to me and I decided I would just have a go.

Starting the process

I had no idea where to start; no technical knowledge of how web pages or blogs were created or worked and as I had just moved to a new city – Bristol – a few months before, I had no one who I felt I could approach to ask for help. But actually, the great thing about blogging for scientists who are interested in trying some form of science communication, is that getting started requires almost no resources at all; just a computer and some time.

Setting up a blog

Even if you know nothing about blogging or programming, it's easy to get up and running quickly using a free blog hosting service. Popular choices include wordpress.com, posterous.com, tumblr.com and blogger.com. Just search for them online and submit your email address and you'll be set to go. You can chose from preloaded 'themes' (blog designs) or get a technologically minded friend to help you design your own. These providers will often automatically synchronise your blog with Twitter (and other social medias), so all your friends will know when you make a post.

In fact setting up a blog is really very easy with sites like wordpress or posterous. These sites offer WYSIWYG (what you see is what you get) editors for writing posts and user-friendly interfaces for controlling the appearance of your site with pre-loaded themes. It's easy to set up a nice looking blog without knowing any programming codes. The hard part is making it your own and the writing itself.

I started writing about topics that I knew a bit about already and could do research on easily to flesh the stories out. For example, being a chemist, I often wrote about interesting molecules in a series of posts which I called 'molecule of the month'. I also wrote about how science influences government policy and law – things which I thought would be relevant to the lay person.

What didn't go as well as expected

One of the problems with a blog is that you do not know who you are writing for, and at the start, I was very much writing for myself. The process was helpful for me and I found it fun. If anyone else read it, that was a bonus. The bloggers I respect like Ed Yong of *Not exactly rocket science* have threads on their blog where they specifically ask their readers to leave comments to explain who they are; this is the only way to find out who reads what you write.

I soon found that to write interesting blog posts, I had to read academic papers, check facts and spend substantial periods of time writing and re-writing pieces. I enjoyed it for sure, but I began to worry if all this was a waste of time. No one really read my blog (one can tell with the in-built hit counters on wordpress) and I wasn't sure if there was any point spending such a large amount of time practicing writing like this when my PhD itself demanded quite a lot of effort.

The positive outcomes

I was encouraged though, when after more than a year of sporadic writing and worrying, I won inclusion in an anthology of the best science writing on blogs in 2009; a book called *OpenLaboratory*, the brainchild of one of the editors of open access journal *Public Library of Science*. This brought more traffic to my site quite quickly and gave me a fuzzy warm feeling of slight success.

With this behind me, I decided to see if I could sell my science writing elsewhere. I eventually secured a slot writing on another, more popular multi-author blog hosted and edited by a writer named Martin Robbins, who would go on to write for *The Guardian* about science.

Over the next year or so I found that the slight success I had achieved led to quite a number of opportunities if I was motivated enough to ask for them; the fact that I had a blog showed dedication, I think, and made people take me a little more seriously. For example, I was awarded a bursary to attend the British Science Association's Science Communication Conference in 2010 where I leant a lot and talked to some very clever people. Then the student newspaper at the University of Bristol agreed to allow me to create and edit a brand new science section in their print paper. These new experiences also gave me interesting material for the blog of course.

People also occasionally write to me and ask me to do science communication things. The most interesting example of this is a man named Tom Marshman who got in touch to ask me to help advise him on a piece of one-man theatre he is doing on the subject of operations under local anaesthetic.

Is blogging beneficial to a mainstream career in science?

I think that getting involved with science communication is something all scientists should do to some extent. It forces you to look at your research from a much wider perspective. Remembering that the general population are in fact probably paying for at least part of the science you do, forces one to be critical of it and makes you think how you justify your research.

Very importantly, it also forces you to focus on presenting your work in an accurate, yet understandable and interesting way. This is difficult to do, but the process of learning how to communicate with different audiences is vital to becoming a good scientist.

Summary of the experience

Writing a personal blog about science may not be the most effective way of engaging the general public with science in itself, mainly because – unless you're quite famous – it seems clear that they may not read it. However, I have found that maintaining a blog is an extremely useful tool for educating oneself in how to do good public engagement. It is a practice ground where written communication skills can be developed and new areas of thought explored. For a young scientist like me, it can open doors to experiences that stretch you and expand your horizons to areas of thought and arenas of discussions which would never have been previously considered. In that respect, maintaining a blog is a valuable and enlightening thing to do.

Finally, on a really personal note, I've found that compiling my blog has definitely helped me in terms of motivation for and direction in my PhD studies. It showed me how other people view what I do and it has taught me how vital story-telling is to science. It's no good just doing endless experiments – even if they're all really good ones. To do high-impact science, experiments must be carefully chosen and the results crafted into a story that other people can easily understand. Like it or not, that's how widely-read scientific journals work, and after all, if we don't communicate the results of research effectively, no one can benefit from our contributions to science.

How do I blog? Josh's top tips

1 Firstly, don't try to write too many posts. It's very obvious on a blog if you haven't written anything for a while – but don't let this pressure you into writing something just for the sake of it.

2 Pick subject matter that you're already pretty clued up about – then it won't take hours of your time to research it.

3 Find a 'hook' if you can – a reason why a particular piece of science is of interest to a general reader.

4 Public engagement is ideally a two-way conversation and writing a blog is only engaging the public if they read it. So amateur science blogging works best if done alongside other activities – school visits, conferences, workshops and festivals. These give you both new material to write about and a chance to promote what you've already written.

5 Lots of blogs offer widgets (basically boxes which you can insert into your posts) which survey your readers on a question of your choice. These can be a good way to begin a conversation with the audience and can provide ideas on how to engage them more effectively.

6 Get yourself added exposure by emailing the authors of more established blogs and asking to do guest posts. Often they'll jump at the chance to get an external expert involved in their project.

8.9.2   Other forms of social networking

Social networking sites have three main components (Baron, 2010):

1 a personal page with details about you;

2 a network of relationships which connects you to others;

3 a message function that allows you to contact your network relationships.

Examples include Facebook and MySpace, with the former being the larger. Most Facebook pages are kept by individuals, but others are created and maintained by other 'groups' such as universities or departments within universities. Facebook is quite high maintenance, as you need to keep the information current. But it is very useful as a medium for keeping in touch with people (e.g. your alumni).

8.9.3 Twitter

Twitter was created in 2006, and has spread rapidly over the past five years and now has millions of users worldwide. Twitter is a micro-blog, where the user is limited to 140 characters, in a post which is called a 'tweet'. A wide variety of people use Twitter, including celebrities, politicians and journalists, and it has the potential to reach across many different boundaries, including that of the scientist and the lay public (Letierce *et al.*, 2010). Within the scientific community, Twitter is commonly used by individual scientists, but also by groups and larger organisations, such as universities and research institutes. It can be used for a variety of purposes:

- research updates by individuals;
- individuals sharing new research papers;
- news updates by institutions;
- following scientific conferences – a conference stream can be created by using a hash-tag.

An interesting phenomenon of social media has been the rapid response to papers published online by journals. Through both blogs and Twitter, scientists have been quick to respond to errors contained within papers. This can sometimes mean withdrawal of the paper or a retraction by the authors. An example was a paper published by scientists at the NASA Astrobiology Institute, who claimed to have found bacteria which use arsenic instead of phosphorous in their DNA backbone. The swift criticism left the authors reeling, and unsure of how to reply. While this type of rapid response is viewed with concern by biologists, mathematicians and physicists have relished the dialogue and indeed, they have been posting draft papers onto a preprint server, arXiv.org for a number of years (Mandavilli, 2011). This rapid response phenomenon can also be used to harness global scientific endeavour. In May 2011, Germany suffered a severe *E.coli* outbreak. Initial attempts to identify the disease source proved difficult. The new speed and lower costs of DNA sequencing suggested a radical new approach to disease source identification. Several sequencing centres sequenced and released outbreak *E.coli* data to the public for Crowd-Sourcing analysis. An unprecedented level of global cooperation through blogs, Twitter and private web pages resulted in scientists sharing data over the internet outside of the peer-review process. CrowdSourcing analysis identified the closest sequenced relative as *E.coli* 55989, isolated in Central Africa several years previously (Crossman, pers comm., 2012).

8.9.4 Websites

There is no doubt that the inception of the World Wide Web and the rapid expansion in global interconnectivity has brought many opportunities to communicate science to a wider audience. Many professional science societies such as the Royal Society of Chemistry (RSC) and well-known science institutes (the Royal Institute) and institutions use their websites to provide information about science and recent scientific breakthroughs. For example,

Reading University recently established a blog on its website, inviting comments upon recently published papers (Box 8.8). Top journals such as *Nature* and *Science* have associated websites that house additional features that highlight the science breakthroughs covered in these journals. These include free access to chosen research papers, summaries of featured science stories, commentaries, podcasts and blogs that invite comment and communication. As scientists it is important that we consider communicating our science more widely by tapping into the potential worldwide audience offered via the internet. There is real opportunity to put content onto a website that will attract thousands of viewers (see Box 8.9 for some suggestions). The arrival of Apple's iTunes store and YouTube as a repository for podcasts and video clips have allowed material to be viewed around the world, especially as the content can be free and easily accessible. Two examples of science communication projects that have successfully tapped into this new media are:

- the *Naked Scientists* podcasts that have been housed on iTunes and have increased the audience figures dramatically.[26]
- the University of Nottingham School of Chemistry set up a collaboration between active researchers and professional film maker Brady Haran. They created a series of short videos, the Periodic Table of Videos, about the elements of the periodic table that have been housed on YouTube and have attracted more than 100 000 viewers.[27]

Box 8.8   The University of Reading's forum

The University of Reading has recently launched a new blog on its website; the Forum. The premise is that each month, two or three articles are published which highlight different research areas within the University. The articles are regularly updated and actively invite comments, viewpoints and discussions to develop two-way conversation about the university's current research. http://blogs.reading.ac.uk/the-forum/

Box 8.9   Making your website more public friendly

If you have your own personal webpage or website, there are some simple steps and changes that you can make to communicate your science with a wider audience. For example consider the following:

- look at the information published on your web page(s) and consider whether you could make it more accessible to a wider audience;
- upload attractive, informative images that illustrate your research and/or the wider scientific area;
- provide links or access to your published articles, or early drafts of your research papers or manuscripts;
- link your personal webpage to your social networking page where you may attract a wider audience;

> - set up a science blog that links to your web page. This can focus on your own research journey, or areas of science that interest you and may interest your audience (Section 8.9.1);
> - upload video clips of you, your research your research laboratory, your research students, or your colleagues chatting about or doing research.

What both the *Naked Scientists* podcasts and the Periodic Table of Videos have in common is that they are both appealing and professional looking products. They are also examples of the idea that people are unlikely to visit your website without a good reason. As an alternative, instead of expecting your audience to come to you, you need to take your content and use different routes to get it in front of as wide an audience as possible. This in turn can bring the web traffic back to your own website. One piece of advice if you are considering developing web content, comes from Dr Chris Smith from the *Naked Scientists*. He suggests that dumbing down the content and making it as simple as possible may not necessarily be the best idea. Instead, he suggests that you 'describe the methods really carefully and simplify the language. People can understand anything if you explain it to them clearly ... we keep as many facts as possible but make sure we don't bamboozle people with jargon' (Broadwith, 2011).

You can also write content for other established websites. Box 8.10 shows how undergraduate students on a science communication course at UEA, got involved with science writing for the web-based ARKive project.

Box 8.10   Writing for ARKive

Students in the School of Biological Sciences, who did a third year module in science communication, have had some of their work published on the ARKive Website (http://www.arkive.org/). ARKive is a project run by the charity Wildscreen, whose patron is Sir David Attenborough. ARKive aims to create the ultimate multimedia guide to the world's endangered animals, plants and fungi.

Our students were invited to write a page each for one species, ranging from the Hildegarde's tomb bat to the Golden Vietnamese cypress tree. The students had to research their chosen species using only peer-reviewed material, and then write the profile in an accessible style for the general public. After submission to ARKive, all the profiles were sent for peer review, and were accepted for publication.

As part of this process, the students had to use a range of research and writing skills which they had developed during their degree programmes. After completion, they had the satisfaction of knowing their work had ultimately reached publication standards through peer review, and that the public were now accessing their written material.

8.9.5   Podcasts

Podcasts are media files, which can be audio and or video (a vodcast). The term is an amalgamation of 'iPod' and 'broadcasting' (Minol *et al.*, 2007). Podcasts are designed to be downloaded onto an MP3 player or a computer. Discussions can be generated through podcasts, which are termed podologues (Birch and Weitkamp, 2010). Many organisations, such as the Royal Society post podcasts onto their websites. Newspapers (e.g. *The Guardian*) also have science podcasts, as does the journal *Nature*. Podcasts are easy and cheap to produce and you don't need any broadcasting experience. You can register your podcast with iTunes, providing that it meets the criteria. If it does, then any iTunes user can then download it. iTunes also provides advice on making podcasts.[28]

8.9.6 Smart phone applications (apps)

With the introduction of smart phones, there are a number of science-related educational apps which can be downloaded. This type of technology is relatively cheap for the end user to purchase, and represents an interesting mechanism for science communication. Science Smart phone apps include an app which downloads images from the Mars rover to your phone, science glossaries, star maps, element periodic tables and wildlife identification guides. Some organisations have their own apps, for example the American Association for the Advancement of Science has 'Science Mobile' which enables you to read science abstracts from papers, get access to podcasts and other media files, and to keep up with the latest science news. You could potentially create your own app, through Appmakr.[29] The citizen science project by Pocock and Evans (Case study 8.4) used an app to help people upload images tagged with a GPS location for their alien moth survey. Apps have been available for a while, but there has been little research into their effectiveness for science communication. One project, which has been evaluated, was at Jacksonville Zoo and Gardens (JZG). In a project named 'Call the Wild', JZG made a smart phone app for the penguin and alligator exhibits which enabled the visitor to:

- find out more information about these species;
- locate specific animals within the exhibit;
- move animals on a screen to match their observed positions in the exhibit.

They found that the app was downloaded and used by the visitors, and younger audiences were particularly captivated. It extended the visit time to the exhibits by a significant margin. The phone was shared between family members, and discussion and observations of the animals within the exhibit took place for longer periods of time compared to exhibits when no app was used (Yocco *et al.*, 2011).

8.9.7 Open Access

A key role of a scientist is to communicate research results to peers and colleagues and one of the major routes for this is publishing in peer-reviewed journals. An effective publishing process allows research findings to be shared, maximising research impact while at the same time ensuring that the quality of published work is maintained through an effective peer-review process. The original publishing process involved production and printing costs that had to be met through various means that included advertising, pay to view

---

[28] http://www.apple.com/itunes/podcasts/specs.html
[29] http://www.appmakr.com/.

costs as well as pay to publish costs. As a result, this publishing process comes with restricted access, to many within the academic community as well as the general public. The arrival of the internet as a facility to deposit, distribute and house primary literature has opened up new possibilities. The emergence of Open Access (OA) allows access to published text because it has removed two key barriers:

1 from the perspective of the consumer, the costs associated with accessing scientific literature have been removed;

2 the growth and expansion of a World Wide Web that is becoming more and more accessible has allowed previously physically unobtainable literature to be accessed at the click of a key.

There is a drive to use the internet to allow Open Access to scientific knowledge. The Berlin Declaration (2003) on OA to Knowledge in the Sciences and Humanities, states that:

> The internet has fundamentally changed the practical and economic realities of distributing scientific knowledge ... the internet now offers the chance to constitute a global and interactive representation of human knowledge.

As scientists, the attraction of OA is that you are expanding your potential audience and thereby the reach and impact of your research; in effect you are potentially communicating your scientific research to a far wider community. So what is Open Access? A definition is:

> 1 The author(s) and copyright holder(s) grant(s) to all users a free, irrevocable, worldwide, perpetual right of access to, and a license to copy, use, distribute, transmit and display the work publicly and to make and distribute derivative works, in any digital medium for any responsible purpose, subject to proper attribution of authorship, as well as the right to make small numbers of printed copies for their personal use.
>
> 2 A complete version of the work and all supplemental materials, including a copy of the permission as stated above, in a suitable standard electronic format is deposited immediately upon initial publication in at least one online repository that is supported by an academic institution, scholarly society, government agency, or other well-established organization that seeks to enable open access, unrestricted distribution.
>
> —Bethseda Statement (2003)

Open Access comes in two main forms:

1 '**green or self-archiving**': allows the author to deposit their publication in an e-archive that is freely accessible to others. For information about the archive or repositories explore The Directory of Open Access Repositories[30] – *Open*DOAR.

2 '**gold or publishing**': the author or the author's institution or grant awarding body pay a fee at the point of publication, so that the published material is freely available to others. For information explore the Directory of Open Access Journals[31].

---

[30] http://www.opendoar.org/
[31] http://www.doaj.org/

Many funding bodies[32] have embraced the philosophy and understood the value of scientific research that is distributed without barriers to a wider audience. For example, the Medical Research Council, the Wellcome Trust, and the National Institute of Health have developed policies that require the public to have access to research that they have funded within a year of publishing or less. They support authors to make their published work freely available by providing the funds to authors to pay the publishing costs, or they provide money to the author's institutions to support the publication of research through OA routes (Suber, 2010).

As a scientist who chooses to publish their research through an OA route you are providing science communication opportunities. However, there are scientists who would choose to use an OA approach to their publishing, but are put off by the costs associated with upfront author fees. It is worth remembering that many journals that are not OA also ask for an upfront publishing fee. Similarly there are OA journals that ask for an upfront fee; however, most OA journals provide free publishing to authors.[33] In addition, if OA journals do ask for a fee, this may be paid by the authors' sponsor, such as their research institution or their research funder.[34] If financial hardship is a real issue, then the fee will be waived. Finally, remember that in the case of an upfront fee paid to an OA journal, you are communicating science by providing Open Access to your scientific research.

8.10 Concluding remarks

Hopefully within chapters 6, 7 and 8 on public science communication, you will find an approach that has caught your imagination and inspired you to go and explore different outlets in more detail. The intention is that these chapters will provide you with a variety of options that will encourage you to find a interest in a science communication activity(ies) that will suit you, your needs and your current career path. We recognise that some of these outlets are not for the fainthearted or the novice, but we have sought to offer alternatives and suggestions that can be undertaken by all scientists with a commitment to widen their science communication activities.

References

Aldiss, B. (2005) Biographical note, in *The War of the Worlds* by H.G. Wells. Penguin Books
Aron, W., Burke, W. and Freeman, M. (2002) Scientists versus whaling: science, advocacy, and errors of judgement. BioScience **52** (12), 1137–1140

---

[32] http://www.mrc.ac.uk/Ourresearch/Ethicsresearchguidance/Openaccesspublishing/index.htm http://www.wellcome.ac.uk/About-us/Policy/Spotlight-issues/Open-access/index.htm http://grants.nih.gov/grants/guide/notice-files/NOT-OD-08-033.html
[33] http://www.earlham.edu/~peters/fos/newsletter/11-02-06.htm#nofee
[34] http://www.wellcome.ac.uk/About-us/Policy/Spotlight-issues/Open-access/Guides/WTX036803.htm

Baron, N. (2010) *Escape from the Ivory Tower: A Guide to Making Your Science Better.* Island Press.

Bennett, J. (2009) From flow to user flows: understanding 'good science' programming in the UK digital television landscape, in *Investigating Science Communication in the Information Age* (eds Holliman, R., Whitelegg, E., Scanlon, E., *et al.*). Oxford University Press

Berlin Declaration (2003) The Berlin Declaration on Open access to Knowledge in the Sciences and Humanities. http://www.zim.mpg.de/openaccess-berlin/berlin_declaration.pdf (accessed 5 May 2012)

Bethesda Statement on Open Access Publishing (2003) http://www.earlham.edu/~peters/fos/bethesda.htm (accessed 5 May 2012)

Bibby, C.J. (2003) Fifty years of bird study. *Bird Study* **50**, 194–210

Bonetta, L. (2007) Scientists enter the blogosphere. *Cell* **129**, 443–445

Boseley, S. (2010) Simon Singh Libel Case Dropped. The Guardian (15 April)

Brier, S. (2006) Ficta – remixing generalised symbolic media in the new scientific novel. *Public Understanding of Science* **15**, 153–174

Broadwith, P. (2011) Reaching out. *Chemistry World* **8**, 42–45

History of the Internet, Internet for Historians (and just about everyone else): http://www.let.leidenuniv.nl/history/ivh/frame_theorie.html

Bucchi, M and Trench, B (2008) Handbook of Public Communication of Science and Technology, Abingdon, Routledge

Christensen, L.L. (2007) The Hands-On Guide for Science Communicators, Springer

Conrad, C.C. and Hilchey, K.G. (2011) A Review of Citizen Science and Community-based environmental monitoring: issues and opportunities. *Environmental Monitoring Assessment*, **176**: 273–291.

Fahy, D. and Nisbet, M.C. (2011) The science journalist online: Shifting roles and emerging practices, Journalism, **12**, 7, 778–793

Fortson, L, Masters, K., Borne, R.N., Edmonson, E., Linott, C., Raddick, J., Schawinski, K. and Wallin, J. (2011) Galaxy Zoo: Morphological Classification and Citizen Science. *Advances in Machine Learning and Data Mining for Astronomy.*

Hand, E. (2010) People Power. *Nature*, **466**, 685–687.

Hanley, B., Bradburn, J. and Barnes, M. (2004) Involving the public in NHS, public health, and social care research: Briefing notes for researchers. Second Edition INVOLVE. http://www.invo.org.uk/posttypepublication/briefing-notes-for-researchers/

Hansen A. (2009) Science, Communication and Media, In: Investigating Science Communication in the information Age, Eds Holliman, R., Whitelegg, E., Scanlon, E., Smidt, S. and Thomas. J. Oxford University Press

Hargreaves, I., Lewis, J. and Speers, T. (2004) Towards a better map: Science, the Pubic and the Media ESRC. London http://www.esrc.ac.uk/_images/towards_a_better_map_tcm8–13558.pdf

Joyce, C (2010) Radio is a Visual Medium, Box 10.1, In: Escape from the Ivory Tower, Author Baron, N., Island Press

Khatib, F., DiMaio, F., Foldit Contenders Group, Foldit Void Crushers Group *et al.* (2011) Crystal Structure of a monomeric retroviral protease solved by protein folding game players. *Nature Structural & Molecular Biology* **18**, 1175–1177

Kouper, I. (2010) Science blogs and public engagement with science: practices, challenges, and opportunities. *Journal of Science Communication* **9** (1), A02

Letierce, J., Passant, A., Decker, S. and Breslin, J.G. (2010) Understanding how Twitter is used to spread scientific messages. *ICWSM* **2010**, 90–97

Mandavilli, A. (2011) Trial by Twitter. *Nature* **469**, 286–287

Markel, H. (2001) Reflections on Sinclair Lewis' *Arrowsmith*: The Great American Novel of Public Health and Medicine. *Public Health Reports* **116**, 371–375

Minol, K., Spelsberg, G., Schulte, W. and Morris, N. (2007) Portals, blogs and co.: the role of the Internet in science communication *Biotechnology Journal* **2**, 1129–1140

Nature (2009) Filling the Void. http://www.nature.com/nature/journal/v458/n7236/full/458260a.html (accessed 20 November 2011)

Nov, O., Arazy, O. and Anderson, D. (2011) Scientists@home and in the backyard: understanding the motivations of contributors to digital citizen science. Working Paper, Ver 1.0. http://faculty.poly.edu/~onov/Citizen%20science%20motivations%20V1.pdf (accessed 15 October 2011)

Ofcom (2004) *Looking to the Future of Public Service Television Programming.* HMSO

Procter, R., Williams, R., Stewart, J., *et al.* (2010) Adoption and use of Web 2.0 in scholarly communications. *Philosophical Transactions of the Royal Society* **368**, 4039–4056

PAS (Public Attitudes to Science) (2011) Ipsos MORI poll for the Department for Business Innovations and Skills. http://www.bis.gov.uk/policies/science/science-and-society/public-attitudes-to-science-2011

Redfern, M (2009) Speaking to the world: radio and other audio, in *Practising Science Communication in the Information Age* (eds Holliman, R., Thomas.J, Smidt, S., *et al.*). Oxford University Press

Rennie, L.J. and Stocklmayer, S.M. (2003) The communication of science and technology: past, present and future agendas. *International Journal of Science Education* **25** (6), 759–773

Rohn, J. (2006) Experimental fiction. *Nature* **439**, 269

Rohn, J. (2010) More lab in the library. *Nature* **465**, 552

Rose, D. (2003) Collaborative research between users and professionals; peaks and pitfalls. *The Psychiatrist* **27**, 404–406

Russell, R. (2010) *Communicating Science; Professional, Popular, Literary.* Cambridge University Press

Sarewitz, D. (2012) Science Advocacy in a Institutional Issue, Not an Individual One. Consortium for Science, Policy and Outcomes, Arizona State University, Workshop on Advocacy in Science, AAAS Scientific Responsibility, Human Rights and Law Programme. http://srhrl.aaas.org/projects/advocacy/workshop/Sarewitz.pdf (accessed 5 May 2012)

Science for Environment Policy (2011) *European Commission DG Environment News Alert Service* (ed. SCU). The University of the West of England, Bristol

Silvertown, J (2011) A new dawn for citizen science. *Trends in Ecology and Evolution* **24** (9), 467–471

Silvertown, J, Cook, L., Cameron, R., *et al.* (2011). Citizen science reveals unexpected continental-scale evolutionary change in a model organism. *PLOS One* **6** (4), 1–8

Smith, C. (2011) The power of the podcast: the naked scientist, in *Successful Science Communication* (eds Bennett, D.J. and Jennings, R.C.). Cambridge University Press

Stilgoe, J. and Wilsdon, J. (2009) The new politics of public engagement with science, in *Investigating Science Communication in the Information Age* (eds Holliman, R., Whitelegg, E., Scanlon, E., *et al.*). Oxford University Press

Stilgoe, J. (2009) *Citizen Scientists Reconnecting Science with Civil Society*, DEMOS, ISBN 978 1 9066930 12 1 http://www.demos.co.uk/files/Citizen_Scientists_-_web.pdf

Suber, P. (First published 2004, revised 2010) Open Access Overview. http://www.earlham.edu/~peters/fos/overview.htm#journals (accessed 5 May 2012)

Terry, S.F., Terry, P.F., Rauen, K.A., *et al.* (2007) Advocacy groups as research organisations: the PXE International Example. *Nature Reviews Genetics* **8**, 157–164

Thompson, J., Popovic, Z., Jaskolski, M. and Baker, D. (2011) Crystal structure of a monomeric retroviral protease solved by protein folding game players. *Nature Structural and Molecular Biology* 1–3.

Wilkins, J.S. (2008) The Roles, Reasons and Restrictions of Science Blogs, Trends in Ecology and Evolution, **23**, 8, 411–413

Wohlsen, M. (2011) *Biopunk DIY Scientists Hack the Software of Life.* Penguin

Yocco, V., Danter, E.D., Heimlich, J.E., *et al.* (2011) Exploring use of new media in environmental education contexts: introducing visitors' technology use in zoos model. Environmental Education Research **17** (6), 801–814

Young, J.R. (2010) Crowd science reaches new heights, in *The Chronicle of Higher Education*. http://chronicle.com/article/The-Rise-of-Crowd-Science/65707/ Date (accessed 14 October 2011)

CHAPTER NINE
Getting Started with Science Communication in Schools

More effective education in science and technology will enable more and more citizens to delight in, and feel a share in, the great human enterprise we call Science.

—Fensham (2008)

9.1 Introduction

In the same way that science societies and organisations have had a major influence on science communication with the public, as outlined in Chapter 1, they have also had a similar influence upon science taught in schools. In the Victorian period, the British Science Association wanted to establish a science curriculum and published a report in 1867 arguing the case for the inclusion of teaching science in schools (Jenkins, 2007). The report suggested that the science curriculum:

- provides mental training and encourages the development of deductive and reasoning skills;
- forms part of a well-rounded education for all students;
- provides a knowledge of science that is important for all citizens as well as for society;
- ensures learning science is pleasurable and useful.

These points still remain remarkably relevant in today's national curriculum, which was introduced in the UK in 1989 and made science a core subject from the ages of 5–16 years (Millar and Osborne, 1998). In terms of historical development of science taught in schools, the focus was initially on the mental training aspect and specifically directed towards the scientific method (Smith, 2010), followed by teaching factual content. The desire to teach the scientific method embedded practical science into schools – the student could rediscover scientific principles through the process of enquiry. Practical science still remains an important component in UK science education and a recent report in 2011 on practical science by the Science Community Representing

Science Communication: A Practical Guide for Scientists, First Edition. Laura Bowater and Kay Yeoman.
© 2013 John Wiley & Sons, Ltd. Published 2013 by John Wiley & Sons, Ltd.

Education (SCORE) suggests that there is more practical science taught in the UK than in most other countries.

This chapter provides:

- background information on the attitude towards science by school pupils in the UK as well as other countries. This will provide evidence for you to justify why you are targeting a school audience in your impact proposals for grant applications;
- tips on how you can begin to get involved with science communication in schools;
- tips on breaking down your research into manageable themes;
- mechanisms for delivering science communication in schools.

9.2 School science education and scientific literacy

The deficit model of science communication has the concept of scientific literacy at its core. This model, described in Section 1.5.1, suggests that citizens do not have enough scientific knowledge and are unable to understand new scientific developments, leading to an anti-science attitude. Although we have tried to move away from this model, with varying degrees of success when communicating science to the public, the idea of science literacy is an important driver in a formal school educational context where the acquisition of knowledge still remains a pertinent issue. Many countries including the UK and the US are concerned about the levels of scientific literacy among citizens and revising school education programmes seem to be the default mechanism for trying to increase literacy levels. The US recognises that very few of its population study science at college and therefore it is clear about the role of its science education, which is to prepare its citizens for making decisions relating to science issues (Dolan, 2008). The US has instigated projects to help improve scientific literacy levels, for example the American Association for the Advancement of Science (AAAS) coordinate Project 2061 which started in 1985, after the publication of the 'Science for all Americans' report. Project 2061 is a long-term programme aimed at helping all Americans become literate in science, mathematics and technology. Initiatives aimed at improving science education have included benchmarking for scientific literacy, that provides specific learning goals which are then used to inform curriculum design (Project 2061, 2011).

Public engagement with science and the ideas behind scientific literacy have influenced and in turn, been influenced by, the science curriculum in schools. There can be tension between the science needed to provide training in preparation for a university science degree and the knowledge needed for a broader social aim of educating the public about science issues (Smith, 2010). The Public Understanding of Science movement has had a clear influence on education reform with the idea of 'science and society' (Donghong and Shunke, 2010). This view is stated clearly in a report in 2008 to the Nuffield Foundation on Science Education in Europe (Osborne and Dillon, 2008).

A science education for all can only be justified if it offers something of universal value
for all rather than the minority who will become future scientists.

—Osborne and Dillon (2008)

There seems to be no clear answer to this problem, but certainly in the
UK there are a plethora of different science courses offered by a variety of
examination boards, which can be taken at school. These courses all have
a different emphasis, ranging from those which have a heavy practical component to those which take a more societal approach. The former include
A-level courses such as Salters' Advanced Chemistry, developed at the Science Education Group at the University of York, and the Applied Science
GCSE and BTEC courses. In 2008, approximately 110 000 students gained an
applied science qualification (Donnelly, 2009), although the statistical data
suggest that these applied courses are taken by students who have not done as
well academically during their preceding school years. Courses which have a
more societal approach include the 21st Century Science GCSE and the Public Understanding of Science AS-level course for 16- to 18-year-olds. These
courses are described by Fensham (2008) as 'Science for Citizenship'. There
are also examples in other countries:
- in the US, the issues-based Science Education for Public Understanding
  Program (SEPUP);
- in The Netherlands a general natural sciences course was designed which is
  compulsory for all students aged 16–17, including those not continuing with
  science education (Osborne and Dillon, 2008). This course is described as
  offering;

all students the chance to develop the scientific literacy that they need to play a full
part in a modern democratic society where science and technology play a key role in
shaping our lives—as active and informed citizens.

—Osborne and Dillon (2008)

The UNESCO Science Education Policy-Making report (Fensham, 2008)
suggests that two courses be offered at schools – one designed for all students as future citizens and another designed for those who wish to pursue a
science-related career. The problem with trying to establish science courses to
fit different needs is that you run the risk of 'science for nobody', fit for neither
academic success, nor preparation for societal issues. However, what is consistent is that most of us receive the majority of our formal science education
in the school environment. Thus it's important that this is an experience that
generates interest and excitement that will have a knock-on effect on future
learning and engagement with science.

The Public Attitudes to Science survey (RCUK and DIUS, 2008) conducted
interviews with 643 young people aged 16–24. When asked about science
education, the respondents thought that it was enjoyable for those who were
good at it, but otherwise seen as hard and not relevant to everyday life; 27%
of young people agreed that 'school put me off science'. Thus scientists,
professional science communicators and specialist science teachers need to

think of more effective ways of bringing science to life for learners in schools. These interventions need not be complex; studies (Porter *et al.*, 2010) have indicated that even brief interventions can have a lasting impact.

9.3   A skills shortage in science

It has been noted that as societies become more reliant on science and technology, schoolchildren are being put off science and fewer are choosing science and technology as a career path (Jenkins and Pell, 2006; Donghong and Shunke, 2010). This is worrying, as recent research by the UK science council suggests that in 2017 over 58% of jobs will require skills in science, technology, engineering and maths (STEM subjects) (SCORE, 2011). Despite initiatives across the Western world, the situation is not improving and a United Nations Educational, Scientific and Cultural Organisation (UNESCO) report on Science Education Policy-Making (Fensham, 2008) indicates that the supply of professionals trained in science and technology is falling short and this needs urgent action. In the UK, the Higher Education Funding Council (HEFCE) has established the National Higher Education STEM Programme to try and encourage more interest in STEM subjects. The programme is being delivered through six universities, with a hub at the University of Birmingham.[1]

In contrast to this negative research, Nathan Green, biostatistican at the University of Manchester, writing for the UK broadsheet newspaper *The Guardian*, noted that A-level entries for maths have risen by 40.2% over five years, with physics and chemistry up by 19.6% and 19.4% respectively over the same time period. Many news stories have put this sudden popularity of maths and science subjects down to the positive influence of physicist and Royal Society Fellow, Professor Brian Cox, whose BBC programmes *The Wonders of the Solar System* and *The Wonders of the Universe*, polled over four million viewers. Nathan Green also points to other possible factors, including the global recession, the increased cost of going to university pushing students towards these more traditional subjects and also better teaching. Interestingly, he also highlights the importance of public outreach programmes, run by government, learned societies, universities and professional institutes which have promoted and encouraged participation in these subjects (Green, 2011).

9.4   Attitudes and knowledge of young people about science

There have been a number of recent key surveys and reports which have investigated both performance in and attitudes towards science by school-age pupils across the world; these have been detailed in Table 9.1. These

---

[1] http://www.hestem.ac.uk/

Table 9.1 Recent key reports in school education covering performance in and attitude towards science.

Report	Date of survey(s)	Date of report	Good information for:	Author	Commissioned by:	Web URL
Top of the Class, High Performers in Science in PISA 2006	2006	2009	Ranking data across different countries in science and maths performance (pupils aged 15). Differences in performance by gender. Extracurricular influences on performance	—	OECD	www.pisa.oecd.org
What Students Know and Can Do – Student Performance in Reading, Mathematics and Science in PISA 2009	2009	2010	See above	—	OECD	www.pisa.oecd.org
Science Education Policy-Making: Eleven Emerging Issues	—	2008	Information about interest in and about science. Scientific literacy. Science education in primary years. Professional development of science teachers	Fensham	UNESCO	http://unesdoc.unesco.org/images/0015/001567/156700e.pdf
Trends in International Mathematics and Science Study (TIMSS)					International Association for the Evaluation of Educational Achievement (IEA)	http://timss.bc.edu/
(a) England's Achievement in TIMSS 2007: National Report for England	2007	2008	Attitudes towards and performance in science of pupils (aged 9–14) in England	Sturman et al		
(b) TIMSS 2007 International Science Report: Findings from IEA's Trends in International Mathematics and Science Study at the Fourth and Eighth Grades	2007	2008	The international report gives trends in performance in and attitude towards science of pupils (aged 9–14)	Michael et al		
The Relevance of Science Education Project (ROSE) in England	—	2006	Provides information about topics in science which boys and girls are, or are not interested in. Provides information on what pupils in England think of school science education and of science in general.	Jenkins and Pell	Supported by The Research Council of Norway, The Ministry of Education in Norway, The University of Oslo and the Norwegian Centre for Science Education	http://roseproject.no/
Attitudes to Science: Survey of 14–16 year olds	2011	2011	UK focused data on attitudes towards science	—	Department for Business, Innovation and Skills (BIS)[1]	www.bis.gov.uk/assets/biscore/science/docs/a/11-p112-attitudes-to-science-14-to-16

[1]Industrialised countries pay for their own survey.

surveys provide very useful statistics which you can use to provide evidence to support your grant impact proposals (see Section 3.5.1). For example:

- if you decide to focus on schools, rather than the lay public, then the reports will provide evidence to support your decision;
- if you decide to target a female school-age audience, through for example Girlguiding organisations, then this choice can be evidenced. If these organisations are located in areas of deprivation, then you have a clearly defined target audience, i.e. girls from a poorer socioeconomic background.

The Department for Business, Innovation and Skills (BIS) have funded public attitudinal surveys about science and technology since 2000 (see Section 6.2). In 2011 they also conducted a junior version of the survey on young people aged 14–16 doing GCSEs (BIS, 2011). BIS commissioned the OpinionPanel (an independent research business) to conduct the survey using *The Learner Panel*. *The Learner Panel* gives access to 10 000 learners from across the UK, who are all involved in education and aged 14 and over. In terms of general attitude, 81% of learners are amazed about the achievements of science. They are interested in science and 16% indicated that it was their favourite subject (science was the top score!). Learners who prefer science are more likely to be male, live in the North West and attend science and engineering clubs. Only 5% of participants indicated that they intended to go into a science-related career but 25% wanted to enter careers which required a STEM subject (e.g. medicine and veterinary science). In terms of the type of science course which 16 year olds are taking in England and Wales, 53% are doing triple science, 25% double and 5% are doing a single award GCSE. There were 17% doing none of these courses.[2] Those doing triple science are more likely to want to pursue a science career (BIS, 2011).

The BIS attitudinal data is specific to the UK, but much larger studies, which focus on knowledge of and about science, have also been undertaken by the Organisation for Economic Co-operation and Development's (OECD[3]) Programme for International Student Assessment (PISA), and also Trends in International Mathematics and Science Study (TIMSS). The PISA studies are carried out every three years, the most recent being in 2009. The 2006 PISA study had a specific focus on science, and involved 400 000 students aged 15 from 57 countries.[4] The PISA studies aim to measure academic levels and attitudes against the characteristics of individual students, country, schools and education systems (OECD, 2006). The studies consist of a 2-hour test of open and multiple choice tasks, which classify students into top performers,

---

[2] This was higher than expected and BIS (2011) suggest that some respondents did not understand the terminology.

[3] The OECD member countries are: Australia, Austria, Belgium, Canada, Chile, the Czech Republic, Denmark, Finland, France, Germany, Greece, Hungary, Iceland, Ireland, Israel, Italy, Japan, Korea, Luxembourg, Mexico, the Netherlands, New Zealand, Norway, Poland, Portugal, the Slovak Republic, Slovenia, Spain, Sweden, Switzerland, Turkey, the United Kingdom and the United States.

[4] The PISA science-focused study will be repeated in 2015.

strong performers, moderate performers and lowest performers. The tasks focus on three areas:

1 using scientific evidence;
2 identifying scientific issues;
3 explaining phenomena scientifically.

For the science-focused PISA study in 2006 (OECD, 2009b), the students also completed a 30-minute questionnaire about themselves and school principals answered a questionnaire about the schools. In addition, parents in 16 countries completed a questionnaire about their role in their children's education as well as their views about science-related careers and scientific issues.

In terms of rank order, in 2006 Finland had the highest number of top performers in science (20%), with the UK ranked eighth with 13.8%. This puts the UK above the US but below Japan, Australia and Canada (OECD, 2009b). In 2009, the PISA data shows that the UK had slipped to twelfth on the rank list with 11.4% of top performers, but still performed significantly above the OECD average (OECD, 2010). In 2006 the UK performed quite badly in the percentages of top performers in mathematics (11.2%). It was ranked 23rd overall, below the OECD average and behind countries such as Iceland, Estonia and Slovenia. Disappointingly, in 2009 the ranking slipped to 31st with an even lower percentage of top performers (9.9%) (OECD, 2010).[5]

In terms of gender, across all countries in the 2006 PISA survey there was no difference in average performance in science between males and females. However, there were significantly more males in the top performers in science in 8 of the 17 OECD countries, with one of these eight countries being the UK (OECD, 2009a). Conversely there was also a higher percentage of males in the lowest performers. Interestingly females performed better on questions which required the identification of scientific issues, while males did better on questions which required a scientific explanation. There was no difference between males and females when answering questions which required using scientific evidence. The 2009 PISA survey (OECD, 2010) also highlighted that in the UK, boys seemed to be performing better than girls in science.

Socioeconomic status is linked to the top performers as more tend to come from advantaged backgrounds. This is not a surprising result. Children coming from advantaged socioeconomic backgrounds are more likely to have parents involved in their education, more educational choice (e.g. public school) and more opportunity to take part in science-related activities outside of school. However, a disadvantaged socioeconomic background was not a barrier to a top performance, on average more than a fifth of the top performers came from a background below the OECD average, in the case of Hong Kong China and Macao-China, 64% and 75% respectively came from a disadvantaged background (OECD, 2009b). This could be due to the value placed on maths and science education in emerging economies.

---

[5] These rankings are based on the percentage of top performers across all countries surveyed, (n = 57 for 2006 and n = 65 for 2009)

In terms of personal attitudes and experiences, what characterised top science performers was their motivation and dedication. They had broad scientific interests, enjoyed science and found it fun, even when it was challenging. The PISA research in 2006, also indicates a very positive general attitude to science, 87% of participants indicated that science was important to society; however, interestingly, only 57% felt that science was relevant to them (OECD, 2009b).

The TIMSS 2007 study (Sturman et al., 2008) involved 425 000 pupils in 59 countries. Year 5 (Grade 4) results came from 36 countries and Year 9 (Grade 8) results came from 49 countries (see Table 9.8 for a breakdown of school years in different countries). Thus the data in this study are from younger students compared to the PISA (2006) study. The other difference between these studies is that TIMSS measures what the pupils know or remember, whereas PISA measures what the pupils can do with the science knowledge that they have (Fensham, 2008). In the TIMSS study, data from England and Scotland were collected separately. England shows a high performance in science and mathematics, compared to most other countries, but a relatively low enjoyment of these subjects.

9.5   The importance of extra-curricular science to achievement

An important part of inspiring science in pupils is providing extra opportunity. The BIS study revealed that only 57% of learners had access to science and engineering clubs at school. Of those who indicated that there were clubs, 60% had never been and only 5% attended on a regular basis. Learners whose favourite subject is science, are significantly more likely to have attended a club. Visitors to school, talking about science and engineering, are also important to make students aware of science outside school; only 7% of respondents said that they had regular visitors (more than one a term) and 53% indicated that they either had never had a visitor or only had one, since they had started school (BIS, 2011).

In the PISA 2006 study (OECD, 2009b), the importance of extracurricular activities was clear. More top performers did science-related activities outside school; this still remained a statistically significant finding, even after an adjustment had been made for socioeconomic status. This characteristic was also longstanding; a significant number of top performers had been attending science-related activities from at least the age of 10, as indicated by the responses to the parental questionnaire. The PISA report comments that:

> ... policy makers may explore ways of encouraging all students to engage in science-related activities outside of school ... in turn improving the average science performance of all students.
>
> —PISA, 2006 (OECD, 2009b)

My own case study 10.3 on the forming of extracurricular science clubs at primary school is an example of an intervention that could lead to more top performers in science in secondary school as their enthusiasm is caught early in their education.

The PISA 2006 study (OECD, 2009b), also noted that only half of the top science performers felt knowledgeable about careers in science. In the Relevance of Science to Education (ROSE) report (Jenkins and Pell, 2006), it was shown that both boys and girls did not feel that science had opened up new and exciting jobs. A recent study on girls' attitudes to education and careers by Girlguiding UK found that girls were far less likely than boys to consider a career in science or engineering. The report also highlighted that girls didn't know about the opportunities within a science and engineering profession (Girlguiding, 2011). This points to a simple, but effective school intervention which you can do (Box 9.1).

Box 9.1   Talk about yourself and your career

The evidence suggests that school pupils are not aware of the exciting career opportunities that studying science can bring. A simple, but effective intervention would be to go into a school and talk about yourself, your career, the people who influenced you and your motivations. This could be done as part of a more general programme about careers, or as a special event. We have provided a PowerPoint Template on the book website to enable you to construct a presentation about your career. We have also provided an example using this template.

... educational excellence goes hand in hand with promoting student engagement and enjoyment of science learning both inside and outside school. The payoff is quite significant: a large and diverse talent pool ready to take up the challenge of a career in science.

—*Top of The class – High Performers in Science, PISA 2006 (OECD, 2009b)*

9.6   Getting started with science communication in schools

In a recent opinion piece in *Nature* (2011) Jon D. Miller states:

This century will bring exciting biomedical advances thanks to stem cells and genetic engineering. If scientists want the public to grasp the meaning of these developments, they need to start getting personally involved in improving the education system.

—*Miller (2011)*

Taking the first initial steps in science communication in schools can feel daunting, but you could consider any of the options in Figure 9.1 to get you started. You will find that as soon as you start, more opportunities will present themselves. It will rapidly become an addictive, satisfying and enjoyable experience.

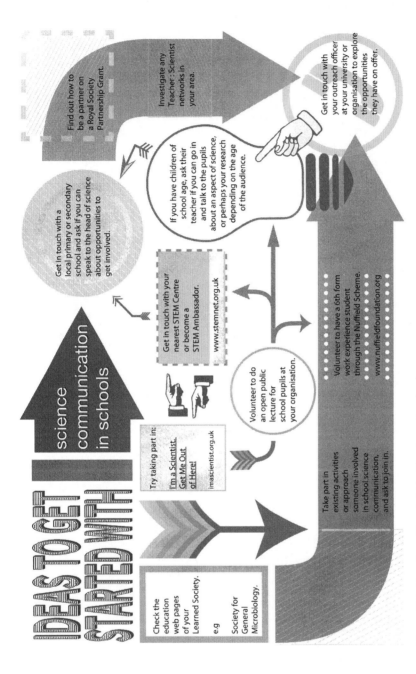

Figure 9.1 Ideas for getting started with science in schools.

Before deciding how to get involved, it's important to consider your own knowledge and examine your interpersonal skills. Ask yourself the question, 'With whom do you wish to work?' You can get involved with any stage of education at schools, but if you have little experience of children, it's quite good to try out a few different year groups to see if you prefer designing and teaching events at primary or secondary level. You may feel that you are better suited to an adult learning audience, or you may have a real desire to work with really young children to inspire them about science early in their education. Check the websites of the organisations listed in Table 9.2, as they contain lots of useful information about science communication with schools.

Table 9.2 Useful organisations in the UK, who support and keep relevant information about science education on their websites. The description has been taken directly from the relevant websites.

Organisation	Description	Website
Association for Science Education	A community of teachers, technicians and other professionals supporting science education.	www.ase. org.uk
Society of Biology	The society of biology advises the government and influences policy. It advances education and professional development.	www.societyofbiology. org
Royal Society	A fellowship of the world's eminent scientists. The aims are to expand the frontiers of knowledge by championing the development and use of science. One of the priorities is to invigorate science and mathematical education.	www.royalsociety.org
Royal Society of Chemistry	The largest organisation in Europe for advancing the chemical sciences. Education activities cater for chemical scientists of all ages.	www.rsc.org
Institute of Physics	A leading scientific society promoting physics research, application and education and brings physicists together for the benefit of all.	www.iop.org
Science Council	A membership organisation which brings together learned societies and professional bodies across science and its applications.	www.sciencecouncil. co.uk
The British Science Association	A charity that aims to advance the public understanding of science. They organise the National Science and Engineering Week.	www.britishscienceassociation. org
Engineering UK	Independent and not-for-profit organisation which promotes the contribution made to society by engineers, engineering and technology.	www.engineeringuk.com
Science Enhancement Programme (SEP)	Established by the Gatsby Charitable Foundation, SEP develops low-cost resources to enhance secondary science education.	www.sep.org.uk

Table 9.3 The percentage of pupils answering the question 'When learning science topics at school, how often do you spend time doing practical experiments?'

	England	Wales	Northern Ireland	Finland[a]	New Zealand	Japan	OECD Average
In all lessons	–	3	2	2	3	3	4
In most lessons	24	17	16	20	18	7	16
In some lessons	62	67	66	52	57	44	43
Never or hardly ever	11	13	16	25	12	45	30

[a]In the 2006 PISA study, Finland had the highest percentage of top performers.
Green crown committee report, as well as PISA 2006 (OECD, 2009b).

9.6.1 Thinking about your planning process

The PISA 2006 study showed that in 12 of the 18 OECD countries, more top performers were exposed to models or applications of science (OECD, 2009b). What is more intriguing is that these top performers had less scientific investigation as part of their science teaching and also lower levels of student interaction. This result seems to run counter to pedagogical research focusing on the benefits of practical investigation. These benefits include acquiring practical and personal skills, a better understanding of concepts and increased levels of motivation (Woodley, 2009). We suggest that the reasons for this may lie in the fact that scientific investigations require significant time investment. Results from investigations can often pose more questions than they answer (as we know is the case for real research), and thus courses which have a very heavy practical component, may not be the best way for students aged 15 to obtain the broad factual knowledge within a formal course at school. This does not diminish the importance of practical investigation, but simply points to the requirement for courses to have a range of teaching methods and to have a practical component of high quality, with clear learning objectives (SCORE, 2011). The PISA 2006 study showed that a varying amount of time was spent doing practical work (Table 9.3). Practical work in schools can be broadly split into three categories, as indicated by Table 9.4 (SCORE,

Table 9.4 Practical science activities in school identified by SCORE (2011).

Core activities	Directly related activities	Complementary activities
Investigations	Designing and planning investigations	Science-related visits
Laboratory procedures and techniques	Data analysis using ICT	Surveys
Fieldwork	Analysing results	Presentations and role play
	Teacher demonstrations	Simulations including use of ICT
	Experiencing phenomena	Models and Modelling
		Group discussion
		Group text-based activities

2011). It is useful to refer to this table if you are intending to have a practical component to your event as it may help you decide on your approach. The SCORE report indicates that practical lessons, and also learning outside the classroom environment, are important for good-quality science education.

Before you begin your planning process, you need to consider the different stages which will have a bearing upon your activity and Figure 9.2 provides a flow chart of what you should be considering. Remember that evaluation is something that should be built into your activity from the start (see Section 5.2) on setting SMART objectives.

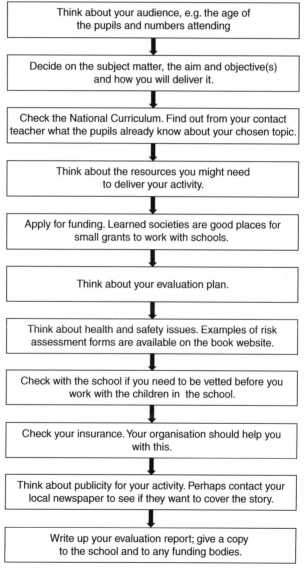

Figure 9.2 What to consider when planning a school science communication event.

To begin, consider the scientific area that you want to deliver and the mechanism you will use to do it. It is always a good idea to start with your own area of expertise, research interest or subject you know well. You could also choose to cover other areas such as:

- the process of science (see Case study 10.1);
- common science misconceptions e.g. GM food (a subject perhaps more suitable for a high school audience);
- scientific controversies in the media, e.g. use of animals in research;
- ethical issues with research, e.g. the use of embryonic stem cells;
- history of science, e.g. the development of vaccines.

Whatever scientific area you choose, you will have to ensure that you tailor the information to suit your audience, and that you build upon their existing knowledge (Section 4.4). To do this, visit websites which give detailed information about the national curriculum[6], and also speak to your contact teacher to find out what the pupils have covered. If you have time at the start of your activity you can always do a quick quiz to gauge the audience's knowledge, or suggest that the teacher does this prior to your activity. Children in primary school enjoy showing their knowledge, but at high school it is much harder to get the pupils to behave in the same way, as they are generally more self-conscious. I like to use personal response systems (PRS), especially with high school audiences as the pupils can take part, but not feel awkward. Many universities have PRS, which they use for undergraduate teaching and you may be able to borrow a set to take out.

9.6.2 The design process

It is often difficult to know how to make our research understandable to a school audience. The following five steps are designed to help you with this process and Figure 9.3 shows how this can be applied in practice:

1 decide on the aim(s) and objective(s) of your activity;
2 break down your research area into broad themes;
3 consider your audience and their prior knowledge;
4 illustrate each of your different themes with either a brief talk or an activity;
5 show how your particular research interest can link the themes together.

During my postdoctoral career I did research into how bacteria acquire their iron, looking particularly at the rhizobium–legume symbiosis. Taking this research area and applying the steps above, you can see how to convey my research to a defined school audience, with a mixture of wet laboratory activity, demonstration, discussion and talking.

You will notice that in this scheme I spent 90% of my time explaining the basic science that underpinned my research. It was only in the last 10% of my time that I explained my particular research interest. This approach, 90% basics, 10% research is *essential* if you are to get your audience to really understand your research.

---

[6] http://curriculum.qcda.gov.uk/

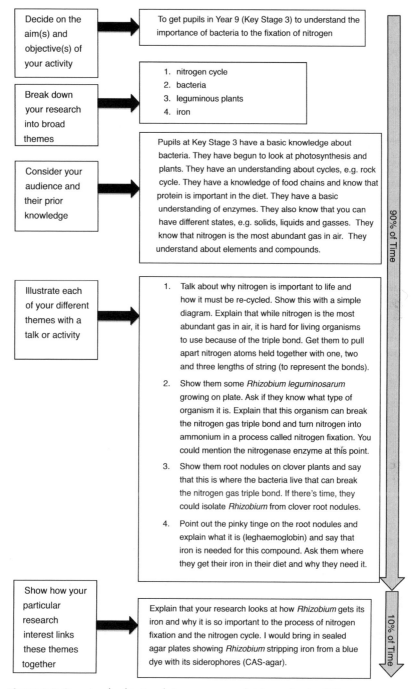

Figure 9.3 Steps involved in translating your research into a school activity.

Table 9.5a The top ten topics chosen by boys and girls as topics they wanted to know more about (Jenkins and Pell, 2006).

Boys	Girls
Explosive chemicals	Why we dream when we are sleeping and what the dreams might mean
How it feels to be weightless in space	Cancer – what we know and how we can treat it
How the atom bomb functions	How to perform first aid and use basic medical equipment
Biological and chemical weapons and what they do to the human body	How to exercise the body to keep fit and strong
Black holes, supernovae and other spectacular objects in outer space	Sexually transmitted diseases and how to be protected against them
How meteors, comets or asteroids may cause disasters on Earth	What we know about HIV/AIDS and how to control it
The possibility of life outside Earth	Life and death and the human soul
How computers work	Biological and human aspects of abortion
The effect of strong electric shocks and lightning on the human body	Eating disorders like anorexia or bulimia
Brutal, dangerous and threatening animals	How alcohol and tobacco might effect the body

If you are confident delivering material not directly related to your research, consider the subjects highlighted in Table 9.5a, that pupils find interesting. The ROSE report for England[7] (Jenkins and Pell, 2006) indicates that there are significant differences between the areas of science which boys and girls wish to know more about. When given a list of 108 topics, there were 80 statistically significant differences. The top ten are give in Table 9.5a. Aspects related to plant sciences tend to be featured in the subjects which both boys and girls are least interested in (Table 9.5b). Given the importance of plant sciences to future food security issues, more scientists active in this area could consider going into schools to encourage interest in these subjects.

9.6.3   Mechanisms used to deliver your science communication activity

Having decided on a scientific area, there are a variety of mechanisms that you can use to deliver your chosen material. A comprehensive list of these mechanisms highlighting their advantages and disadvantages are provided in Table 9.6; many more exciting and innovative ways are also described in the case studies. These types of activities can be mapped onto the graph (Figure 9.4) produced by Dolan (2008), which shows how the potential impact of your science communication activity can be measured against the number of pupils reached. Time-intensive activities such as work experience

---

[7] The ROSE study for England (Jenkins and Pell, 2006) had 1284 responses from 34 schools.

Table 9.5b The ten topics chosen by boys and girls as topics they least wanted to know more about (Jenkins and Pell, 2006).

Boys	Girls
Alternative therapies	Benefits and possible hazards of modern farming
Benefits and possible hazards of modern farming	Plants in my area
Famous scientists and their lives	Organic and ecological farming
Organic and ecological farming	How technology helps us handle waste, garbage and sewage
How plants grow and reproduce	Atoms and molecules
Plants in my area	How petrol and diesel engines work
How crude oil is converted to other materials	How a nuclear power plant functions
Detergents and soaps	Famous scientists and their lives
Lotions, creams and the skin	Symmetries and patterns in leaves
Symmetries and patterns in leaves	How crude oil is converted to other materials

placements, although reaching fewer pupils, have a potentially larger impact than a guest lecture which, although reaching a much larger audience, has less measurable impact.

Activities which occur outside of school can be very valuable learning experiences, but they require:

- a greater degree of planning by the school;
- the cost of transport to be taken into consideration;
- teacher cover.

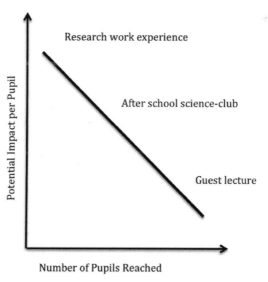

Figure 9.4 Potential impact of different activities. Reproduced from Dolan E. (2008) Education Outreach and Public Engagement, with kind permission from Springer Science+Business Media B.V.

Table 9.6 Different methods of delivering school science communication.

Type of activity/event	Advantages	Disadvantages
Talk (typically 50 min)	Quick to deliver Can fit into a normal school lesson Cheap Easy to organise Not much experience needed	Can be 'dry' unless you make it interactive Little opportunity for more in depth discussion Little opportunity for follow-up
Activity day (typically 9am-3pm). Scientist going to the school	More opportunity to deliver hands-on science More opportunity to stretch the pupils Can be very inspirational School can follow up on the work done during the activity day	Can be expensive Needs more organisation Will have to be arranged some months in advance to fit with the school calendar
Workshops (typically 1–2 sessions lasting 50–120 min). School based, but they can also occur within your organisation	Can fit into a normal school lesson, perhaps on consecutive days or in a double period Can be a mix of dry and wet laboratory activities Good opportunity for discussion	You need to carefully design the event to fit into the time available Can be expensive Needs to be organised with the school in advance Your planned activity may need to match with their curriculum
Science Clubs (can vary but meetings tend to be once a week over a period of months). Clubs tend to be located in the school but are extracurricular	Excellent for scientific investigation; a project can be carried out over a longer period of time Enables pupils to explore the process of science Builds real commitment and enthusiasm for the subject	Requires extensive planning Can be expensive You will need help in delivering the sessions Clubs run outside of school hours and teachers may not be available to help
Visit to a place of work. Teachers will bring pupils to visit you in your workplace	Very good for showing how scientists work in the real world Shows real science in action Can be aspirational	School will need to organise permission from parents, transport and supply teacher cover
Work placement – many pupils in Year 10 go on a work placement and there are also sixth form bursary schemes, e.g. Nuffield Foundation	Very good for showing how scientists work in the real world Shows real science in action Can be aspirational	Time-the pupil will need supervision Organisation-you will need to inform your HR department Can be expensive if there is no funding for the pupil

A report from a project funded by AimHigher[8] on raising aspirations in rural Norfolk secondary schools by Yeoman and Jones, showed that when asked if an event at their school or an event at UEA would be more helpful in raising aspirations, all teachers felt that scientists coming to the school was a better approach (Yeoman, K. H. and Jones, H., Report for AimHigher, pers. comm. 2006).

One mechanism of delivery is the demonstration lecture. This works equally well with a school audience either inside or outside the school environment. The case study by Tim Harrison and Professor Dudley Shallcross shows how a demonstration lecture can really help to illustrate scientific concepts. The lecture has been adapted to suit different aged school pupils and those with different abilities. The design allows other scientists to be trained to give the talk, providing sustainability for the initiative.

Case Study 9.1

A Pollutants Tale

Tim Harrison and Dudley Shallcross

Background

A Pollutant's Tale (APT) was created in 2005 by Dudley Shallcross, the Professor of Atmospheric Chemistry and Climate Change and Tim Harrison, the school teacher fellow at the School of Chemistry, Bristol University. It was designed to teach atmospheric chemistry and climate change to school pupils. This engagement activity became an element in the outreach portfolio of Bristol ChemLabS, the UK's Centre for Excellence in Teaching and Learning (CETL) of Practical Chemistry. The lecture can be delivered in lecture theatres, school halls, school laboratories, science festivals and under shade-netting in tropical school playgrounds, to audiences ranging from 20 to 1000 at a time. The demonstration lecture combines experiments, lecture and audience participation and is enjoyed as much by teachers and members of the public as it is the school pupils.

Design of A Pollutants Tale

APT performances are not 'magic shows'. The experiments highlight aspects of the talk and the theory behind the practical demonstrations is provided; they are not simply a series of chemistry demonstrations strung together without explanation. We have adapted this talk to provide different versions of APT for different age and ability groups: the version

---

[8] The AimHigher programme closed in 2011. It funded and delivered a wide range of activities to engage and motivate leaners in schools and colleges who had the potential to enter higher education, but who were under-achieving, undecided or lacking in confidence (http://www.aimhigher.ac.uk/sites/practitioner/home/).

for 4- to 11-year-olds is called 'The Gases in the Air' and lasts 40 minutes, whilst the full performance for pre-university students lasts 75 minutes. With pupils aged 11–16 the chemistry of the atmosphere is always presented but the degree of discussion of climate change varies according to their prior knowledge.

Stages of APT talk for secondary school pupils and teachers

1 The first stage of APT starts off by examining the composition of the gases in the atmospheres of other planets in our solar system. The audiences' attention is grabbed by the first demonstrations which involve the combustion of 4- to 5-litre balloons filled with helium or hydrogen gases. The chemistry of the main components of the Earth's atmosphere, nitrogen and oxygen are discussed with examples of experiments with liquid nitrogen and in the decomposition of hydrogen peroxide into oxygen foam (the elephant's toothpaste experiment). Liquid nitrogen is a material that stimulates the imagination of all who see it, irrespective of age!

2 The talk then examines the temperature profile of the Earth's atmosphere with an explanation of the temperature rise in the stratosphere, given in terms of ozone reactions for those older pupils.

3 In the second stage, the talk moves onto the troposphere and talks about the contributions of anthropogenic and biotic contributors to the volatile organic compound (VOC) load of the troposphere. The audiences are asked if they know what VOCs are – they usually don't as they are not expressly covered by the schools curricula. Examples of traditionally derived plant and animal VOCs are given out to the audience on colour coded perfumers' 'smell sticks', which are slithers of thickened blotting paper. The VOCs themselves are nature identical fragrance components (10% in ethanol). The two animal derived fragrances are ambergris (the audience is told it's whale vomit) and civet, which is originally extracted from the anal glands of the civet cat. The audience are surprised by the pleasant odour of the first and repelled by the second. The audience is questioned about where these odours/VOCs go, as the troposphere does not retain them and they are not soluble enough to be washed out by rain.

4 Stage three is an excuse for looking at incomplete and complete combustion of VOCs. Ethyne (acetylene) gas is first synthesised from calcium carbide and water. Detergent is added and the acetylene is trapped as foam that burns with a yellow smoky flame. After the incomplete combustion demonstration, we show the complete combustion of methanol vapour which is ignited in a dry 18 litre plastic water container (the whoosh bottle experiment). The residual water is shown. Again according to audience prior knowledge, the true mechanism of photochemical decomposition via the hydroxyl free radical is described in a simplified way.

5 The fourth stage goes on to look at what the climate is and explores some of the data on climate change.

6 The penultimate demonstrations use dry ice, another material that captures the imagination of school pupils. The pupils are shown that dry ice (solid CO_2) does not melt but sublimes. Small pieces of it are tied into latex gloves and warmed by the heat of a hand and passed to the audience so that they can see there is no liquid in the fingers of the gloves, that by now resemble small cows' udders. A chemical property of the dry ice is then demonstrated by putting small pieces into large beakers of water that have been made alkali with dilute sodium hydroxide to which universal indicator has been added. The 'Hollywood film directors' special effect' of colour changes and cloud formation is explained in terms of the acidic nature of the carbon dioxide.

7 The fifth and final stage of all the talks aimed at secondary age pupils, involves the technologies that already exist that can be used to offset the problems of climate change if there is sufficient economic, political and social willpower to do so. The very last demonstration returns to the gases hydrogen and helium. If time allows, members of the audience are called up on stage to ignite the balloons (sometimes in the dark if facilities allow), both of which are hydrogen as the performance must always end on a bang!

APT for primary schools pupils and teachers

The primary version of the talk 'The Gases in the Atmosphere' uses many of the same experiments described above but concentrates on;

- what is a gas;
- what is and isn't an irreversible chemical reaction;
- changes of state.

The performance lasts approximately 40 minutes but has been known to go for 80 minutes at the teachers' request. With younger pupils the VOC section is always omitted as is the PowerPoint presentation. The audience members can vary between 4 and 12 years of age. As a performer, it is often noticeable that teachers as well as their pupils are sat watching with their mouths open!

Health and safety considerations

- In the APT talk small pieces of dry ice are tied into latex gloves and warmed by the heat of a hand and passed to the audience, a health and safety warning is applied about not touching the ice for too long to avoid frost bite. Latex allergy can also be a problem, so nitrile gloves can be used instead.
- Typically with a primary school audience, the children are sat crossed legged on the floor of the school hall so care must be taken when pouring liquid nitrogen, to make sure it does not spill. In such performances we suggest that the younger children sit in the middle of the school hall as some get a little anxious by the odd explosion. Younger children migrating to the apparent safety of their teachers is often seen.

Taking A Pollutants Tale to South Africa

APT is being delivered in two regions in South Africa by two separate organisations. APT was first performed at Scifest Africa 2008 by Harrison and Shallcross with sponsorship from the British Council. As part of the promotion of chemistry outreach, the Chemistry Department at Rhodes University in the Eastern Cape adopted the talk after having been trained by the Bristol Chemists. Final year honours undergraduates and research students at Rhodes University regularly give performances to the township schools in their region and even much further afield as and when they can obtain sponsorship. Postdoctoral chemists from Rhodes have also performed APT at Scifest Africa 2011 in place of Harrison and Shallcross, showing sustainability of the project.

Several staff at the Sci-Bono Discovery Centre in Johannesburg have also been trained to deliver the talk. The city-based science centre had a new lecture theatre in 2010 accommodating 200+ students. Thanks to sponsorship from the local branch of the Royal Society of Chemistry, Tim Harrison was able to spend a week with staff training them in this performance, and other outreach activities. The staff members from the Sci-Bono Discovery Centre deliver the lecture in local dialect as well as in English and can take the lecture out to the township schools. By the end of 2010, the lecture demonstration had been seen by an estimated 25 000 people in South Africa.

Future plans for A Pollutants Tale

The end is definitely not in sight as far as this demonstration talk is concerned. APT is regularly given as part of outreach activities for pupils engaged in other events, whether they are at workshops at Bristol ChemLabS, part of a summer school at Trinity College Dublin or a winter chemistry school at the National University of Singapore, as well as in its own right as a single component event. There are still pupils and teachers who have not been excited by the sight of liquid nitrogen and dry ice even in the immediate region around Bristol University, let alone much further afield. Even local teachers who have seen the demonstration four or five times still bring new pupils into the department to see it again and again. There is a major demand without even considering those pupils in other countries.

Sponsorship is currently being sought to take APT on the road to more remote areas of southern Africa. The education authorities in Namibia and universities in Botswana and northern South Africa are supporting a concerted effort to not only allow more school pupils, teachers and undergraduates to see APT but also to train postgraduate chemists and academics how to deliver the talks for themselves whilst *en route* in late 2011. The concerted outreach event will be delivered jointly by Bristol and Rhodes Universities. A film crew from Rhodes Journalism School will make a documentary about the engagement, and the data collected about the impact will be analysed by a future researcher in chemistry outreach.

There is also an expectation that postgraduate chemists in several universities in Malaysia will be taught how to deliver the talk to senior school pupils in local dialects, Arabic and English as part of the International Year of Chemistry and beyond.

9.6.4   Hard to reach school audiences

Section 6.2.2 described the different public attitudinal groups including the disengaged-sceptics (PAS, 2011). A school audience is a good opportunity to engage with similar groups. Case study 9.2 by Dr Tristan Bunn is a perfect example of how you can reach disengaged students, by carefully considering the design of an event or activity.

Case Study 9.2

Taste and Flavour
Tristan Bunn

Background
Engaging young people is always a challenge. How to capture their attention and truly interest them in science is often a daunting prospect. This is especially the case when trying to engage pupils who have challenging behaviours, or who are not engaging with mainstream education. I was approached by the Local Education Authority's Advisor on Improving Behaviour and Attendance at Secondary School to deliver enrichment activities for 11- to 13-year-olds that would both promote healthy and sustainable lifestyles, as well as spark an enthusiasm for science. It is an interesting challenge to create something that will get teenagers to sit up and pay attention when the topics are healthy living and science.

Activity design
I started with a simple activity that had been used by scientists from the Institute of Food Research with a general public audience: Taste and Flavour. In order to make it more suited for pupils aged 11–13, I added a variety of quick investigations to stimulate the senses and provide more of a 'shock factor', as well as the opportunity to investigate their own genetics and develop their numeracy.

The activities used were modular, enabling them to be shortened or lengthened according to the time available and the setting. They can also be delivered to a wide age range, as they simply require the common experience of eating as the starting point for discussing the science. Participants can learn about everything from their own basic senses of taste to the molecular and genetic basis of taste and flavour. You can link the activities to chemistry, biology and food science courses, covering topics such as the nervous system, healthy eating, genetics, anatomy, molecular biology and organic chemistry.

There are six key activities.

1 The basic tastes – what are basic tastes and can you recognise them?
2 Taste and flavour – what is the difference between taste and flavour?
3 Taste and colour – how does colour affect your perception of taste?
4 Taste and genetics – how do your genetics affect your perception of taste?
5 Taste and temperature – what are the spicy and minty sensations?
6 Taste buds – what do your taste buds look like?

What you need to run this activity session

- Basic taste solutions – salt, sour, bitter, sweet, umami
- Drinking water
- Diagram of tongue taste map
- Flavoured sweets – parma violets are particularly effective
- Clear still flavoured drinks or jelly and food dye
- Genetic taste strips – phenylthiocarbamide (PTC) strips are available from Blades Biological
- Tabasco or chilli powder
- Mints
- Blue food dye
- Cotton buds
- Disposable cups
- Mirror
- Flexi-camera and monitor or projector (optional)

Running the session

Part one: the basic tastes

Setting the scene for the first activity involves introducing the concept of 'Basic tastes' (sweet, salty, sour, umami and bitter). The pupils are challenged to identify the basic tastes by tasting five clear solutions. This enables them to get started testing out their senses straightaway. At this stage you can elicit their understanding and preconceptions regarding tastes and flavours. The taste solutions used are sugar, salt, citric acid, monosodium glutamate and flat tonic water, tested by dipping cotton buds in prepared solutions and dabbing on the tongue. People often react differently to umami; some may hate it, others may not be able to taste it at all and the taste will range from salty, meaty, fishy or vinegary and several other things too!

At first some pupils are reluctant to try some of the tastes, especially if they see other people screwing up their faces in reaction to sour or sticking their tongues out when tasting umami, but this soon changes and I found that curiosity takes over with normally reluctant and sullen pupils eager to join in. It isn't long before pupils are asking questions and describing their feelings. Most notably, the pupils displayed considerable empathy and interest in each other's perceptions of the tastes, discussing how it tastes and expressing whether they like it or not. They soon come to realise how different people have different likes and dislikes to the same things and that these differences are intriguing and perfectly normal.

Once pupils have familiarised themselves with the tastes they can conduct an investigation into the taste map. Pupils investigate their own taste 'zones' and compare their sense experiences to the taste map to see if it is accurate.

It is a commonly held myth that there are specific areas on your tongue for the separate taste buds and most people have seen diagrams showing the bitter, sweet, salty and sour zones of the tongue. While the taste map is partially true it is now known that the sensation of taste is far more complicated with different taste receptor cells, variation amongst people and taste buds that change with age. All taste buds detect all tastes though there are different types of taste buds located around the oral cavity.

The map was familiar to some pupils and new to others. Interestingly while some pupils believed in 'the map' because it seemed authoritative, others were instantly sceptical drawing on their own experiences to discount it.

Part two: taste and flavour

The sensations of taste and flavour are synonymous for most people. To encourage pupils to think about this, you can ask them to consider the question 'What is the difference between taste and flavour?' Pupils were asked to hold their nose, then provided with a parma violet (or a similar fragrant sugary sweet) to suck. Whilst still holding their nose, pupils described what they could taste (the activity alone produces a sense of amusement for all involved). The typical response was that they could taste something sweet, which is the sugar in the parma violet. When they let go of their nose they got a sudden rush of the flavour as the smell permeated their nasal cavities. The response seemed to be a mixture of surprise and wonder, at the difference in sensation, along with recognition of the flavour of a parma violet. This demonstrates that it is smell which produces a sense of flavour. Just like when you have a cold, food has little flavour.

The basic tastes do not depend on smell, so pupils can now go back and identify the basic tastes just as easily with or without holding their nose.

Part three: taste and colour

The colour of foods has a strong influence on our likes and dislikes, due to our personal experiences and human evolution. We have developed protective responses to avoid poisoning and to seek out nutritious food. The pupils were provided with a selection of drinks with added food dye and the challenge was to identify the flavour of the drink. The activity can be carried out using clear, still flavoured drinks that can be found in most supermarkets along with a few drops of food dye. Carbonated drinks are more acidic due to the dissolved carbon dioxide and less effective for the activity. Alternatively lemon jelly and varying amounts of red food dye can be used to produce yellow, orange and red jellies. Commonly pupils will be strongly influenced by the flavours suggested by their peers. Black food colouring tends to generate a strong negative response that may be conserved in evolution in order to avoid rotten food. Despite this, Cola drinks are hugely popular and the power of branding, advertising and expectations can be discussed with pupils.

Part four: taste and genetics

The next activity involves testing the genetic basis of taste and comparing the numbers of 'tasters, non-tasters or super-tasters'. Pupils are provided with PTC taste test papers and are classified according to their ability to taste the PTC. Some pupils react strongly, moderately or have a delay in their response, while others cannot taste anything. You can classify the pupils according to the strength of their reaction. The results of the group are collated and percentage proportions of each category calculated by the pupils and compared to the general population. Being able to taste PTC is a dominant trait. About 66–75% of people in Western cultures are able to taste it. Individuals who are homozygous for the dominant allele are commonly described as 'supertasters'.

Part five: taste and temperature

After conducting the genetic taste test, the concept of hot and cold can be introduced. Spicy foods have become hugely popular in the UK and some pupils are often keen to demonstrate their fondness for hot foods. Pupils can taste a dilute solution of Tabasco sauce, as they did with the basic tastes, and are subsequently very eager to get the cool sensation of a mint. Due to the strong sensations and subsequent dulling of the taste buds, this activity is usually reserved for the end of the tasting experiences. The sensation of hot food is caused by substances such as chilli and capsaicin activating a receptor that is also activated by heat. The receptor is connected to the trigeminal nerve in the tongue, mouth and nose that is stimulated by a variety of chemical 'irritants' such as onions. The cold minty sensation is also detected by receptors in the trigeminal nerve that can detect changes in temperature.

Part six: taste buds

The last activity involves applying blue food colouring to the tongue to reveal the 'taste buds', which can be viewed in a mirror or on a TV monitor using a flexi-cam. The food colouring leaves the 'taste buds' as little pink dots while dyeing the rest of the tongue bright blue. This is a great visual ending enabling the pupils to actually see what they have been investigating and providing a reminder and interesting talking point they can bring up with friends and family once the activity is over.

One girl, who had been somewhat grumpy during the previous activities, when persuaded to have her tongue dyed really opened up and displayed her vivacious personality, revelling in the attention and eagerly encouraging her teacher to take a picture of the taste buds on her tongue with her mobile phone.

Reflection

In retrospect, the link between our sense of taste and flavour and how they shape our preferences for healthy or unhealthy lifestyles could have been more explicit. The inclusion of tastings of foods such as brussels' sprouts or broccoli to demonstrate the bitterness experienced by supertasters and their aversion to some healthy foods is the sort of activity that would have provided a clearer link between the science of taste and dietary choices. In future activities I plan to let pupils taste Beneforté broccoli, developed by scientists at the Institute of Food Research and John Innes Centre to have higher levels of a beneficial chemical called Glucoraphanin that lowers rates of heart disease and some forms of cancer as well as boosting the body's antioxidant enzyme levels.

Throughout the session the activities provoked many questions and had otherwise disengaged young people eagerly taking part and becoming more open and communicative with their peers and staff. In the evaluation they said 'All of us found science the most interesting . . . if the chance came up again we would love to be invited back.' But it was the changes that took place in the young people, the personal development and emotional shift that was the most rewarding outcome. When asked his name at the start of the day by the Local Education Authority's Advisor, one pupil's defensive reply was 'Why do you want to know?', but by the end of the day he had opened up and revealed he had been an idiot to have been excluded from his school, and realised how much he has missed out on. He made eye contact and was prepared to show his vulnerable side.

Tips to engage with a disengaged student audience
1 Provide short, quick activities that enable pupils to take control of their own learning.
2 Explain the purpose of the activity and find out what they already know.
3 Be friendly and laid back and remain unconcerned if they do not get involved in the activities straightaway.
4 Provide thinking time after questions and 'acclimatisation' time for pupils to engage, as this ensures meaningful interactions that pupils choose to be a part of.
5 Add an element of competition and an emotional component that provides meaning to the activities.
6 Provide interesting and surprising facts that they can relate to.
7 Smile, make it a fun experience and enjoy yourself; it is entirely possible that they may want to take part.

9.7 Think about your resources, consumables and equipment

After you have decided on the design and delivery mechanism of your activity, you now need to think about resources, consumables and perhaps equipment. Some funding sources will allow you to buy equipment and this is very useful if you are trying to build up a suite of equipment for science communication activities (see Section 6.8). Building up a stock of equipment can be very useful, as you may find that taking it out from your research laboratory may not be a suitable strategy.[9]

For pre-prepared resources there are a number of excellent organisations where you can download suitable material (see Table 6.7). There are also several companies where you can buy good quality, but reasonably priced consumables, kits or pieces of equipment for school projects and places where you can obtain them are provided in Table 9.7.

Your resources also include people to help you run your activity or event. This is equally applicable to both school and public science communication. If you work at a university, you have at your disposal a potential army of helpers in the shape of both undergraduate and postgraduate students. These young people studying science are a remarkable resource, full of energy and enthusiasm. They can often inspire school-age pupils as they are closer to their own age. I have also found my students give me a limitless supply of new ideas for activities, many of which I use repeatedly. This 'human resource' is sustainable, as there is a constant supply of fresh students passing through departments. One way I involve students is in an undergraduate module in science communication, during which the students design and deliver an activity as part of a themed public event held during Science Week. I gain valuable help and the students gain a range of transferable skills and credit towards their degree programme (Yeoman *et al.*, 2011; Case study 10.3).

9.8 School years and qualifications

9.8.1 School years

Quite a few of the case studies throughout this book refer to school years. School years differ from country to country and also sometimes within countries. The information in Table 9.8 will broadly help you relate the case study information to your own country's school structure.

9.8.2 Qualifications obtained in schools

Qualifications are very diverse and vary considerably, not just between countries, but also within countries. The following information is intended to be a broad guide to the most common qualifications. Currently pupils in:

---

[9] There may be an excess on the insurance or it may be needed in the laboratory.

Table 9.7 Organisations which offer resources for teaching and companies where you can buy educational products across a wide range of scientific disciplines.

Organisation/company	Practical kits/resources/ equipment and activities	Website
National Centre for Biotechnology Education (NCBE)	Kits for DNA activities, cheap equipment, bioreactors, bacterial strains	www.ncbe.reading.ac.uk
Science Museum	Classroom resources, good activities for Key Stage 3 and 4	www.sciencemuseum.org.uk
British Science Association	Crest Awards; bronze, silver and gold. There are a variety of activities which are detailed for different age groups. There are also packs for National Science and Engineering week	www.britishscienceassociation. org
Practical Chemistry, Physics and Biology	Practical experiments and a wide range of resources for teaching chemistry, physics and biology	www.practicalchemistry.org www.practicalphysics.org www.practicalbiology.org
Science and Plants for Schools (SAPS)	Primary and secondary school resources for plant education	www-saps.plantsci.cam.ac.uk
Edvotek	The Biotechnology Education Company have a range of kits and equipment	http://edvotek.co.uk/
BioRad	A wide range of kits and equipment for the classroom	www.bio-rad.com
Carolina Biological (a US-based company-to order Carolina Biological products go through Blades Biological)	A wide range of kits, resources and equipment covering all science areas	www.carolina.com www.blades-bio.co.uk
ESPO	The best place for cheap art materials, they also have some school science equipment as well	www.espo.org
Wards Natural Science (a US based company)	Wards has some excellent kits, models and equipment. The are based in the US but have reasonable shipping costs	http://wardsci.com/

- England, Wales and Northern Ireland do GCSEs typically at age 15–16, then AS levels at 16–17 and A2 ('A' levels) at 17–18. Some schools also offer the International Baccalaureate (IB);
- Scotland do the Scottish Qualifications Certificate (SQC) followed by Higher Grade Examinations, typically at ages 14–15 and 16–18 respectively;
- Republic of Ireland, do junior certificate examination and then the leaving certificate examinations typically at ages 15–16 and 17–18 respectively;

Table 9.8 School years comparison between different countries.

Age	US	England and Wales	Scotland	Northern Ireland	Ireland	Australia
4–5	Pre-kindergarten	Reception	Primary 1	Primary 1	Junior Infants	Pre-School
5–6	Kindergarten	Year 1	Primary 2	Primary 2	Senior Infants	Kindergarten
6–7	1st Grade	Year 2	Primary 3	Primary 3	First Class	Year 1
7–8	2nd Grade	Year 3	Primary 4	Primary 4	Second Class	Year 2
8–9	3rd Grade	Year 4	Primary 5	Primary 5	Third Class	Year 3
9–10	4th Grade	Year 5	Primary 6	Primary 6	Fourth Class	Year 4
10–11	5th Grade	Year 6	Primary 7	Primary 7	Fifth Class	Year 5
11–12	6th Grade	Year 7	S1	Year 8	Sixth Class	Year 6
12–13	7th Grade	Year 8	S2	Year 9	First Year	Year 7
13–14	8th Grade	Year 9	S3	Year 10	Second Year	Year 8
14–15	9th Grade	Year 10	S4	Year 11	Third Year	Year 9
15–16	10th Grade	Year 11	S5	Year 12	Transition Year	Year 10
16–17	11th Grade	Year 12	S6	Year 13	Fifth Year	Year 11
17–18	12th Grade	Year 13	a	Year 14	Sixth Year	Year 12

Years involved in primary/elementary education

Years involved in Middle school

Years involved in secondary education/high school

[a]Note in Scotland students can enter higher education at 17.

- European schools study for a Baccalaureate, at ages 17–19;
- The US and Canada leave high school with a High School Diploma typically at ages 17–18, although the International Baccalaureate is also becoming increasingly popular;
- Australia study for the Secondary School Certificate of Education (SSCE), which they do typically between the ages of 16–18. The SSCE is required for entry into university.

Many of the case studies which take place in England refer to Key Stages within the National Curriculum, which are detailed in Table 9.9.

Table 9.9 The National Curriculum is split into four Key Stages (England, Wales and Northern Ireland).

Key Stage	Ages	School Years	Qualification
1	5–7	1 and 2	—
2	7–11	3, 4, 5 and 6	—
3	11–14	7, 8 and 9	—
4	14–16	10 and 11	GCSE

9.9 Concluding remarks

This chapter has outlined some of the reasons why it's important to target school-aged pupils in your science communication activities. It has hopefully given you a brief overview of the current attitudes towards and performance of pupils in science across different countries. It has introduced you to the different types of school qualifications in science currently being offered in schools. Finally we have given you advice on how to start building up your science communication portfolio within a school environment.

References

BIS (2011) Attitudes to Science: Survey of 14–16 year olds. http://www.bis.gov.uk/assets/biscore/science/docs/a/11-p112-attitudes-to-science-14-to-16 (accessed 2 August 2011).

Dolan, E.L. (2008) *Education Outreach and Public Engagement. Mentoring in Academia and Industry*. (ed. Bell, J.E.). Springer.

Donghong, C. and Shunke, S. (2010) The more, the earlier, the better: science communication supports science education, in *Communicating Science in Social Contexts* (eds Cheng, D., Claessens, M., Gascoigne, T., *et al.*). Springer.

Donnelly, J. (2009) An Invisible Revolution? Applied Science in the 14–19 Curriculum. A Report to the Nuffield Foundation. http://www.nuffieldfoundation.org/applied-learning-14-19-science-education (accessed 5 May 2012)

Fensham, P.J. (2008) Science Education Policy-Making: Eleven Emerging Issues. Commissioned by UNESCO, Section for Science, Technical and Vocational Education

Green, N. (2011) A-level results Send Science and Mathematics to the Top of the Class, The Guardian (18 August). http://www.guardian.co.uk/science/blog/2011/aug/18/a-level-results-science-mathematics (accessed 5 May 2012)

House of Commons Science and Technology Committee (2011) Practical Experiments in School Science Lessons and Science Field Trips. Ninth Report of Session 2010–12, Volume 11.

Jenkins, E. (2007) School Science: a questionable construct? *Journal of Curriculum Studies* **39** (3), 265–282

Jenkins, E.W. and Pell, R.G. (2006) The Relevance of Science Education Project (ROSE) in England: a summary of findings. http://roseproject.no/network/countries/uk-england/rose-report-eng.pdf (accessed 5 May 2012)

Michael O.M., Mullis, I.V.S. and Foy, P. (2008) TIMSS 2007 International Science Report: Findings from IEA's Trends in International Mathematics and Science Study at the Fourth and Eighth Grades

Millar, R. and Osborne, J. (1998) Beyond 2000: Science Education for the Future. The Report of a Seminar Series Funded by the Nuffield Foundation. http://www.kcl.ac.uk/content/1/c6/01/32/03/b2000.pdf (accessed 5 May 2012)

Miller, J.D. (2011) To Improve Science Literacy, Researchers should run for School Board. Opinion. *Nature* **17** (1), 21

Osborne, J. and Dillon, J. (2008) Science Education in Europe: Critical Reflections. A report to the Nuffield Foundation

OECD (2006) *Assessing Scientific, Mathematical and Reading Literacy: A Framework for PISA 2006.* OECD, Paris

OECD (2009a) Equally Prepared for Life? How 15-Year-Old Boys and Girls Perform in School. http://www.oecd.org/dataoecd/59/50/42843625.pdf (accessed 5 May 2012)

OECD (2009b) Top of the Class, High Performers in Science in PISA 2006. http://www.oecd.org/dataoecd/44/17/42645389.pdf (accessed 5 May 2012)

OECD (2010) PISA 2009 Results: What Students Know and Can Do – Student Performance in Reading, Mathematics and Science (Volume I). http://dx.doi.org/10.1787/9789264091450-en (accessed 5 May 2012)

Porter, C., Parvin, J. and Lee, J. (2010) Children challenging industry: all regions study of the effects of industry-based science activities on the views of primary school children and their teachers. http://www.cciproject.org/research/documents/CCI_National06–08.pdf (accessed 15 October 2011)

Project 2061 (2011) http://www.project2061.org/ (accessed 6 May 2012)

RCUK and DIUS (2008) Public Attitudes to Science, Report prepared for Research Councils UK and the Department for Innovation, Universities and Skills by People Sciences & Policy Ltd

SCORE (2011) Practical Work in Science: A Report and Proposal for a Strategic Framework. http://www.score-education.org/media/3668/report.pdf

Smith, E. (2010) Is there a crisis in school science education in the UK? *Educational Review* **62** (2), 189–202.

Sturman, L., Ruddock, G., Burge, B., *et al.* (2008). *England's Achievement in TIMSS 2007: National Report for England.* NFER, Slough

Woodley, E. (2009) Practical work in school science –why is it important? *School Science Review* **19** (335), 49–51

Yeoman, K.H., James, H.A. and Bowater, L. (2011) Development and Evaluation of an Undergraduate Science Communication Module. beej, 17-7 http://www.bioscience.heacademy.ac.uk/journal/vol17/beej-17-7.pdf (accessed 5 May 2012)

Useful websites

CLEAPSS – www.cleapss.org.uk/
Department of Education in NI – www.deni.gov.uk/
National STEM Centre – www.nationalstemcentre.org.uk
National HE STEM Programme – www.hestem.ac.uk/
Nuffield Foundation – http://www.nuffieldfoundation.org/education
Primary Connections in Australia – http://www.science.org.au/ primaryconnections/
Science Learning Centres – www.sciencelearningcentres.org.uk/
The Science Lab – www.sciencelab.org.uk/

CHAPTER TEN

Demonstrating Interactions between Scientists and Schools

In order that the relations between science and the age may be what they ought to be, the world at large must be made to feel that science is, in the fullest sense, a ministry of good to all, not the private possession and luxury of a few, that it is the best expression of human intelligence and not the abracadabra of a school, that it is a guiding light and not a dazzling fog.

—William Jay Youmans, Hindrances to scientific progress, *The Popular Science Monthly* (1890 [Nov] 38, p.121)

10.1 Introduction

Working with schools can be an incredibly rewarding experience for the scientist and provide both professional and personal satisfaction. Chapter 9 provided information about performance in and attitudes towards science education in the UK and beyond. It also gave advice on getting started with science communication in schools. Scientists and professional science communicators engage with schools in a number of different ways:

1 enhancing the curriculum at primary and secondary levels in the school environment;
2 enhancing the curriculum with activities with a school audience, but outside the school environment;
3 developing cross-curricular activities for both primary and secondary schools;
4 influencing curriculum change;
5 embedding scientists into schools;
6 training teachers and technical staff.

Figure 10.1 maps these different mechanisms against whether they are teacher or pupil focused as well as fact or policy focused. Most of the mechanisms scientists can use to engage with schools are biased towards fact and are more obviously pupil centred.

The case studies presented in this chapter aim to highlight these different scientist and school interactions and provide ideas on activities that can be

Science Communication: A Practical Guide for Scientists, First Edition. Laura Bowater and Kay Yeoman.
© 2013 John Wiley & Sons, Ltd. Published 2013 by John Wiley & Sons, Ltd.

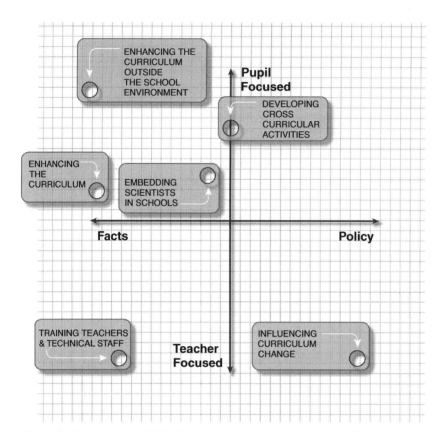

Figure 10.1 Mechanisms which scientists can use to become involved with school education.

done with schools. They also provide background information, methodology, potential pitfalls, specific safety advice, top tips and where to go to for resources to enable you to replicate and or adapt the activities presented. The case studies we have chosen to include in this chapter are those that contain ideas for approaches and activities which we felt you could use directly with little modification. There are other examples of activities provided in boxes, and more information about these can be found on the book website.

10.2 Enhancing the curriculum within the school environment

Enhancing the curriculum within the school environment includes activities within school time and out of school time. These can range from talks by scientists about their careers to full science activity days and science clubs, the latter run over a longer period of time. The advantage is that pupils remain within their own environment and as a consequence feel more confident. There is less organisation involved, for example the school need not provide additional teacher cover or arrange transport. The case study examples in this section are split into initiatives for primary school (Section 10.2.1) and secondary schools (Section 10.2.3).

10.2.1 Primary school curriculum enhancement

Primary schools face particular problems when teaching science (Fensham, 2008). Primary teachers, although very skilful in pedagogy having trained as generalist teachers, are not usually trained in science and can lack confidence. The TIMSS study showed that only 16% of primary school teachers had any specialist primary science training (Sturman *et al.*, 2008). A study in Australia has shown that as little as 2.7% of classroom time is spent on science (Angus *et al.*, 2004). Australia has recognised the need to improve primary science education and established Primary Connections[1], run by the Australian Academy of Science. Primary Connections supports teachers through professional workshops and curriculum resources (Hackling *et al.*, 2007).

The built school environment is also a problem; there are no specialist science laboratories, no chemical storage facilities and no technical staff. However, this need not be a barrier to school science communication. The 2011 SCORE Practical Science report indicates that teachers in primary school value practical science, which develops scientific enquiry and enables an understanding of investigative processes. In science courses, the understanding of the scientific method is an important but difficult concept to teach. Case study 10.1 by Dr Niamh Ní Bhriain shows how the scientific method and the importance of writing scientifically can be taught to a class of primary schoolchildren, which involves no wet laboratory experimentation and is rooted in the pupil's existing experience and understanding. The activity has been well designed; it is resource-light and could be repeated by someone with limited school communication experience. The approach taken in this case study could easily be adapted to an older high school audience.

Case Study 10.1

Is Chocolate Dangerous in a Slang-Free Zone and How Do *You* Make a Jam Sandwich?

Niamh Ní Bhriain

Background

I work with pupils in the eighth (and final) year of the primary cycle in the Republic of Ireland. Most will be 11–13 years old. I teach ten classes in a normal school environment. The programme is ambitious. Using biology and microbiology as exemplar branches of science, we look at how scientific understanding developed in Europe from the time of ancient Greece

---

[1] http://www.science.org.au/primaryconnections/

to today. We consider how science has developed and evolved arm in arm with human understanding and technology – for example, without the development of the microscope (by people such as van Leeuwenhoek and Hooke), 'microbes' would not have been discovered: Pasteur would not have made the discoveries that prompted him to hypothesise the 'germ theory' of disease and Koch would not have proved that hypothesis. We look at some of the problems individuals and societies faced and how they were solved. We see how Aristotle, Redi, Hooke, Semmelweiss and Pasteur, to name but a few, got some things right, other things wrong and were often frustrated by not being able to convince people of the truth as they saw it.

Why have such a programme?

At the moment science is not a mandatory subject in secondary school in the Republic of Ireland. It is therefore formally possible for students to complete their education without taking any science subjects beyond primary level. My personal aim is to ensure that even if the students I teach do not study science after primary level they will understand that science and scientists are relevant to their lives. I am also keen that those who *do* choose to study science will realise that the real challenge of the subject is not simply to memorise the topics and facts that feature on the examination curriculum, but rather to develop and refine a certain way of thinking – the scientific method.

This case study describes the structure of the first session in which I have a number of objectives. The first, and most immediate, is obviously to get to know the class and become familiar with its dynamics. The class teachers will 'flag up' any unusual challenges in terms of individual academic ability and/or behaviour, but I need to get to know the class as a whole. This is achieved by working with the class, asking the children about their understanding of science, scientists and how they work, and then engaging with them in the exercises described below. This interaction provides the information necessary to ensure that the subsequent classes can run smoothly – or at least as smoothly as is possible in any such classroom situation!

There are three additional aims:

1 to dispel the myth that scientists 'do' experiments that have 'right' or 'wrong' answers;
2 to communicate the fact that science writing should be slang-free;
3 to make the case for presenting comprehensive and faithful records of experimental protocols and results.

Aim One: To dispel the myth that scientists 'do' experiments that have 'right' or 'wrong' answers

The misapprehension that scientists do experiments that have right or wrong answers seems to be an unfortunate consequence of studying science as part of a National Curriculum. Classes have specific learning objectives. Teachers may have limited scientific training and be working outside their areas of expertise. There is a natural reliance on experiments which have been designed and selected to give reproducible and unambiguous results. This is probably necessary in the classroom situation but has the major drawback that students come to think that in science – just like in primary level mathematics for example – there is always a 'right' answer. By extension, they see alternative answers as 'wrong'. However much a hard-pressed teacher tries to change this way of thinking, the idea that 'my experiment didn't work/I got the wrong answer' prevails.

> I need to communicate the fact that it is *information* provided by experiments and not a 'right' or 'wrong' answer.

Exercise for teaching that a result is not 'right' or 'wrong'

I have found that what really helps is to introduce the term 'hypothesis' – a special word for an idea that can be tested. Discussion of how ideas are tested by doing experiments, allows the introduction of the idea of the scientific method.

> (*My contribution* / class response)
>
> Me: *I think that all children love sport. Is this a hypothesis – can I test the idea?*
> Class: Yes.
> Me: *How?*
> Class: Ask us?
> Me: *Hands up if you love sport.*
>
> The result will be yes, no or maybe. Some will ask 'exactly what do you mean by 'sport'?' which allows discussion of 'good' vs 'bad' questions and 'good' *vs* 'bad' experiments.

We go through the statements given in Case study Box 10.1.1 using simple examples such as:

Case Study Box 10.1.1

The **scientist** comes up with an **idea** – the **hypothesis**.

You can *only* use the term **hypothesis** for an idea if you can do **tests** to see whether of not it's correct.

These **tests** are called **experiments.**

You have to do experiments more than once to be sure that the result you get is reliable.

Experiments can give **three** kinds of answers – **yes, maybe** and **no**.

If you get a **yes** answer this means that your hypothesis **might be** correct,

If you get a **maybe** answer you might need **to do the experiment again** or to **design a slightly different one** to get a clear answer.

If you get a **no** answer then there is something not right with the hypothesis and you might need to think about it some more.

This way of doing things is called the Scientific Method.

If lots of experiments suggest that a hypothesis is correct it may become widely accepted as a **theory**.

If experiments are done properly the results generate useful information/data/results.

If this is **new** information it is a **discovery**.

> Class: What do you mean by sport?
> Me: What does your answer mean for my hypothesis?
> Class: 'It might be right'. 'It's wrong.' 'It can't become a theory.' 'You can't be sure.'
> Me: What experiment should I do to get more information?
> Class: Ask the other classes the same question and see what they say.

After some more simple examples like this it's time for a more taxing exercise. The following works really well.

My Hypothesis
Children will do better in their Friday morning spelling tests if they feel happy and have had a chocolate treat beforehand.

Me: Is this a Hypothesis?
Class: Yes.
Me: Why?
Class: Because you can test it.
Me: How can I test it?
Class: Give us chocolate before our spelling test.
Me: What will happen?
Class: If the hypothesis is correct we'll all get good marks.
Me: But maybe you all get good marks all the time.
Class: We don't!
Me: But the spellings might just have been easy . . .

Pause for thought

Class: What you need to do is just give chocolate to some of us.
Me: *Ok, so how will I do that?*
Class: Divide the class in half and give chocolate to one half.
Me: *What will happen?*
Class: If you're right the half that had chocolate will get better marks than the half that didn't.
Me: *But maybe I just happened to give chocolate to people that are good at spellings.*
Class: That's easy to check. Next time swap it around and only give the chocolate to the half that didn't get it the first day and then do the spelling test.
Me: *What will happen?*
Class: You should get the opposite – the ones who did worse the first time should do better.
Me: *What will that tell me?*
Class: You'll have a yes answer so your hypothesis might be right.
Me: *What can I do to be more sure?*
Class: Do it lots of times.
Me: *Anything else?*
Class: You could do the same test/experiment in all the classes.
Me: *Would that be a good idea?*
Class: Yes because the more yes answers you get the more you can be sure you're right.

Pause for thought.

By now the class has a much improved and practical grasp of scientific method and experimental design.

Short-term reinforcement

In the last five minutes of class time the students are given a worksheet to evaluate what they have learnt (see worksheet on book website). Most complete it quickly and correctly. The few who struggle can be given individual attention while the rest are busy. Once the sheets are graded I ask the class how they would have coped with the worksheet before the activity. They are stunned to see how much they have learned and respond enthusiastically to the instruction to take the worksheets home and 'show off' the new words they have learnt to others in their families!

Aim Two: To encourage slang-free scientific writing

Scientific writing should be slang-free. This age group is often unaware that the words and phrases that they use in everyday speech are either incomprehensible to, or mean something very different to, those outside of their immediate peer group. Relaxed, idiomatic expression has a place in self expression but is unhelpful in scientific writing.

Exercise to make the case for slang free scientific writing

English-speaking children take their command of the language for granted. Their experiences of holidays, television news reports etc. tell them that English is widely spoken and it is easy for them to fall into the trap of believing that English is universally understood. I use a very easy method to rattle this complacency.

I ask for three volunteers who enjoy reading aloud. Two are told that they will be asked to read from popular novels (I use books from the *Harry Potter* and *Artemis Fowl* series) and the third will read from a scientific journal, any journal will do but the longer and more impressive the name the better. The first volunteer is given a book and asked to read … silence! Why? The text of the 'familiar' title is in French. Laughs and apologies ensue and the second volunteer opens the next book. There is a pause followed by careful and very hesitant progress through the opening sentences. This volume is in Irish – the language is familiar but the text is difficult. The final text is a selected passage from the journal of choice which is read with fluency and confidence. The reader is asked if they understood the content. They say No, and when pressed explains they could read the words because they were in English but they still struggled to understand them. The class agrees.

> The class is then asked:
>
> Me: *How would you find out what the difficult words mean?*
> Class: Use a dictionary (or an on-line translator).

The scene is now set for the second lesson of the day – that when writing for a scientific journal you have to choose carefully the words you use.

I exploit the fact that the word 'deadly' is used in common parlance in Dublin to mean 'really good'/ 'excellent'/ 'completely desirable'.

We return to the 'chocolate hypothesis' discussed earlier in the class.

> Me: *So now imagine that I've written up my experiment. I'm really excited about the '… We examined the effect of a pre-test chocolate treat on improving pupils' performance in spelling tests. The results were deadly…'*

Then I describe the results.

The paper gets published.

Far away in a remote village in Siberia Ivan Ivanovitch is using his computer. He is a teacher and his pupils are having serious problems learning their spellings. He's really worried because the inspector is coming soon and he needs the children to do well. He's hoping to find some helpful ideas so he checks his Russian/English dictionary and searches 'improvement, spelling tests, results'. He finds my paper and reads it, using his dictionary to help him translate because his English is not very good.

> Me: *Will he give his class chocolate before the inspector arrives?*
> Class: Yes.
> Me: *Are you sure?*
> Class: Yes.
> Me: Are you absolutely sure?

Pause and some grins appear.

Class: No he won't. Unlikely. It depends on how much he hates his class!
Me: *Why won't he?*
Class: Because he'll think that they'll die.
Me: *Why?*
Class: Because you said that the results were deadly.
Me: *And?*
Class: If he looked in the dictionary he'd think that deadly means it would kill them.
Me: *Is that what deadly means?*
Class: Yes, in the dictionary.
(It may be necessary at this point to allow someone to confirm this fact in the dictionary in order to convince the sceptics that this is true.)
Me: *Should I have said "wicked" or "sweet" instead?* (Either of these words would be appropriate in the idiomatic parlance.)
Class: No they're not what you want to say either.
Me: *What is wrong with them?*
Class: They're slang. . .

Point made!

It is important to ask for volunteers for this exercise. Those with the confidence to volunteer to read in public are unlikely to feel embarrassed or humiliated when challenged by the unfamiliar texts. The choice of 'non-English' texts can obviously be varied depending on the local situation.

Aim Three; Science write-ups should be faithful and comprehensive accounts of what has been done

For the third aim, presenting comprehensive and faithful records of experimental protocols and results, I need to remind the class that just because it is obvious to them why or how they did part of an experiment they cannot assume that it is equally obvious to the person reading their account. They **must** 'state the obvious' and give an accurate and detailed account of what they have done.

When writing an account of an experiment performed in class the pupils are actually doing two things – they are:
1 reporting what they have done;
2 trying to comprehend and explain the new information.

I wanted to design an exercise that does not require additional tasks so that assessment can be based solely on the clarity of expression.

Exercise for writing a comprehensive account of how to execute a simple and familiar task

The first homework task I set my classes is to: write out instructions on how to make a jam sandwich in such a way that anyone could follow them and make a sandwich exactly the way you like to eat it.

This exercise may not initially look like 'science' but it has a number of merits:
1 all class members, regardless of academic ability, can tackle this task with some degree of confidence;
2 there is no 'wrong' answer. Specific recommendations for improvement can be made to each individual but there is always something that can be praised, however weak or incomplete the answer may be;
3 a general discussion of class results, helps reinforce the need for clarity and attention to detail in experimental writing but since nobody has got the exercise 'wrong' the 'criticism' is general and constructive.

Sometimes a pupil will include a 'diagram' – a drawing of either the preparative steps and/or the completed sandwich. I single these out for particular praise and explain how much information can be put in a picture and that it saves writing lots of words. The idea that scientists like to save effort – are as lazy as the rest of the population – goes down well and reinforces the idea that you don't have to be 'special' or 'different' to be a scientist.

Highlights of the course

The exercises I have described above can readily be adapted for different age groups and environments. Apart from what they teach the class about the scientific process and its communication, they are invaluable for establishing a rapport with the class members. They allow me to engage with each group of children in a structured but relatively relaxed way. Pupils who have formed the idea that 'science is hard' or 'I can't do science' find themselves fully engaged. I get an enormous amount of in-class feedback and I get a sense of the class group dynamics that is hugely useful in deciding how subsequent formal classes should be conducted. Above all, they are FUN!

Top tips

1 Warn the class in advance that you are going to teach them some big words (e.g. hypothesis), that are *really* useful in word games like 'hangman' and then write big words up slowly, letter by letter, on the black/white board. This age group loves sounding out the syllables as they appear and trying to predict the completed word.

2 Get some local knowledge – ask the class teacher if there is a particular time of the week at which tests are scheduled. This allows the scene for the 'chocolate hypothesis' to be set in a way that really draws the children in.

3 If there are children in the class for whom English is not their first language, ask them to share examples of slang words that they have found difficult or amusing.

4 Ask for a volunteer to describe how to make a cup of tea. The conflicting – and often unsolicited (!) – 'advice' offered by the rest of the class sets the scene nicely for the homework task.

It is vitally important that an enthusiasm for science is nurtured at an early age. The 2006 PISA study (OECD, 2009) shows that early engagement with science produced more 'top performers,' and the TIMSS study showed that only 59% of Year 5 pupils in England had a positive attitude to science (Sturman et al., 2008). The attitudinal study by Porter and Parvin (2008) on behalf of the Chemical Industry Education Centre at the University of York suggests that it's very important to engage with primary school children before their ideas about science become fixed. Universities have a tendency to concentrate on secondary school pupils in terms of their outreach strategy, but studies have shown that interventions aimed at primary school are not too early, and that children remember and remain influenced by the experiences they have had (Porter et al., 2010). Thus scientists have a real opportunity to make a difference to enrich the curriculum, provide inspiration and excitement and to make science fun. Case study 10.2 by Dr Alison Ashby, 'Fun with Fungi' is an excellent example of curriculum enhancement in a primary school. It shows how scientists can influence what is taught in schools by taking part in these initiatives. It also shows how existing resources available on a variety of websites can be used to design an effective lesson, tackling an area of science which is poorly covered by the National Curriculum. Alison has broken down the subject of mycology into three broad areas of interest:

1 how fungi differ from plants and animals;
2 fungi in the environment;
3 fungi in our lives.

Fun with Fungi for Primary School Children: a Curriculum Enhancement Project

Alison Ashby

Background

Fungi are the forgotten kingdom, yet they are omnipresent and play a vital role on our planet. It is important that we encourage an early engagement with the Fungal Kingdom and further develop this across the full academic arena. With support from the British Mycological Society (BMS), the Department of Plant Sciences in Cambridge and the Royal Society, a set of primary school resources were brought together and trialled in both the primary school environment and at science communication events.

The aim was to provide a platform from which children can explore and engage with the 'Fungal Kingdom'. The resources comprise a set of 'kitchen science' and 'art and craft' based activities designed for use by educators of children at Key Stage 1 and 2 (between the ages of 5 and 11 years old) and are currently hosted on the BMS primary schools web pages (see Case study table 10.2.1 on useful websites). The primary school resources can be used independently in a science lesson or grouped together in a carousel of activities as part of a 'Fun with Fungi' day.

The project

Initially to get the project running, contact was made through teachers at my children's schools. Further contact was made with teachers at Teachers Forums (formal and informal groups where discussion on teaching issues occurs).

A 'Fun with Fungi' day was trialled at St Faith's school, Cambridge, which gave us the opportunity to test the primary school activities in a school environment. The day formed part of an enrichment week of activities during the summer term at the school. Subsequently the activities were delivered to other primary schools in the local area.

The key objectives were to:
• have an enjoyable and informative day;
• describe what fungi are and how they differ from plants and animals;
• discover what fungi can do for us and our environment.

The day began with a general introduction to fungi, where the whole year group of approximately 50 children learnt the fundamentally important message that fungi were not plants or animals but had their very own kingdom. The children learnt:
• how fungi range in size from microscopic to the largest living organism on Earth
• that you can find fungi just about everywhere on Earth;
• that they have been around on Earth far longer than the dinosaurs!

The children pretended to be fungi by using string to make mycelium (the body of the fungus) and digesting dead organic matter (bread sticks were substituted for dead tree branches). After a brief rain shower (from water sprayers), fruit bodies (cocktail umbrellas) were produced and fungal spores (like the seeds of flowering plants and in this case, glitter!) were released. A fungal slide show was presented illustrating the diversity in colour, size and shape of different fungal fruit bodies. Props were used, including rotten strawberries, penicillin, compost heaps, chocolate and a basket full of groceries demonstrating the importance of fungi to us and to our environment. The year group was then split into four and the remainder of the day focused on four parallel 'hands-on' activities that were performed in a carousel.

The day was drawn to a close with a 'Questions and Answers' session and a glimpse at some amazing fungus facts. At the end of the activity day in schools, each child went home with a 'mushroom spots' book (supplied by the BMS free of charge), a spotty red balloon, fungus models made from modelling clay, a spore print, gill and spore photograph, a certificate and a chocolate bar (because fungi help to put the flavour in chocolate!).

In October of the same year we were able to trial the same activities plus others at a science communication event held at the Royal Botanic Gardens in Edinburgh, hosted by the BMS. The activities were also run in the public arena at the Cambridge Science Festival weekends held in March each year.

Resources and equipment

Using funding from the Royal Society, five small portable stereo microscopes were purchased from Brunel Microscopes. These were ideal for viewing the gills and other structures comprising the fruit body. The Department of Plant Sciences kindly supplied a compound microscope, camera and TV allowing children to view the spores produced by a mushroom which had been deposited onto a microscope slide. The BMS provided *How the Mushroom got its Spots* booklet for each child and loaned some of their many mushroom models. A local mushroom supplier kindly donated sufficient Portobello mushrooms for spore prints and a range of other edible exotic mushrooms for the children to investigate.

The details and web resources which accompany the activities are shown in Case study tables 10.2.1 and 10.2.2.

Case study table 10.2.1 Activities and web resources for 'Fun with Fungi'.

Activity	Web resources
Make your favourite fungus	http://www.britmycolsoc.org.uk/education/resources-and-materials/primary-school-resources/introduction-to-fungi/activity-1-make-your-favourite-fungus/
Make a woodland collage to show the importance of fungi in the woodland habitat	http://www.britmycolsoc.org.uk/education/resources-and-materials/primary-school-resources/introduction-to-fungi/activity-5-making-a-collage-of-fungi-in-the-woodland-habitat/
Fungus detectives involving using microscopes and magnifying glasses to observe different fungal fruit bodies, making spore prints and looking at fungal spores.	http://www.britmycolsoc.org.uk/education/resources-and-materials/primary-school-resources/introduction-to-fungi/activity-4-mushroom-detectives/ · The mushrooms were all edible species and were donated by Produce Global Solutions Ltd
Watch a fungus fizz	http://www.britmycolsoc.org.uk/education/resources-and-materials/primary-school-resources/introduction-to-fungi/activity-3-watch-a-fungus-fizz/
How the mushroom got its spots	http://www.britmycolsoc.org.uk/education/resources-and-materials/primary-school-resources/introduction-to-fungi/activity-2-how-the-mushroom-got-its-spots/

I developed the mushroom detectives activity using resources from the BMS (spore print worksheet, mushroom spots booklet). I provided mushroom books, BMS mushroom posters and a set of fungus fruit bodies from the BMS gallery of images. Watch a fungus fizz was modified from the *Naked Scientists* experiment. The balloon spots activity was developed as a BMS resource and is published in the BMS *How the Mushroom got its Spots* booklet.

Case study table 10.2.2 Useful websites.

Website	URL address
British Mycological Society Education Resources	http://www.britmycolsoc.org.uk/index.php/education/ http://www.britmycolsoc.org.uk/education/resources-and-materials/primary-school-resources/introduction-to-fungi/
NCBE Microbiology resources, growing oyster mushrooms	http://www.ncbe.reading.ac.uk/NCBE/MATERIALS/MICROBIOLOGY/oyster.html
Mushroom masks	http://www.planet-science.com/outthere/lifemasks/fungi/toadstool/toadstool_bw.pdf
Badges	http://www.fungi4schools.org/GBF_web/Props&images/Props_Badges.pdf
Brunel Microscopes	http://www.brunelmicroscopes.co.uk/

This was modified to include a schematic of the lifecycle of fly agaric and some coloured photos of the activity. The 'make a woodland collage' activity brings together information from previous BMS resources such as 'Appendix 14 Fungi4schools: Understanding fungi in the forest' and 'mushroom murder mystery' which is published in *How the Mushroom got its Spots*; however, for younger children the activity is simply 'art and craft' based.

Project highlights
One of the most rewarding experiences was to watch the faces of children as we performed the mushroom spots activity. They were truly amazed! Making mycelium and pretending to be fungi are best carried out in larger groups for the best overall effect.

Project extensions
The 'Fun with Fungi' days are intense and involve several days of planning. It is a pity that at present these days are run as part of the enrichment programme in schools rather than as a curriculum topic area. As an enrichment exercise there is little time or scope for further expansion. However, I was delighted that in one school, children were encouraged to spend a further two science lessons developing a booklet on everything that they had learned about the Fungal Kingdom. In another, children were asked to write a fungus fact on a white circle which then formed the spots of the fly agaric mushroom that was mounted on the wall in the classroom. Some schoolchildren were allowed to grow their own oyster mushrooms following the protocol and using oyster culture supplied by NCBE Reading. In one year, Produce Global Solutions (PGS, an edible mushroom supplier in Ely, Cambridgeshire), supplied shiitake mushroom growing kits for a whole year group. The children loved to grow their own fungi and many were very successful.

Personal gain
Because fungi do not feature on the Science Curriculum, most teachers know very little about them and what they do for us and our planet. One of the most rewarding things for me as an academic was to be able to enlighten the teachers with a little fungal knowledge. Teachers enjoy the challenge of learning too.

Specific safety considerations for these activities
1 Mushrooms are either supermarket bought or edible species supplied by a mushroom grower so there is no risk of poisoning.
2 Mushrooms can produce spores, and if fruit bodies are kept in a room without ventilation for long periods of time, there could be a potential risk to individuals who are sensitive to fungal spores. If mushrooms are stored overnight in a refrigerator and are disposed of in black bag waste or onto a compost heap at the end of the activity day, this risk is reduced.
For other general health and safety considerations see Section 6.10.

Top tips
1 Plan everything in advance.
2 Make sure that the teachers are confident with the science activities by providing them with sufficient background information beforehand. This will help them to re-run the activity without you.
3 Make sure teachers know if the activity is going to generate mess.
4 Ensure that the children can take home something, e.g. art work or a certificate.
5 Think about extension activities to allow the teachers to expand their teaching in this area.
6 Think about using materials or equipment available from different organisations.
7 You can use the same activities in a school or public environment.
8 Get your developed resources onto a website for wider dissemination.
9 Make the activities fun as well as learning experiences

10.2.2 Extra-curricular activity within the school environment

The importance of extra-curricular activity has been well documented (OECD, 2009). It provides gifted pupils an opportunity to be stretched and struggling pupils an opportunity to achieve (Mahoney *et al.*, 2003; Turner, 2010). School science clubs, although time intensive, have the potential for lasting impact, although research by Aschbacher *et al.* (2010) has shown that science club experiences may not be enough to make up for poor school science teaching. I ran my first science club with the aid of a Royal Society Partnership grant (see Section 6.8). These first pupils are now of university age and although I didn't follow the pupils in any formal longitudinal study, I know from having taught their brothers and sisters in subsequent science clubs, that at least two are studying science at university. Science clubs have since formed a highly rewarding part of my engagement portfolio and they are an ideal training ground for both undergraduate and postgraduate students who are considering a teaching career (Yeoman *et al.*, 2011). Case study 10.3 highlights the science clubs I have developed from a Peoples Award from the Wellcome Trust.

Case Study 10.3

Design and Delivery of Extra-curricular Science Clubs
Kay Yeoman

Background
I developed this project because of my belief that it is important to engage children by sharing enthusiasm for science. In 2007 I received a Wellcome Trust Peoples Award to fund a 3-year project aimed to bring biomedically related science to different audiences in Norfolk, England:

the 'Mobile Family Science Laboratory'. As part of the project I developed extra-curricular science clubs to be held in primary schools after the school day for year 6 children (10–11 year olds) at Key Stage 2.The education literature has shown that extra-curricular activities are known to have a positive effect, both on high and low achieving pupils.

The project

The science clubs take place in the school after school has finished. They last five weeks, with 2 hours contact time per week. Children attending the club are selected for their enthusiasm, not their academic ability. Club themes have included 'Cells to tissues to organs', 'My visible and invisible body' and 'Totally amazing me!' The science clubs have explored the workings of the human body, in a way that caters for different learning styles and different academic abilities. We have cultured microorganisms on our hands and tested the hypothesis about whether washing removes them. Pupils have learnt about the different cells of the human body – where they are and what they do. They have learnt about the sub-structure of cells and how to extract DNA. Club members have measured brain waves and heart activity, and learnt about the effect of exercise on the body. I was amazed to see how quickly the children learnt, not only to use quite sophisticated equipment, but to think about the results they generated. With some of the clubs I have enrolled the pupils onto the Bronze Crest Award Scheme run through the British Science Association (see useful websites case study table 10.3.2).

Equipment and resources

I was lucky to have a budget from my Wellcome Trust Peoples Award, enabling me to buy equipment. I bought a BioPac (see useful websites) which enables measuring of physiological functions, such as brain and heart activity. I also bought other equipment such as cell models, a camera which could be attached to microscopes and also compendium sets of pre-prepared microscope slides, covering human physiology, parasites and microbiology. These were expensive items, but could be used for many other school and public events. In terms of consumable money, the clubs can be run quite cheaply (£20 per pupil per club). I like to buy the children a laboratory coat each as this makes them feel like scientists and they get to take the coat home with them when the club ends. One of the important design elements to my clubs is that the children get to take something home with them after every session (e.g. organ T-shirt, or a DNA necklace or model). This allows them to show off their work and acts as a discussion point within the family.

Project highlights

One of the best things I did was to involve some of my third year undergraduate students who had enrolled on a science communication module. As part of their module they helped run the clubs. Each student took ownership of one of the sessions and came up with the ideas to engage the children. The students were brilliant at coming up with new ideas for activities within the club, trying them out and then delivering them to the pupils. Seeing young men and women (the students) being enthusiastic about science was an unforeseen benefit for the children and these positive role models were fantastic at breaking down the stereotypical image of a scientist.

The design of the activities was crucial for the success of the club. People learn differently, and it's important to think about how the material can be delivered in as many different ways as possible to cater for these learning styles, through visual and written material, auditory information and hands-on learning (kinaesthetic). For example, in a club session on the human body, the children were briefly introduced to the roles of the major organs through verbal communication, we found out what they already knew through discussion and the use of an organ tunic (see useful websites). They then matched the function of organs to pictures using a worksheet. They were then able to create their own organ T-shirts by placing organ templates on the T-shirt, drawing round them, and then colouring in. The heart was then looked at in more detail, by measuring heart activity on a BioPac.

What didn't go quite so well as expected?

The clubs I have run have been very intense experiences and you need to know how confident you feel about controlling a group of excited 10–11 year olds (or younger). Some of the behavioural issues were a surprise, although these tended to arise from excitement. You have to be prepared for the unexpected; for example, at one session the taps were not working in the sink used for the disposal of chemicals, and on another occasion the data projector had been removed as it had broken down during the week.

Project evaluation

The project was funded by the Wellcome Trust and an important part of grant conditions was the evaluation. I used a simple questionnaire to evaluate how the children felt about the science clubs and the things they learnt during it.

The children were asked four questions:

1 What did you think of our activities?
   fantastic   ☐   good   ☐   alright boring   ☐   terrible   ☐
   (This Likert scale was converted to numerals 1–5 respectively for the data analysis.)
2 Do you think science is fun?
   yes   ☐   no   ☐   don't know   ☐
3 Has today made you think about science differently?
   yes   ☐   no   ☐   don't know   ☐
4 What are the three most amazing things you learnt?
   free text response
   We also conducted a parent questionnaire consisting of the following questions:
1 Did your child have fun in the club?
   yes   ☐   no   ☐   don't know   ☐
2 At home did they talk about the club?
   regularly   ☐   occasionally   ☐   no   ☐   don't know   ☐
3 Do you think the club helped with your child's understanding of science?
   a lot   ☐   a little   ☐   no   ☐   don't know   ☐
4 Do you think the club made a difference to your child's enjoyment of science?
   a lot   ☐   a little   ☐   no   ☐   don't know   ☐
5 Do you think your child will go on to study a science subject at 'A' level?
   yes   ☐   no   ☐   maybe   ☐   don't know   ☐

The questionnaires were useful to find out what the children had learnt and we were pleased and surprised by the level of knowledge they had remembered (see Case study table 10.3.1 and case study Figure 10.3.1).

Case study table 10.3.1 The most amazing things you learnt.

What are the three most amazing things you learnt?

That you get diseases if you don't have a balanced diet
If one DNA base changed the whole body could go wrong
We learnt where the organs were and what they looked like
How bacteria grow and expand
How UV can be harmful to DNA
The sun cream blocks UV light
That you can take DNA out of the body
That you should only use antibiotics when you need to
That DNA is stored in the nucleus
How easily bacteria can spread
That DNA can be stored on a chip
There are thousands of cells multiplying all the time
That bacteria have different shapes
I found that there were lots of different cells everywhere in the body
There is a cell which can change into any cell
So many different blood cells in your body

Case study figure 10.3.1 Wordle of the free text in Case study table 10.3.1.

Personal gain

There is an incredible sense of achievement in getting children excited about doing science. I feel that science clubs expand their knowledge, but more importantly their enthusiasm. All the children receive certificates and medals to celebrate their success. These are usually presented at an assembly in front of parents, their peer group and other teachers. This is always a proud moment for all involved in the club. The goal of the clubs is not to get children to study science, although it can certainly help, but to give them confidence and the feeling that you don't have to be special to do and enjoy science.

Safety considerations specific to these activities

1 The parents are informed by letter exactly what the children are going to do and permission is sought from parents for their child to attend the club. This is an essential part of any outreach event to a school.
2 If you intend to use your own vehicle to transport equipment and consumables from your place of work to the school, check you have the correct insurance.

Case study table 10.3.2 Useful websites.

Website	URL address
The BioPac supplied by Linton Instruments	http://www.lintoninst.co.uk/mp100_main.htm
Organ tunic	http://www.rapidonline.com/1/1/470-organ-tunic.html
Children's Laboratory Coats	http://www.biz-e-kidz.co.uk/acatalog/Kid_s_Lab_Coat.html
British Science Association Crest Awards	http://www.britishscienceassociation.org/web/ccaf/CREST/

Top tips for running primary school science clubs

1 It can often be difficult to contact schools to find out if they would like to have a science club. Most universities will have an outreach office, which should be able to help you get started. I initiated my first contact through my son's school and then built up other contacts through running public events, which primary school teachers attended.
2 Be very clear about your expectations of the role of the school in delivering the club and also what the school expects from you. If you require teacher assistance in running the club, then make sure that this has been agreed with the school in advance. This is particularly important, when running a club outside of normal school hours.
3 Be very clear about what the club will deliver in terms of wet laboratory activities. There are certain experiments, for example, taste with salt, sugar and lemon, which will require parental permission for younger pupils. One of the activities I do on a regular basis is extraction of DNA from cheek cells. I ensure that the parents know that the DNA remains anonymous and that any DNA not taken home is destroyed at the school at the end of the club session.

4 You are not allowed to take pictures of children engaging in your science club, unless permission has been obtained, the school will be able to help you with this but it is important that you receive permission for this in advance.

5 You need to know what to do in case of an emergency and who to contact within the school.

6 Discuss with the school how the children will be selected for the club, I usually like a first-come-first-serve, as I want enthusiastic children, rather than those selected on ability; however, be guided by the schools judgement.

7 Make sure that the school sends you a list of the children attending the club, so you can take a register. This is also useful if you then intend to give certificates. Be aware that some of the children may have special learning needs, but this might not be disclosed. If this worries you, always ask the teacher involved.

8 Be well prepared for the activities taking place and know how long they will last. A detailed lesson plan of each session is crucial in the smooth running of the club. Always make sure you have a back-up activity up your sleeve for those children who finish quickly.

9 Get help to run your sessions; undergraduate and postgraduate students are perfect and generally full of energy.

10 When running a club for younger children (ages 5–7) be aware of their capabilities; for example, some will have problems using scissors.

11 It is always a good idea to go and see the place where you will be teaching ahead of time. This will give you an idea of how to lay the equipment out, how much room you have and also if there are any stairs to get equipment up and down.

10.2.3   Curriculum enhancement in secondary school

At secondary school, science should be taught by specialist science teachers in physics, chemistry and biology. It has been recorded however that only 14% of science teachers have a physics degree, 22% have a chemistry degree and 44% have a biology degree[2]. This means that subjects like physics, where there is a shortage of teachers are being taught by teachers without a physics degree. The lack of specialist teachers can lead to lower levels of achievement by pupils (Dolan, 2008). To reverse this, the profession needs to attract more top quality graduates from across the sciences and to provide incentives for STEM subject teachers. Thus programmes which allow both undergraduate and postgraduate students access to working with schools are valuable (Yeoman *et al.*, 2011). Case study 10.7, by Sarah Holmes, an undergraduate biological sciences student, shows how the opportunity given to her to design and deliver a gifted and talented event, through a final year science communication module (Yeoman *et al.*, 2011) confirmed her desire to pursue a teaching career.

The benefits of doing practical science in schools has been widely reported, but there are several barriers to increasing practical work.

• Many schools have improved their laboratory space, but the cost of equipment and consumables to run practical sessions can still limit the amount of practical work done in the classroom. Schools must balance their budgets and the cost of laboratory work has to be weighed against other demands.

---

[2] School Workforce Census http://www.education.gov.uk/schools/adminandfinance/school admin/ims/datacollections/swf

- Insufficient technical support for practicals can also be a problem. Technical staff within secondary schools need to be more valued and also given opportunity for further Continuing Professional Development (CPD) training (House of Commons Science and Technology Committee, 2011).
- There are time constraints in the school timetables, many techniques in molecular biology for example, cannot easily be carried out in a 50-minute lesson.
- Many science teachers still lack confidence when it comes to laboratory work, as few would have had extensive training, unless qualified to a PhD level. Science can move rapidly, so science teachers need to keep up-to-date with new advances; however, this is time consuming. Teachers and technicians need access to good CPD courses. Such courses do exist through CLEAPSS and the Science Learning Centres, but the evidence suggests that teachers don't seem to attend, or are not allowed to attend such courses (House of Commons Science and Technology Committee, 2011).

Good examples of curriculum enhancement at secondary school are Case study 10.4 by Drs Laura Bowater and Colwyn Thomas on 'Genes and Health', and the BioPunks science club (Box 10.1).

Box 10.1   BioPunks – a synthetic biology club

Kay Yeoman

In the academic year 2011–12, I started to run a Year 9 science club (pupils aged 13–14) on synthetic biology at local secondary school. The club is funded by Cue East, UEA's Beacon for Public Engagement. I decided to call the club 'BioPunks' in reference to the movement in the US where citizens were conducting their own synthetic biology research. The pupils were selected to take part on the basis of their science attainment. They are working towards their British Science Association Silver Crest Award, which requires a minimum of 40 hours laboratory work. The aim of the club to create a 'biosensor' for caffeine is very ambitious. To achieve this goal, the pupils have undertaken a series of laboratory work which has taught them the skills associated with molecular biology, including transformation, DNA isolation, PCR and cloning. The pupils are also developing additional skills such as scientific writing and presentation.

Case Study 10.4

Genes and Health

Laura Bowater and Colwyn Thomas

Background

This event was a workshop, designed to deliver a combination of a hands-on science activity and a led discussion on an issue relevant to society. The workshop was held in schools which had low aspiration towards higher education amongst their pupils. We decided to engage with pupils at Key Stage 4, designing an event that focused on the impact of the huge expansion of our knowledge about the human genome and its consequences for our health and society. We made this decision because all pupils have to study science at this stage of the school curriculum, but the activity would be experienced by pupils before they actively choose the subjects that they want to study in the sixth form.

Event design

The event was designed to fit within a double science session for pupils (which varied between an 80 and 100 minute time slot). It took place in the school environment over a period of six weeks. Six different schools were targeted from Norfolk and each session was delivered to a class of up to 30 pupils. The event was designed to include several discrete activity blocks that included:

- an initial session that described the structure of DNA and the process of genetic inheritance;
- a core laboratory experiment undertaken by all pupils who isolated DNA from their own cheek cells. This activity highlighted the ease of isolating DNA that can subsequently be used for genetic profiling;
- a complementary class discussion; pupils were introduced to the advances in genetics and biotechnology and asked to discuss their views on the possible beneficial or detrimental impacts that this might bring to society;
- a complementary, group-based activity undertaken by all pupils facilitated by us but led by pupils. A hypothetical case study was provided that illustrated some of the ethical problems associated with screening for genetic conditions. Pupils were encouraged to discuss some of the ethical issues and to consider potential solutions.

Points we considered

- Contacting schools, with details of the event that we were offering, and inviting the schools to take part was time consuming. Taking advantage of a known point of contact within each school proved hugely useful at this stage. We were also able to use contacts that had already been developed by outreach teams within our University.
- Taking the event into schools had the advantage of not having to remove pupils from other lessons that they were scheduled to attend on the day. Schools did not have to find money to pay for transporting pupils to a venue or additional costs of providing cover at the school for any accompanying teacher. However, there were our travel costs to consider and the costs of materials had to be covered.
- We secured funding from the university as part of their Outreach Opportunity Fund to help us cover the costs of the project.
- We took care to ensure that the project dovetailed with the school curriculum and it also prepared pupils for parts of the curriculum that they were due to cover in the future.
- The activities that we chose to deliver had to be portable and undertaken by all pupils in a school laboratory.
- We realised that there may be some pupils who were interested in science but may not enjoy laboratory work. In order to engage these pupils, we felt that it was important to provide an activity that actively encouraged them to consider science within society. The group discussions that centred on the ethical scenarios that had science at the heart of the activity were designed to allow this to take place.
- When a school agreed to take part, they were provided with the lesson plan in advance. This also included a written risk assessment form that contained instructions on the safe handling of all equipment and potentially hazardous material. We supplied safety equipment such as gloves and goggles and we requested that a teacher remain in the classroom throughout the session to take responsibility for class management.

What worked well?

1 Dividing the event into discrete activities was very effective and ensured that pupils were able to remain focused throughout the session. It was really interesting to see that different

pupils enjoyed different aspects of the event with some pupils enjoying the ethical group discussions whereas other pupils enjoyed the experimental sections. 'DNA is so interesting as was learning about the ethical implications of science.'

2 The ethical discussion worked because pupils were reassured that in ethics (unlike their perception of science) there is no right or wrong answer and everyone's opinion is valid and should be considered. This clearly encouraged almost all students to take an active role in the group work.

3 Teachers also commented how pupils who were normally reticent with the science content of the course took an effective role in the ethical aspects and the group work. Several teachers also commented that they would use parts of the event in future lessons.

What we would change in the future

1 Setting up this event took a lot of investment in time and resources and we only delivered it to six different schools. In retrospect, we could have spent more time contacting additional schools and delivered the event to more students and teachers, which would have meant that we had a better 'return for our initial investment'.

2 The event was run during term time, which was key for running an event in a school, but it also meant that it clashed with term time for our undergraduate students. This meant that we could not use our undergraduate students to help deliver these sessions. I feel that this was a lost opportunity for school pupils to interact with undergraduates and to share experiences, hopes and concerns. Pupils would also have benefitted from an opportunity to talk to undergraduates about why they chose to become involved in science after leaving school. Finally, we also failed to provide an opportunity for undergraduate students to become involved in a science communication event.

The full resources for this event are provided in the book website.

Citizen science (see Section 8.7) can also be done within a school environment. Case study 10.5 by Dr Adam Hart on the Bee Guardian Foundation is an excellent example of how pupils can be engaged early with the concept of citizen science. The students who took part in the project designed and collected data from an experiment and further acted as 'advisors' to the Bee Guardian Foundation.

Case Study 10.5

Investigating the Nesting Preferences of the Red Mason Bee: a School Citizen Science Project

Adam Hart

Background

The Bee Guardian Foundation (BGF) is an educational conservation organisation, which raises awareness of the need to conserve bees with the creation of 'bee guardians'. I got involved as a scientist with the BGF because of my research involvement with bees. However, it became very clear that the activities, workshops and general attitude towards public engagement that the BGF already had would fit in very well with my outreach activities. The BGF were already extremely good at engaging youngsters and families with bee conservation, but they were keen to develop a more scientific side to their activities.

There are more than 250 different types of bee in the UK and most of these bees do not live in big colonies like honeybees and bumblebees. The so-called solitary bees do things on their own. The females of many species seek out cavities that they subdivide into 'cells' and fill with nectar and pollen to nurture the larvae that develop from eggs that they lay. Despite the solitary bees being incredibly important pollinators, and garden centres being filled with 'mason bee nests' and 'solitary bee houses', we know very little about what sort of nest type they prefer (e.g. bee houses can have nest tubes made from cardboard, bamboo or drilled wood) and whether the bees thrive in artificial nest boxes. The BGF and I decided to fill this gap in our knowledge by developing a school-based scientific research project to investigate the nesting preferences of a particular species of solitary bee, the red mason bee. If sufficient data were collected then the aim was to write up a scientific paper, with the pupils as named co-authors.

The project

The Stroud High School and the BGF put together an application for the Royal Society's Partnership Scheme to fund the design, development and operation of experimental bee houses, laid out in the school's grounds in an experiment that would be run by the pupils. I would provide expert help with bee biology, experimental design and analysis, while a mixed group of 14- to 17-year-old pupils would develop designs for the experimental bee house that would allow us to determine which nest tube type (cardboard, bamboo or drilled wood) was preferred by bees, and what effect this has on the number of bees that are reared. They would also act as consultant advisors to the BGF, advising on bee nest construction for future BGF projects. This gave the project a 'real world' feel and made the pupils act as professional consultants as well as scientists.

The pupils split into groups and came up with designs for the experimental bee house, based on some fundamental design constraints and scientific criteria that we developed together. All of the six groups produced workable designs, but what was exceptional was that all six had some element of their design that was novel and useful. After their presentations, the judging panel found it very hard to decide which design to go for, and in the end, at the suggestion of the pupils, the final design was an elegant mixture of elements of all designs.

The construction phase was remarkable – pupils and staff alike gave up time during half term to make the bee houses and the final product was beautifully constructed and perfect for the scientific needs of the project.

Once the bee houses were put out into the grounds, the pupils devised a monitoring strategy to visit all boxes regularly and inspect for activity. The idea was to monitor which of the nest tubes (paper, bamboo, drilled holes) were being used by bees and to analyse the data statistically to determine whether bees showed a preference for a particular type. Extension work was to involve monitoring the success of bee nests in different nest tube materials. Unfortunately, the abundance of bees on the school grounds was low and this meant that we did not get sufficient bees using the nest boxes to get any statistically meaningful data. However, this did mean that the pupils had to think on their feet, and already plans are in place to plant more bee-friendly trees and plants to attract bees to the school grounds. Also, this 'failure' allowed for considerable reflection on the problems of 'real world' science. Hopefully the changes made will mean that useful data can be gathered in future years (the project is ongoing).

Project highlights

The Royal Society selected this project to represent the Partnership Scheme at the 350th Anniversary Summer Science Festival in London. As well as being a considerable honour for the school and the BGF, it gave the pupils a chance to demonstrate and talk about their project and bee biology to several thousand people. The scientific development that students gain through these projects is only part of the picture, and the chance to be taken seriously at such a prestigious science event gave them a fantastic confidence boost.

Safety considerations

1 Use of tools in the construction stage of the project, as the specialist tools, e.g. laser cutters and computer controlled routers, required specialist support
2 The school grounds are quite large and the experimental bee houses were quite widely dispersed. Any concerns I had about the pupils' safety in checking these were quickly allayed by the experienced school staff, who made sure pupils were in groups and that they let people know where they were.
3 A CRB check is useful for school work even if it isn't technically necessary. I got mine, for free, through my involvement with the Science and Engineering Ambassadors Scheme (see useful websites for URL).

Personal outcomes

Outreach work is often seen as a one-way street – all give and no take. This certainly wasn't the case here and I got a great deal from my involvement with this project. As well as providing data for my ongoing bee studies, the project helped to build publicity for the Bee Guardian Foundation, both locally and nationally through the Royal Society's 350th Anniversary exhibition.

Top tips

1 Foster a good relationship with the teachers you will be working with to make sure from the start that they are as engaged with the project as you are.
2 Having an involvement with another external group, in this case the BGF, is a huge advantage. Not only does it spread out responsibility, it also provides additional expertise and shows the pupils how collaborative real world projects and ventures tend to be.
3 Think about project location, being based in a school and able to use all the facilities within in a school (we were based in Biology but needed Design and Technology) was a huge advantage to the success of the project.
4 Try and make the project appeal to a wide range of pupils, being able to run a project that involved design and technology as well as science meant that pupils were engaged who were not normally 'sciencey'.
5 A CRB check is needed for school work.
6 Think about specific safety issues; for example, in the data collection aspect of the project the pupils needed to check the bee boxes. Sharing my experiences running adult field courses with them has usually been a good ice-breaker if I have had any safety concerns.

Useful websites

Stemnet
www.stemnet.org.uk/
Bee Guardian Foundation
www.beeguardianfoundation.org/
Royal Society Partnership Grants
http://royalsociety.org/education/partnership/

10.3 Developing cross-curricular activities for primary and secondary schools

Cross-curricular education breaks down the barrier of individual subjects as the same material is taught from a variety of different subject perspectives. For example, schools who chose to cover the 200th anniversary of the birth of Charles Darwin and the 150th anniversary of the publication of his book *On the Origin of Species*, could study the material from a:

- **historical perspective** – learning about Victorian Britain, its science and scientists;
- **geographical perspective** – mapping the voyage of *The Beagle* and investigating the geology and the climate conditions of the places he visited;
- **scientific perspective** – exploring the theory of evolution by natural selection by looking at the different plants and animals that he found.

There are distinct advantages to a cross-curricular approach as identified by Johnston (2011):

1 makes learning relevant and motivating;
2 enables children to see how different subjects relate to each other;
3 enables wider coverage of the curriculum.

Cross-curricular learning is more akin to real life experiences and promotes creativity (NESTA, 2005). The philosophy of cross-curricular learning does not seek to dilute each individual subject, but to create a system of learning which is greater than the sum of the parts. It also promotes better communication between teachers in different departments, and a greater understanding of each other's subject areas. It allows the opportunity to develop dynamic resources and schemes of work which emphasise the acquisition of generic skills, such as problem solving, interpretation, evaluation, team work and communication (Brodie and Thompson, 2009). Box 10.2 provides an example of cross-curricular projects between maths and history. Other successful cross-curricular projects include partnering science with the arts. This is beautifully illustrated in Case study 10.6 by Professor Anne Osbourn and Dr Jenni Rant, which demonstrates the powerful links which can be made between art and science.

Box 10.2   Millenium Maths Project

The Millenium Maths Project (**http://mmp.maths.org/**) is based at Cambridge University and is for children aged between 5 and 19 years, as well as the general public. The aim is to help with maths education, but also to help everyone 'share in the excitement and understand the importance of mathematics'. Examples of cross-curricular projects include 'Babylonian Maths' and 'Maths and Our Health'. 'Babylonian Maths' for pupils who are at the Key Stage 2/3 transition, explores what maths children learnt 4000 years ago. They also learn about mathematical history and archaeology through the objects people left behind.

'Maths and Our Health' explores the practical aspects of mathematics. It is designed for Key Stages 3 and 4 and examines probability and statistics associated with medical research. For example, the link between eating bacon sandwiches and the risk of developing bowel cancer is examined.

The SAW Trust
Jenni Rant and Anne Osbourn

Background

Science, Art and Writing (SAW) is a creative science education initiative that dissolves the barriers between science and the arts. Learning and creativity are stimulated by visually striking scientific images and through practical science experimentation. Children do not normally think in 'boxes', i.e. divide their thinking between science and the arts. However, it is often easier for the curriculum to be delivered in a subject-specific way. Best practice developed during the past six years of SAW projects has enabled the delivery of contemporary science activities to children of all ages by professional scientists, artists and writers. These projects aim to break down the barriers which seem to exist between art and science and provide a cross-curricular bridge to allow powerful links to be made.

SAW projects are flexible in that they are suitable for all ages and abilities and can run for as little as one day or for a whole week depending on the requirements and budget. Projects centre on a scientific theme supplemented by a set of intriguing scientific images used to weave together practical science experimentation with creative writing and art. Schools can use this philosophy to run 'in-house' projects or can enrich projects by inviting professional scientists, artists and writers to work with them. Projects have run in over 70 schools in Norfolk and in several other regions of the UK on a diverse range of themes, some addressing parts of the national curriculum, others on topics chosen by the school (for example, relevant to the schools geographical environment) and other projects on cutting-edge research topics. For this case study, one project will be used to give an example of the nuts and bolts of how SAW works as a guide to the reader. This project was one of three themed on plant natural products, designed and delivered as the formal outreach mechanism for 'Smartcell', a major European science consortium (detailed lesson plans for the science components of all three projects can be found on the SAW Trust website).

The colour pigment project

The colour pigment SAW project was run in a primary school with a group of 60 children aged 5–6 on the theme of plant natural products. Initially the scientist met with the teacher to select and shortlist scientific images that would be used to encourage curiosity and creativity. The final set of images is provided in the colour plates of this book. Some are from an online database, the Science Photo Library (http://sciencephoto.com) that schools are able to use at a concessionary rate and others were produced in-house. It is important to note that images sourced externally may be subject to copyright protection and therefore permission may need to be obtained prior to their use.

A local artist and writer were invited to join the project and were given a brief introduction to the theme and copies of the images to work with. As this was a 1-day project, the day was divided into sections, beginning with science until morning break, writing until lunchtime and then art for the afternoon session. The scientist, artist and writer then worked independently on the design of their respective sessions, practising and timing practical activities whilst considering appropriate dialogue to use when communicating with the target age range. After a few weeks planning time, the SAW team met and went into school to discuss the session content with the teacher, look at the available space and resources and discuss any possible health and safety issues. This is an extremely valuable part of the planning process as it ensures the project runs as smoothly as possible and it gives the teacher the opportunity to include some activities in class leading up to the day of the project to maximise the learning potential. Schools are also required to sign an agreement prior to the start of a SAW project covering all risk assessment and copyright issues.

The timing and the way images are introduced vary depending on the age of the audience. It is not necessary to reveal the identity of all the images as this leaves room for creative exploration. As the children in the class were very young, some of the images were introduced before the practical science activity.

The science

The project took place in the school hall with 60 children arranged around ten tables. Each table of six children had one adult to help (the teacher, scientist, artist, writer and two classroom assistants) and the children at each table worked in two groups of three. Children set to work extracting colour pigments from a plant sample that was placed on their table, so as a class, we had ten samples. A simple protocol was placed on each table for the adult to follow and to guide the children. The school provided aprons and goggles for the children to wear. Pupils extracted pigments by grinding plant samples in 75% (v/v) ethanol. Liquid samples were then transferred to vials with a plastic pipette. Another table was set up in the middle of the hall as a 'chromatography station' that the children could take their extracts to for loading by an adult. I had a colleague to help do this, but older children may be able to load their own samples. The children then explored the effect of pH on colour by adding acid (lemon) or alkali (bicarbonate) solutions to their samples, which they really enjoyed. Some of the samples that changed colour were also loaded onto the chromatography paper. Samples were swapped between tables so that everyone got to explore a variety of pigments. The chromatography was left to run while the children helped to tidy up and during a short break, after which we all looked at the results together. They were surprised to see that some colours were made of more than one ingredient!

The poetry

A set of images had been placed on each table and during the science activity the children were keen to talk about them, so when it came to the writing session they were encouraged to participate in a group discussion on what the images looked like. The children became very excited when they were asked to give their opinions and it is amazing to hear how the images are perceived through their eyes. The writer asked them to pick one image to write about. A lack of inspiration for creative writing is not a problem at this age but many of the children had a very limited vocabulary, couldn't spell and couldn't read all that well either! This wasn't a problem as there were no great expectations at this level to produce more than a sentence or two with the help of the adults. What was important was the dialogue between children, and with the circulating adults. Whilst some of the conversations allowed more scientific questioning, all of them led to creativity and built confidence. Some of the children read out their poems to the class.

The art

The artist used the concept of colour pigments travelling at different speeds through paper using pots of coloured powder called 'Brusho®' to create flower pictures to display in the school hall.

The children used wax crayons to draw patterned flowers onto filter paper, and then wet the paper and dropped small amounts of different coloured Brusho® onto it. The colours spread out at different rates and were resisted by the wax. They decorated paper straws to use as stalks and when the flowers were dry the shapes were cut out and assembled onto the wall for the rest of the school to see.

Plenary session

At the end of the day we asked the children if they had enjoyed the day and how many of them wanted to be scientists, artists or writers when they grew up. They all wanted to be all of those things! With older children we sometimes ask them to fill in a short questionnaire to get a snapshot of how inclusive the project is and the value of the learning process. We covered a large number of topics during the day and the children were all able to be very involved in every aspect. The thought of going into a class of five and six year olds to teach them that plants have the ability to make different coloured compounds would be a scary one if it were to be approached in a less creative way!

SAW has received very favourable national and international coverage and has been acclaimed by those involved in projects: scientists, artists, writers, teachers, and – in particular – the children (see www.sawtrust.org for example projects and evaluations).

Top tips for a successful SAW project

- The SAW project begins with a teacher and a scientist working together to select a theme and high-quality scientific images on which to build.
- The design of the sessions and the level of detail are then developed to suit the age range.
- Practical activities should be practised and timed beforehand and, most importantly, at least one meeting with the teacher and the SAW team is needed to discuss the project. Look at the space and resources available and to discuss health and safety issues.
- More information and assistance on setting up a SAW project can be accessed via the website.
  www.sawtrust.org

Acknowledgements

Funding for the project was supplied by the European Commission FP7 Smartcell consortium, work package9.

Poetry session designed and delivered by Mike O'Driscoll.

Art session designed and delivered by Chris Hann.

Alien World

Sticky trees protecting the pumpkin.

Alien world with a spaceship landed.

Freddy (aged 6)

A Green Alien

A big alien walking on skin,

It sucks blood,

It has spikes on it.

Its got red eyes,

Its skin is green,

It has a yellow poisoner.

Its got six legs

And its antennae are green and mighty.

Emmanuel (aged 6)

Hedgehog

Little nectar flecter on the flowers,

Hedgehogs are spiky,

And frighty,

Hairy microphone looking at the flower!
Regan (aged 6)

My Flower
My flower is good medicine,
The middle reminds me of nettle,
Violet and yellow, those are the colours
Of my flower.
Caty (aged 6)

Cross-curricular education can be used with a wide variety of ages. The example provided in Box 10.3 is with a group of 'AS'-level students, who had to combine their scientific knowledge with business skills. In developing a biotechnology company they had to think of both economic and ethical considerations and hone their presentation skills.

Box 10.3 Natural Products Biotechnology Business Competition

Kay Yeoman (UEA) and Rachel Jarrold (Wymondham High School)

A biotechnology business competition for pupils who had completed their first year of A-levels (AS level) was designed with funding from Cue East (UEA Beacon of Public Engagement). The format was a two-day workshop at the School of Biological Sciences, which consisted of laboratory practical sessions (using educational kits supplied by BioRad), lectures on natural products, business, marketing and ethics.

The workshop was based around the scenario of a plant found in a remote South American village in the rain forest, traditionally used as a 'cure-all' treatment for many different diseases. Using World Health Organisation data, the pupils had to decide which disease would be the most profitable for them to develop a product for. They then had to form a biotechnology company – including company name, logo and name of product and consider how they would develop the company, taking into account the different costs associated with research and development. They needed to research any drug competitors that were already in the market place for their target disease. They also needed to consider the ethics of their actions. The students had to 'pitch' their company and product to a panel of scientists acting as 'business angels' at the end of the two-day workshop. The panel then picked the winning company who presented the best investment opportunity.

For more detailed information on the Biotechnology Business Competition visit the book website.

10.4 Enhancing the curriculum with activities with a school audience but outside the school environment

The BIS 2011 attitudinal study (on 14–16 year olds) showed that learners hear about science in a variety of places, with TV news being the most common (35%) followed by books (25%) and other TV programmes (23%). The most

Table 10.1 Background of learners who visit different venues and events (SEG is Socioeconomic Group).[a]

Background	Differences
Gender	Females are significantly more likely than males to have been to an art gallery, a science or discovery centre, another type of museum, live concert or zoo.
Age	Younger learners are significantly more likely than older learners to have visited an event during NSEW. Older learners are more likely to have been to a live concert or lecture/talk on a science related subject outside of school.
Ethnicity	White learners attend more events and visited more places than non-white learners. In particular they are significantly more likely to have been to an art gallery or other type of museum or live concert.
SEG	Leaners from higher income backgrounds have attended more events, and visited more places than learners from lower income backgrounds. They are significantly more likely to have been to a science and discovery centre, art gallery or live concert.
Favourite Subject	Learners whose favourite subject is science are more likely to have attended a science related event or been to science related venues.
Attendance of clubs	Leaners who have attended science or engineering clubs have on average visited more places, and attended more events than those who have never been to a science and engineering club. In particular they are more likely to have attended science related events and venues.

[a]Reproduced with permission from the BIS 2001 Attitudes to Science: Survey of 14- to 16-year-olds.

popular events or places to go are theme parks (57%) and live concerts (53%). The most popular science-related activity/event outside of the classroom was one associated with the school, community centre or university. In Table 10.1 the background of different learners is linked to where they are more likely to go in terms of visits. If you are trying to reach a particular audience, then this may help you decide on where an event should be held, or not held. Learners seem to go on educational trips, for example, to museums and galleries with their families. This indicates the importance of parental/guardian influence in getting learners more engaged with science. They are more likely to go to specific science events with their school, for example a lecture or talk, or a science-related activity at a university. Parental involvement has been shown to be of importance to pupils' learning and eventual attainment, not only by providing a stable environment, but also providing a stimulating atmosphere where there is discussion, good role models and high aspiration. The level of parental involvement is influenced by social class and, interestingly, the level of education achieved by mothers (Desforges and Abouchaar, 2003).

Research has shown that extra-curricular activities are associated with many positive outcomes, including increased academic achievement, greater life aspirations and higher levels of self-esteem. It also seems to stop pupils from dropping out of school at an early age (Mahoney et al., 2003; Turner, 2010).

Your school event may involve different organisations, your own, the school and perhaps other providers. Each organisation is a stakeholder in your project and potentially will require different outcomes from the event. Case study 10.7 by Sarah Holmes shows how an activity can be organised with several different stakeholders.

Case Study 10.7

Living or Lifeless: a Workshop on the Ethics of Natural History Collections for Gifted and Talented Year 10 Pupils

Sarah Holmes

Background

Extra-curricular activities are an important part of communicating with young people, especially those in the Gifted and Talented (G&T) programme. Research has shown that participating in extra-curricular activities has a positive effect on young people (Holland & Andre, 1987). Pupils deemed to be G&T are considered to have the potential to develop to a higher standard than the rest of their year group (Directgov, 2010). This G&T programme was responsible for providing support for 4- to 19-year-old G&T pupils; their learner academy provided opportunities for pupils to meet and participate in activities in order to challenge and stretch them (Directgov, 2010). However in 2010, the responsibility for G&T young people was placed with Capita, the same company responsible for the National Strategies (Department for Education, 2011). Since June 2011, the National Strategies has also been scrapped. There is now no external funding available for excellence hubs, and schools are expected to provide full support for their G&T pupils, but without any 'ring-fenced funding' (Henry, 2010).

The project

The project was a 2-day event workshop called 'Living and Lifeless' aimed at G&T year 10 pupils from a range of secondary schools across East Anglia, and was carried out at the Norwich Castle Museum. The aim of the event was for pupils to create their own documentaries about the ethics of natural history collections. The project was arranged through Excellence East, based at the University of East Anglia (UEA) and we also worked with BBC Voices, based at The Forum in Norwich. BBC Voices is a media and production unit, who offer film making and radio editing workshops to individuals and groups free of charge (The Forum, 2011). They offer these services to the community, helping the public to make their own films and audio features (The Forum Trust, 2010). They also played a large part in our project, providing camera training and equipment alongside editing capabilities and help. Our event was therefore carried out partly at The Forum in Norwich, where they are based, and partly at the Norwich Castle Museum.

Event design

The Norwich Castle Museum had recently refurbished their natural history collection in order to provide a display that better reflected the era in which most of the specimens were collected, rather than a layout of specimens based solely on taxonomy. The event was designed to explore the ethics behind the collected specimens rather than just the museum displays themselves. It was during the second day of the event that the pupils filmed material for their documentaries.

The pupils found the filming and editing parts of the event the most enjoyable, but they were enthusiastic and engaged throughout the two days. We did find it surprising, however, that most of the pupils had come to the event with the preconceived idea that natural history collections were unethical, a viewpoint which was a difficult barrier to break down.

Event highlights

The best part about the event was that all pupils enjoyed it! They were excited and engaged by the filming and seemed to enjoy learning about natural history collections, even if they did find it difficult to present a balanced view.

One of the main things that helped the event run smoothly was to liaise personally with all three of the organisations involved, Excellence East, BBC Voices and the Castle Museum. This meant that even when personal contacts changed (as they did with Excellence East), we were still up to date with where everything was in terms of planning. Keeping in regular contact with all three organisations meant that we were able to minimise any problems or misunderstandings.

Involving outside organisations such as BBC Voices and the Castle Museum was beneficial as we would not have been able to run such a high-quality event without their help and equipment. The museum staff also gave a lecture style talk and a question and answer session about taxidermy. This was really effective, as they were able to present this information in a far better way than we would have managed within the given time frame. It is definitely a good idea to get the experts involved where possible.

Improving the event

I would structure the two days slightly differently, although this can be a challenge when trying to meet the needs of several different organisations. I would address the potential misconceptions held by the pupils first, before presenting new information. I would then give time and opportunity to explore new ideas and concepts.

Although we were able to film in the museum before general opening hours, it would be beneficial to complete filming without the crowds, especially during half term. It was difficult to keep track of all groups during filming to ensure guidelines were being followed.

Working with Excellence East, BBC Voices and the Norwich Castle Museum for one event also proved a challenge. When liaising with three organisations such as this, it can be difficult to make sure all of their needs are being fulfilled and any necessary compromises organised in advance. For example, BBC Voices required filming to take place with specimens not being behind glass, and the museum staff were happy to accommodate this by opening the cabinets for the pupils to film inside them.

The main difficulty seemed to be advertisement. The event was advertised by Excellence East who contacted schools in Norfolk with the poster designed for our event. This was then extended to contacting schools in the whole of East Anglia, which eventually resulted in 13 pupils attending the event. We had initially wanted to have 20 spaces, but reduced this to 12 based on advice from BBC Voices.

Personal gain

I had a great sense of achievement seeing the pupils engage with the activities throughout the 2-day event. Gifted and Talented pupils can disengage easily from activities if they are not set at the right 'intellectual' level. It was a pleasing reflection on the structure and content of the event that all of the pupils who returned the evaluation form stated that the event was the right level for them.

I very much enjoyed working on the event. Not only was it great fun and a good experience but confirmed my decision to pursue a teaching career and provided me with skills that I have subsequently used as a teacher. The event introduced me to working with G&T pupils and vastly added to my experience of working with this group. It taught me valuable communication skills, not only when teaching and talking to children, but also to other adults when organising the event and diffusing situations.

Specific safety considerations

When filming, we had to ensure that the pupils did not film any children in the museum due to child protection and privacy issues, which proved difficult when it was busy. We overcame this by the museum allowing early entry so that we could complete the majority of the filming without the general public being in the museum.

Top tips

1 Research your audience, especially if you have not worked with them before. This way you are more likely to pitch your activity at the right level and know what to expect from them.

2 Look at what is included in the National Curriculum for school age pupils to ensure these G&T pupils knowledge and understanding is being extended rather than reinforcing knowledge they may already have from school. A good place to start is the National Curriculum website (National Curriculum Key Stages 3 &4, 2007). It may also be useful to look at the specifications from the various GSCE exam boards: AQA (2011; The Science Lab, 2011), OCR (OCR GCSE Sciences, 2011) and Edexcel (Edexcel GCSE Science, 2011).

3 When running an event that includes several different organisations, make sure you double check all details and plan everything as far in advance as possible. It can be difficult to suit the needs of everyone involved, so advanced planning is essential.

4 Get experts involved where possible.

10.5 Influencing curriculum change

It is not always obvious how scientists can influence the content of science curricula. In the UK there are a number of different examination bodies that provide both GCSE and A-level courses, for example OCR, AQA and Edexcel.

Currently, the UK government is carrying out a review of the national curriculum and the first consultation has already taken place. Learned societies such as the SGM responded to this via the Society of Biology, which sent in a joint response with the Institute of Physics and the Royal Society of Chemistry through SCORE. They are now reviewing the evidence before a second consultation which will take place in 2012. To get involved in curriculum change, the best way is to engage with your learned society and work through them. The Royal Society recently announced a new project aimed at producing a vision for the future of science and mathematics education for pupils aged 5–19.[3] In January 2012, the project asked for views through an online questionnaire. The project will then investigate five specific areas;

---

[3] http://royalsociety.org/education/policy/vision/

1 teachers and the wider workforce;
2 leadership and ethos;
3 infrastructure;
4 skills, curriculum and assessment;
5 accountability.

10.6   Embedding scientists into schools

There are a number of national and international schemes aimed at getting scientists into schools; some of these are outlined in Table 10.2.

Table 10.2 Organisations across the world aimed at partnering scientists with schools.

Organisation	Location	Description	Website
Scientists in Schools	Australia	A national programme that creates and supports long term partnerships between teachers and scientists	www.scientistsinschools. edu.au
Scientists in Schools (SiS)	Canada	SiS is a charity aimed at getting children interested in science	www.scientistsinschool.ca
Science and Health Education Partnership (SEP)	US	Promotes partnerships between scientists and educators for K-12 students	http://biochemistry.ucsf.edu/ programs/sep/about-sep. html
Science Learning Centre North East	UK (North East)	The teacher scientist network puts scientist working at universities in the region with teachers. There is also a 'scientist is residence' scheme where scientists are seconded to work with pupils on science projects	www.slcne.org.uk/tsn/
Bristol Teacher-Scientist Network	Bristol, England	Develops working partnerships between teachers and scientists	www.clifton-scientific. org/partnership.htm
National Science Foundation's GK-12 Program	US	Provides funding for graduate students in STEM subjects to acquire additional skills by partnerships in K-12 education programmes	www.gk12.org
Teacher Scientist Network (TSN)	UK (East of England)	Organisation which links the scientific community of the Norwich Research Park (NRP) with teachers of science in the East of England	www.tsn.org.uk

The partnerships developed between schools and scientists are mutually beneficial.

- Pupils gain by having access to scientists who work in a real context. This can provide positive role models and enables pupils to gain an understanding about different career paths within science. Pupils develop skills and gain access to specialised knowledge as well as equipment.
- Teachers gain knowledge and more confidence about the process of science. Partnerships may well also open up more career development opportunities.
- Scientists can become re-enthused about their subject. Seeing a new audience being fascinated by your research can really help to invigorate you. It can also potentially lead to new lines of enquiry, due to the unique view often taken by pupils (Dolan, 2008). Scientists can also gain better communication skills which can improve their own teaching to undergraduate and postgraduate students. Younger scientists can get experience about a teaching career, and some will choose to develop this further and do a teaching qualification when their higher research degree or first post-doctoral contracts have been completed.
- Successful partnerships can lead to more outreach on the part of the scientist.

Case study 10.8 by Dr Phil Smith MBE, outlines the development of the teacher scientist network which runs from the Norwich Research Park (NRP).

Case Study 10.8

The Teacher Scientist Network – Partnerships in Practice
Phil Smith

Background

In 1994, quite by chance, Professor Keith Roberts MBE, then Head of the Cell Biology Department at the John Innes Centre, was introduced to Frank Chennell, a local (Norfolk Education Authority) science adviser. The National Curriculum had begun a staged introduction beginning in 1989, meaning that for the first time all primary and secondary pupils had a common entitlement to study science. Teachers across the age groups were nervous – the changes were revolutionary in terms of teaching practice and expectation and the era of assessment (with an initial 17 attainment targets) and league tables was one step closer. Primary teachers would now *have* to teach science, as science was now a core subject. Many teachers of primary science did not possess a science degree, their last formal qualification in science being their own 'O' or 'A' level. This would inevitably lead to under-confidence in their ability to teach the sciences effectively – a situation that has changed little in the intervening 20 years (House of Commons Education Committee, 1995; Wellcome Trust, 2005). Concern was not just limited to primary school teachers; high school teachers were also worried that they were out of date both academically and practically, lacking many of the hands-on skills themselves. In reality, despite much progress and positive change in education, these challenges remain within science education. The academic world was changing too – an era of public understanding of science, and a culture in which engagement between academics and the public was just beginning, with schools being one of those 'publics' (Bodmer, 1985).

The development of the Teacher Scientist Network (TSN)

In answering the challenge, a precedent existed already – the Science and Health Partnerships Scheme initiated by a colleague of Keith's in California – Professor Bruce Alberts (http://biochemistry.ucsf.edu/programs/sep/). Keith shared his knowledge of this programme with Frank and quickly a group formed to assess its viability for Norfolk. Importantly, from the outset, this group was strongly represented by teachers. This ethos remains at the heart of what Norfolk's TSN does today – teacher-led and responsive to teacher-need. In less than five months, an abstract idea was transformed into a reality. The concept was a partnership scheme between teachers and scientists. A Steering Group of six people was formed, funding was sought and an arbitrary, yet aspirational, target of 30 partnerships between teachers and scientists was set.

Keith and Frank, strong voices within the scientific and educational communities respectively, championed the idea widely. At a meeting on the 30th June 1994, the network was launched and a staggering 50 scientists were linked with 50 teachers, who started to plan how they would work together.

The Gatsby Charitable Foundation (www.gatsby.org.uk/) supported the programme with an initial grant of £20 000 over two years. The trustees continued to support the programme generously over the next 10 years and this period saw the organisation formalised into the Teacher Scientist Network (TSN). TSN agreed a mission statement, formulated a logo, and employed Frank as its part-time coordinator. Ideas from teachers continued to form the basis of what TSN offered.

> TSN's mission is to enhance science in schools across Norfolk and North Suffolk through the supportive involvement of the local science community in science education:
> - by encouraging activities of mutual benefit to both scientific and educational communities;
> - by providing support, advice and resources to the teachers and scientists involved in the network;
> - by regularly reviewing, adapting and updating activities in accordance with the changing needs of the network's members. www.tsn.org.uk/index.htm

Of course it has not been plain sailing, but TSN benefitted enormously in its first year from the US experience documented in the publication *Science Education Partnerships, Manual for Scientists and K-12 Teachers* (Sussman,1993) in which the authors identified a 10-step recipe for starting a partnership programme. These experiences, TSN's early years and a visit to the US in spring 1996 to see first-hand the American experience, has from the start helped TSN to learn much about what makes a successful partnership after throwing together 50 pairs of teachers and scientists in the first gathering at the John Innes Centre in 1994.

How TSN works

Initial matching

Careful matching of partners and induction for the partnership are crucial. One of the biggest lessons has been the need to balance the expectations of partners and ensure an understanding from both sides of their respective responsibilities. Taking time to highlight to both partners the importance of communication and how and when to contact each other can be hugely valuable. With these ideas in mind, TSN quickly adopted an induction approach for each individual partnership. So the first meeting between teacher and scientist is a three-way meeting involving the TSN coordinator who brokers the meeting, ensuring the expectations of each partnership is voiced and an understanding reached. By joint planning of each activity, partners should both have some ownership of the session.

Roles and responsibilities

The dictionary definition of a partnership is one in which two or more people meet to achieve a common goal sharing designated roles in order to achieve that goal. And so it is with our partnerships – the common goal is the enhanced science experience for the teacher (and pupil). But what are designated roles for each person within the partnership? The roles of teacher and scientist are outlined below.

- **Classroom management**. This remains the responsibility of the teacher. Your scientist partner is not here to teach your lesson for you and give you a coffee break! Scientists will often work with small groups within the classroom setting, but you should never be far away. Most commonly partners tend to team teach and this works better the more times you work together.
- **Classroom language.** Teachers need to guide their partners on the most appropriate language to use with their pupils. This does not mean 'dumbing down' the science; it's all about making it accessible to pupils, qualifying any scientific jargon that is used and guiding the scientist on the relative abilities of the different age-groups and different abilities that you teach. It is good that pupils extend their scientific vocabulary.
- **Classroom and curriculum context**. You've agreed with your scientist that they will contribute to a session on 'ocean acidification' by CO_2, but were does this fit in the curriculum? What other areas does it link to? Perhaps acids/alkalis, climate change, etc. Giving this context to your scientist will help their planning and allow them to see the bigger picture. This context may enable them to bring in other relevant perspectives, enhancing the session further.
- **Scientific knowledge and updating current science.** Some teachers may not have a post-16 qualification in science, others might have a doctorate, each will have different motivations for wanting to work with scientists. Practising scientists will have knowledge of recent work in the subject, current techniques and commonly greater scientific knowledge (at least in a specific area of science). It is rare that even highly qualified teachers are able to 'keep up' with developments in science; this is often one of the motivations for a scientist taking part in TSN. To be able to share that knowledge with teachers and their pupils, when done in the right way, is a real asset to the partnership. TSN partnerships are much more than an extra pair of hands in the classroom.
- **Planning experiments.** We have found that teachers and their pupils (particularly in the primary classroom) find planning robust, reliable and well-planned scientific investigations a real challenge. Working with scientists who do this every week, can be very valuable and far more instructive than scientific inquiry from a textbook – after all science is a truly practical subject. Likewise scientists can learn about well-planned lessons from the teachers.

Sustaining the partnership

The partnerships tend to self-manage themselves after the initial induction. Time for partners to plan at the start of their partnership is critical and TSN provides half-a-day supply cover to facilitate time away from school for a second meeting between teacher and scientist. At this meeting, the partnership plans how and when they might work together in the future. For a successful partnership, the scientist's input must address the school's existing science curriculum (not least because of time restrictions in the classroom), and the teacher must use the scientist's expertise appropriately.

Teacher–scientist partnerships imply genuine collaboration in planning in-school activity. We have found that it is the combination of the teacher's skills and the scientist's skills that make partnership activity so productive, and which give students the opportunity to reshape their attitudes to science and scientists.

Range of TSN activities

1 The main focus of TSN is the long-term partnership, between teachers and scientists. The focus is providing a support to teachers. Other schemes exist but they are not teacher focused and are more geared towards short-term encounters between scientists and pupils. Many partnerships last in excess of three years.

2 After a series of popular and relatively successful 'quick-response' talks, a wider-reaching Master Class programme was begun in 1997, again in response to teacher need. Master Classes are day-long events providing high school teachers with the latest ideas and understanding in the field, enriching the teachers subject knowledge some years before such ideas would reach a textbook. TSN's first Master Class, addressing 'Climate Change' saw several leading scientists from the School of Environmental Science at UEA present talks and lead activities around climate change and modelling. As its Master Class programme developed, TSN is now able to support science in the broadest sense across all age groups.

3 Uniquely TSN recognised that time out of the classroom was a barrier to many teachers attending these valuable subject-knowledge focused events. So TSN offers every teacher attending a master class, a contribution towards their supply cover as well as half-a-day supply cover for those teachers developing partnerships.

4 Demand for scientist partners from teachers continues to out-pace supply. In 1999 the suggestion was made to offer some kits for schools to borrow. The kits needed to either fill gaps in curriculum materials or encourage a hands-on investigative approach to science practicals. Supported by science technician John Mallott, an initial six kits were offered – covering aspects of microscopy and primary physics. Booking is by phone or email and the responsibility rests with the teacher to collect and return the kits they borrow (www.tsn.org.uk/KitClub.htm). The kits were also a way to counter-balance the life-science bias of the region's scientists. With over 11 000 people, the Norwich Research Park (NRP) has one of Europe's largest single-site concentrations of research in Health, Food and Environmental Sciences (www.nrp.org.uk).

5 TSN has supported the development of similar networks (in Cambridge, Cardiff and Durham), but at the time of writing only the TSN North-East remains viable (http://www.slcne.org.uk/tsn). Norfolk's TSN location at the centre of a 'hot-spot' of science and the support it has received from the organisations of the NRP have certainly contributed to its longevity.

TSN continues to be led by a teacher-dominated steering group including representatives from the main scientific organisations from across the NRP. This truly 'bottom-up' approach in which we respond to teacher need rather than 'deciding what teachers need' is the key to TSN's success. All of the core activities that TSN now supports – partnerships, Kit Club and continued professional development opportunities via the Master Class programme and Primary Science Workshops, have evolved in response to teacher feedback. Additionally these activities support each other with the kits providing an ideal 'prop' for a scientist to work with during an early visit into school.

Benefits to pupils, teachers and scientists

For the benefits to pupils, teachers and scientists of TSN, please refer to Case study 3.3.

Further detail about the activities of the TSN can be found on our website at www.tsn.org.uk.

Acknowledgements

The author of this case study, the present coordinator of the Teacher Scientist Network, wishes to gratefully acknowledge the vision and drive shown by Professor Keith Roberts MBE, and Mr Frank Chennell in establishing the network, the input of all steering group members from 1994 to the present day and the long-term funding offered by the trustees of Gatsby Charitable Foundation and GTEP programmes.

10.7   Training teachers

Many studies have identified the need for science teachers to update their subject specific knowledge through Continuing Professional Development (CPD) courses. Getting Practical[4] is a programme run by a consortium of different organisations, but the main lead is the Association for Science Education. Getting Practical provides CPD in the area of practical science for primary, secondary and post-16 education.

A government funded organisation called Myscience was established in 2005 by the White Rose University Consortium[5]. Myscience runs the UK Science Learning Centres[6], which were also established to provide CPD for both teachers and school technicians. The Science Learning Centres host residential programmes, which are run by practitioners who are experts in science pedagogy. A report by ENTHUSE[7] in 2009–10 has shown that these CPD programmes have helped teachers improve their classroom teaching and pupils have benefitted as a result, with higher levels of motivation and attainment. Scientists can also get involved in both designing and delivering CPD programmes and Box 10.4 provides an example of a CPD programme run by scientists for teachers in the area of genes and health.

Box 10.4   Genes and Health: the Role of Science in Society

Laura Bowater

An all-day workshop, supported by funding from the UEA's Outreach Opportunity Fund, that mapped onto Key Stage 4 (GCSE) was organised for secondary school science teachers and classroom assistants from local schools. After initial discussions with secondary school teachers to establish a successful recruitment strategy, the workshop took place at the University of East Anglia and it was held after the GCSE and A-level exam period but before teachers broke up for the summer holiday. We decided that a focus for our workshop would be the advancing knowledge in the areas of human genetics and biotechnology and, in particular, how genes affect our health. In addition, we wanted to explore the possibilities of genetic profiling reducing the risk of developing human diseases as well as leading to more effective treatment strategies including 'personalised medicine'. We were keen to include activities that highlighted the ethical issues associated with these technological advances and how this growth in genetic information should be used in our society. The workshop was designed to include the following components:

- an introduction to the theory of inheritance and human genetic diseases;
- stimulating practical, laboratory based activities that demonstrate the theory or technology that underpin genetic profiling;

---

[4] http://www.gettingpractical.org.uk/

[5] the consortium consists of the universities of Leeds, Sheffield, Sheffield Hallam and York

[6] https://www.sciencelearningcentres.org.uk/curriculum/secondary/contemporary-science/fresh-science-ecpd

[7] A partnership supported by the Wellcome Trust, the Department for Education, AstraZeneca, AstraZeneca Science Teaching Trust, BAE Systems, BP, General Electric Foundation, GlaxoSmithKline, Rolls-Royce, Vodafone and Vodafone Group Foundation.

- a discussion of the ethical issues raised by genetic profiling through interactive seminars and workshops.

The workshop attracted 36 school staff from 21 different secondary schools. Prior to attending the session, teachers were provided with a programme for the workshop as well as written instructions in the safe handling of all equipment and potentially hazardous materials and asked to observe good laboratory practice. The activities were designed to incorporate different learning styles with a didactic lecture, an active laboratory based session, a group based session focused around an ethical dilemma and a wrap up session at the end. Evaluation of the workshop revealed that the teachers had enjoyed the mixture of learning styles, and the mixture of ethics and science. They reported that the workshop mapped onto the curriculum but also included extension topics that could be explored back in the classroom. In addition 90% of teachers reported that they recognised that the topics covered were contemporary and that the session was very useful. All teachers reported that they had found something within the workshop that they could undertake or use back in school. This included 'using the ethical problems', 'updating the cancer lessons' and 'feeling brave enough to use the school DNA electrophoresis kit to look at DNA'. After the workshop, several requests were received from teachers for additional learning materials including access to PowerPoint presentation, clear instructions for the laboratory based experiments and lesson plans for the ethical discussion sessions. These additional materials were produced and sent to email addresses supplied by teachers who had requested them. An added bonus is that these materials have been used in subsequent workshops.

10.8   Concluding remarks

Interacting with schools can result in long-term beneficial impacts for both pupils and scientists. We feel that our work with schools and the wider community has had a real and tangible effect upon individuals. Maybe these young people will not go on to be scientists (for us, this outcome is welcome but not essential), but they will have gained an appreciation and excitement of science and an understanding that scientists are not special, nor super-intelligent, but just ordinary people who have curiosity – just like them. Our parting piece of advice would be just to give school engagement a go. You don't have anything to lose. You might try it and discover it's not for you, and that's fine. But if you do enjoy it, you may open up a wealth of new possibilities, which could form a whole new set of interactions and relationships and it could even enhance your research and teaching career.

References

Angus, M., Olney, H., Ainley, J., et al. (2004). *The Sufficiency of Resources for Australian Primary Schools*. Canberra: DEST

Aschbacher, P.R., Li, E. and Roth, E.J. (2010) Is Science Me? High School Students' Identities, Participation and Aspirations in Science, Engineering, and Medicine, *Journal of Research in Science Teaching*, 47 (5), 564–582

AQA (2011) http://web.aqa.org.uk/subjects/science.php (accessed 5 May 2012)

BIS (2011) Attitudes to Science: Survey of 14–16 year olds. http://www.bis.gov.uk/assets/biscore/science/docs/a/11-p112-attitudes-to-science-14-to-16 (accessed 2 August 2011)

Bodmer, W. (1985). The Public Understanding of Science. Report of a Royal Society ad hoc group endorsed by the Council of the Royal Society, ISBN 0854032576

Brodie, E. and Thompson, M. (2009) Double crossed: exploring science and history through cross-curricular teaching. *School Science Review*, 90 (332), 4752

Department for Education (2011) http://www.education.gov.uk/schools/toolsandinitiatives/nationalstrategies (accessed 30 July 2011)

Desforges, C. and Abouchar, A (2003) The Impact of Parental Involvement, Parental Support and Family Education on Pupil Achievements and Adjustment: A Literature Review. Department for Education and Skills. Research Report RR433.

Directgov (2010) http://www.direct.gov.uk/en/Parents/Schoolslearninganddevelopment/

Dolan, E.L. (2008) *Education Outreach and Public Engagement. Mentoring in Academia and Industry* (ed Bell, J.E.). Springer

Edexcel GCSE Science 2011 (2011) http://www.edexcel.com/quals/gcse/GCSE-science-2011/Pages/default.aspx (accessed 30 July 2011)

ENTHUSE (2009–10) National Science Learning Centre, Report on Impact. https://www.sciencelearningcentres.org.uk/research-and-impact/enthuseimpactreport.pdf (accessed 5 May 2012)

Fensham, P.J. (2008) *Science Education Policy-Making: Eleven Emerging Issues*. Commissioned by UNESCO, Section for Science, Technical and Vocational Education

Hackling, M., Peers, S. and Prain, V. (2007) Primary connections: reforming science teaching in Australian primary schools, *Teaching Science*, 53 (3), 12–16

Henry, J. (2010) Ministers Pull the Plug on Gifted and Talented Academy, The Telegraph (23 January). http://www.telegraph.co.uk/education/educationnews/7062061/Ministers-pull-the-plug-on-gifted-and-talented-academy.html (accessed 30 July 2011)

Holland, A. and Andre, T. (1987) Participation in extracurricular activities in secondary school: what is known, what needs to be known? *Review of Educational Research* 57, 437–466

House of Commons Education Committee (1995) Fourth Report, Science and Technology in Schools, Vol. 1. Report together with Proceedings of the Committee, Session 1994–95 (published 5 July)

House of Commons Science and Technology Committee (2011) Practical Experiments in School Science Lessons and Science Field Trips. Ninth Report of Session 2010–12, Vol. 11

IET (2008) Studying STEM: what are the barriers? A factfile produced by the Institution of Engineering and Technology. http://www.theiet.org/factfiles/education/stem-report-page.cfm (accessed 5 May 2012)

Johnston, J. (2011) The cross-curricular approach in Key Stage 1, In: *Cross-Curricular Teaching in the Primary School* (eds Kerry, T.). Routledge

Mahoney, J.L., Cairns, B.D. and Farmer, T.W. (2003) Promoting interpersonal competence and educational success through extracurricular activity participation. *Journal of Educational Psychology* 95 (2), 409–418

National Curriculum Key Stages 3 & 4 (2007) http://curriculum.qcda.gov.uk/key-stages-3-and-4/index.aspx. (accessed 5 May 2012)

NESTA (2005) Real Science, Encouraging Experimentation and Investigation in School Science Learning. http://www.nesta.org.uk/library/documents/RealScienceFullReport1.pdf (accessed 5 May 2012)

OCR GCSE Sciences (2011) http://www.ocr.org.uk/qualifications/type/gcse_2011/index.aspx (accessed 5 May 2012)

OECD (2006) *Assessing Scientific, Mathematical and Reading Literacy*: A framework for PISA 2006. Paris: OECD

OECD (2009) Top of the Class, High Performers in Science in PISA 2006. http://www.oecd.org/dataoecd/44/17/42645389.pdf (accessed 5 May 2012)

Porter, C. and Parvin, J. (2008) Learning to Love Science: Harnessing Children's Scientific Imagination. A report from the Chemical Industry Education Centre, University of

York. Shell Education Services. http://www.ciec.org.uk/resources/ses_report.pdf (accessed 16 October 2011)

Porter, C., Parvin, J. and Lee, J. (2010) Children Challenging Industry: All Regions Study of the Effects of Industry-Based Science Activities on the Views of Primary School Children and Their Teachers. http://www.cciproject.org/research/documents/cci_national06-08.pdf (accessed 15 October 2011)

SCORE (2011) Practical Work in Science: A Report and Proposal for a Strategic Framework. http://www.score-education.org/media/3668/report.pdf (accessed 5 May 2012)

Sussma, A. (ed.) (2003) *Science Education Partnerships: Manual for Scientists and K-12 Teachers*. University of California

Sturman, L., Ruddock, G., Burge, B. *et al.* (2008). England's Achievement in TIMSS 2007: National Report for England. Slough: NFER

Turner, S (2010) The benefit of extracurricular activities in high school: involvement enhances academic achievement and the way forward, academic leadership, *The Online Journal*. http://www.academicleadership.org/508/the_benefit_of_extracurricular_activities_in_high_school/ (accessed 5 May 2012)

Wellcome Trust (2005) Primary Horizons: starting out in science. http://www.wellcome.ac.uk/About-us/Publications/Reports/Education/WTX026627.htm (accessed 5 May 2012)

Yeoman, K.H., James, H.A. and Bowater, L. (2011) Development and Evaluation of an Undergraduate Science Communication Module, beej, 17-7 http://www.bioscience.heacademy.ac.uk/journal/vol17/beej-17-7.pdf (accessed 5 May 2012)

Epilogue

This book has benefitted greatly from the depth and breadth of case studies written by scientists, teachers and professional science communicators. It could not have been written without their valuable contributions, which have highlighted the many different ways that we communicate with diverse audiences. Practical science communication – whether done face-to-face or remotely – is out there and growing year on year, but this is perhaps not reflected in the science communication literature, which tends to focus more on the process rather than practice. The case studies presented in this book have provided a rich resource of evidence that enhances the literature for the effective practice of science communication.

As we began writing this book and contacting science communicators working both nationally and internationally, we became increasingly aware of the diverse, innovative and exciting ways that scientists were engaging with the public. It is also clear that the message of 'more dialogue' has reached scientists, who instinctively adopt this approach – we like discussing our science. Many of the events and activities described within the case studies have adopted a mixed approach, offering facts and dialogue combined together to great effect. However, note that what social scientists really mean by dialogue – resulting in a more empowered citizen and a more democratic science culture – is not so easy to translate into practice. However, it would be a misconception to assume on the behalf of scientists that the will for dialogue is not there.

One of the aims of this book was to bridge the gap between the scientists 'doing' science communication and the social scientists who study the process and effectiveness of what we do. This book provides the scientist with an overview of how science communication has developed and of the models used to explain the processes involved in communication and learning. We have outlined the benefits of and the barriers to getting started with science communication or becoming more involved in this activity. We have tried to provide ideas on getting started with both school and public science communication, and the case studies have illustrated the many different and innovative ways that projects can develop over time.

Science Communication: A Practical Guide for Scientists, First Edition. Laura Bowater and Kay Yeoman.
© 2013 John Wiley & Sons, Ltd. Published 2013 by John Wiley & Sons, Ltd.

We hope that this book has inspired you to try your hand at science communication, or perhaps to develop what you currently do. We also hope that we have provided you with the practical steps needed to make this happen. We can only re-state what a powerful and personally transformative experience practical science communication can be, and that no scientist has anything to lose and everything to gain from taking part.

Good luck!

Abbreviations and Acronyms

AAAS	American Association for the Advancement of Science
BBSRC	Biotechnology and Biological Sciences Research Council (UK)
BAAS	British Association for the Advancement of Science
BAYS	British Association of Young Scientists
BSE	Bovine Spongiform Encephalopathy
CLEAPSS	Consortium of Local Education Authorities for the Provision of Science Services
CSA	Citizen Science Alliance
COPUS	Coalition on the Public Understanding of Science (US)
COPUS	Committee on the Public Understanding of Science (UK)
CoSHH	Control of Substances Hazardous to Health
CRB	Criminal Records Bureau
CUE EAST	University of East Anglia Beacon of Public Engagement
EPSRC	Engineering and Physical Sciences Research Council
EUSCEA	The European Science Events Association
FEC	Full Economic Costing
GMAG	Genetic Manipulation Advisory group
GM	Genetic Modification, Genetically Modified
HEFCE	Higher Education Funding Council for England
HEI	Higher Education Institute
KISS	Keep it short and simple
KS	Key Stage
NCBE	National Centre for Biotechnology Education
NCCPE	National Co-ordinating Centre for Public Engagement
NSEW	National Science and Engineering (UK)
OECD	Organisation for Economic Co-operation and Development
OPAL	Open Air Laboratories
PAS	Public Appreciation of Science
PAS	Public Attitudinal Survey
PAT	Portable Appliance Testing
PCST	Public Communication of Science and Technology
PEST	Public Engagement with Science and Technology
PUS	Public Understanding of Science
QR	Quality Related Funding

Science Communication: A Practical Guide for Scientists, First Edition. Laura Bowater and Kay Yeoman.
© 2013 John Wiley & Sons, Ltd. Published 2013 by John Wiley & Sons, Ltd.

RAE	Research Assessment Exercise
REF	Research Excellence Framework
RCUK	Research Councils UK (UK)
RI	Royal Institution
ROSE	Relevance of Science in England
SCOPE	Science Outreach and Public Engagement
SCORE	Science Community Representing Education
SAW	Science Art and Writing
SGM	Society for General Microbiology (UK)
STEM	Science, Technology, Engineering and Maths
TIMSS	Trends in International Mathematics and Science Study
UEA	University of East Anglia
UK	United Kingdom
US	United States
VARK	Visual, Audio, Read/Write, Kinaesthetic

Index

Note: Page numbers followed by b, f or t refer to Boxes, Figures, or Tables.

Science Communication: A Practical Guide for Scientists, First Edition. Laura Bowater and Kay Yeoman.
© 2013 John Wiley & Sons, Ltd. Published 2013 by John Wiley & Sons, Ltd.

Printed and bound by CPI Group (UK) Ltd, Croydon, CR0 4YY